FOOD WARS

Michael Heasman dedicates this book to Mum, Dad, Susan, Colin and Jason

Tim Lang dedicates this book to Anna and Alfie and their generation's future

Food Wars

The Global Battle for Mouths, Minds and Markets

Tim Lang and Michael Heasman

London • Sterling, VA

First published by Earthscan in the UK and USA in 2004

Reprinted 2004, 2005, 2006

ISBN-10: 1-85383-702-4
ISBN-13: 978-1-85383-702-9

Typesetting by JS Typesetting Ltd, Wellingborough, Northants
Printed and bound in the UK by CPI Bath
Cover design by Declan Buckley from a painting by William Crozier
(Joie de Vivre, private collection)

For a full list of publications please contact:

Earthscan
8–12 Camden High Street, London, NW1 0JH, UK
Tel: +44 (0)20 7387 8558
Fax: +44 (0)20 7387 8998
Email: earthinfo@earthscan.co.uk
Web: **www.earthscan.co.uk**

22883 Quicksilver Drive, Sterling, VA 20166-2012, USA

A catalogue record for this book is available from the British Library

Library of Congress Cataloging-in-Publication Data

Lang, Tim.
Food wars : the global battle for minds, mouths, and markets / Tim Lang and
Michael Heasman.
p. cm.
Includes bibliographical references and index.
ISBN 1-85383-701-6 (hardback : alk. paper) — ISBN 1-85383-702-4
(pbk. : alk. paper)
1. Nutrition policy. 2. Food supply.
[DNLM: 1. Nutrition Policy. 2. Diet. 3. Environmental Health. 4. Food
Industry. 5. Food Supply. 6. Nutrition Disorders. WA 695 L271f 2004]
I. Heasman, M. A. (Michael Anthony) II. Title
TX359.L36 2004
363.8–dc22
2003022771

Earthscan is an imprint of James and James (Science Publishers) Ltd and
publishes in association with the International Institute for Environment and
Development

Printed on elemental chlorine-free paper

CONTENTS

List of Figures, Tables and Boxes

Figures

Tables

Box

Acknowledgements

This book has been a long time in gestation. We began to talk seriously about it in the run-up to the December 1999 Seattle World Trade Organization meeting, and an early version of our thinking was launched there by the International Forum on Globalization. The book is being published when there is a full-blown debate about obesity and the cost of health care associated with it. Back in 1999, the UN system's World Health Organization prepared a draft strategy on tackling the epidemic of diet-related disease sweeping our world, which received a hostile reaction from sections of the food industry. We had realized that, in the very welcome and rising debate about globalization, the vital area of food and public health was somehow being marginalized or perceived as being limited to a few issues such as food safety and GM foods for example. As we show in this book, food and health issues go far wider than that, and include large issues such as the health impact of the spread of Western diets to the developing world.

While environmentalists and citizens groups had well-developed debates underway about the cultural and political transition (and about the need to reform government and policies in the pursuit of the public, not just the corporate, interest), the food and public health movement appeared to have been left on the sidelines – ironically, in the face of the evidence supplied by epidemiologists and nutritionists arguing for policy change. We decided that we had to set down our arguments and thoughts. The process took longer than we expected, as it required us to enter areas and review data which are themselves immensely complex and require labyrinthine understanding. The book underwent an iterative process of being written, read by specialists and friends, criticized, wholly rewritten and round again.

We therefore want to pay tribute to our many friends and colleagues who have encouraged and helped us in this process. It began for both of us on two fronts. First, from involvement in the public policy debate throughout the 1990s, in the course of which a coalition of interests came together to monitor, engage with and lobby on the arcane area of international trade regimes and regulations. Second, we were both involved in following the global food industry as it struggled to integrate a 'health' agenda into its new product development and marketing strategies.

We felt that health had to be a significant feature in the re-alignment of food with society. Yet if a good understanding of food and health issues was poorly represented outside the decision-making process, discussion about their relevance would inevitably, it seemed, be left to officials or the industrial interests about which we were nervous. Our view was that a central role for food and public health policy was a critical test for sound policy-making which should be based on best evidence and best practice.

We have been privileged to be part of that growing debate and of our own self-education process. So our first debt is to all those who, over the last 12 years or more, have been prepared to discuss issues, respond to ideas, ask us to write and present papers, support as well as curtail our enthusiasms, point out errors, and do all such helpful things that friends and colleagues do.

We therefore pay heartfelt tribute, for helping to form the ideas and thinking in this book, to the following: Annie Anderson, John Ashton, Carlos Alvarez-Dardet, David Barling, Fran Baum, Robert Beaglehole, Warren Bell, David Buffin, Colin Butler, Geoffrey Cannon, Martin Caraher, Mickey Chopra, Charlie Clutterbuck, John Connor, Dick Copeman, John Coveney, Michael Crawford, John Cubbin, George Davey-Smith, Barbara Dinham, Liz Dowler, Anna Ferro-Luzzi, Ben Fine, Ken Fox, Yiannis Gabriel, Susan George, Edward Goldsmith, C Gopalan, Jeya Henry, Brian Halweil, Spencer Henson, Ildefonso Hernández, Nick Hildyard, Colin Hines, Vicki Hird, Dinghua Hu, Mika Iba, Michael Jacobsen, Phil James, Jean James, Jørgen Højmark Jensen, Marco Jermini, Andy Jones, Ingrid Keller, Cecile Knai, Mustafa Koc, Al Krebs, Lyndon Kurth, Ron Labonte, Felicity Lawrence, Mark Lawrence, Kelley Lee, Rod Leonard, Tim Lobstein, Jeanette Longfield, David Ludwig, Jerry Mander, John Manoocheri, Barrie Margetts, Karen McColl, Martin McKee, Tony McMichael, Philip McMichael, Margaret Mellon, Erik Millstone, Sid Mintz, Monica Moore, Marion Nestle, Chizuru Nishida,

Aleck Ostry, Roland Petchey, Miquel Porta, David Porter, Barry Popkin, Kaisa Poutanen, Jules Pretty, Bill Pritchard, Pekka Puska, Geof Rayner, Mike Rayner, Tom Reardon, Michael Redclift, Sarojini Rengam, Neville Rigby, Mark Ritchie, Aileen Robertson, Peter Rosset, Sam Selikowitz, Aubrey Sheiham, Prakash Shetty, Mira Shiva, Vandana Shiva, Bruce Silverglade, Boyd Swinburn, Steve Suppan, Geoff Tansey, David Thomas, Peter Timmer, Antonia Trichopoulou, Colin Tudge, Flavio Luiz Schieck Valente, Bill Vorley, Lori Wallach, David Wallinga, Kevin Watkins, Amalia Waxman, Julius Weinberg, John Wilkinson, Martin Wiseman, Derek Yach, Taka Yagi, and Richard Young, and all those in the international food industries and agencies who shared their insights and thoughts over the years about the business side of health. All have provided us with encouragement, thoughts, criticism and advice on the complex range of issues that we feature in this book.

For permission to use data reproduced in this book, we thank the World Cancer Research Fund, the ACC-SCN Expert Committee and Barry Popkin; as ever, all three were inspirations. For permission to quote work, we thank Rita Clifton of Interbrand, and the BHF Health Promotion Research Group at Oxford University.

Many people helped us with the laborious practical process that is book writing. From 1998 to 2000 Heena Vithlani, Jenny Lord and Kelly Andrews took turns as PAs to Tim Lang; in 1999, Pirkko Heasman and Jillian Pitt gave invaluable help on mapping for us the nutritional and public health aspects of the globalization of diet; and in 2002, Yannick Borin and Sylvie Fritche, our inestimable French duo, gave wonderful help with data-gathering, sorting and tables.

We want also to thank the many people who have commented on ideas we presented when teaching or at seminars in Australia, Brazil, Canada, Croatia, Denmark, Finland, India, Italy, Japan, Kenya, Korea, Latvia, Mexico, The Netherlands, New Zealand, Russia, Slovenia, Spain, Sweden, Switzerland, the UK and the US over the last ten years. There have been moments when it has seemed that there is nowhere we will not go to research food supply. Our energy-burning, carbon-burning environmental 'credits' have no doubt been all used up, as we jetted about, at other people's expense, to meetings and conferences where we could explore, observe and pronounce on the world's changing supply chain. At those many meetings, we would try out many of the ideas and data given in this book. We are truly grateful to those who came to discuss with us and give feedback; we hope they think their efforts were worthwhile.

We also want to thank the many organizations who nurtured our thoughts, including friends and colleagues at the International Forum on Globalization, Pesticides Action Network, Sustain (created by the merger of the National Food Alliance and the Sustainable Agriculture, Food and Environment Alliance), the UK Public Health Association, the *Journal of Epidemiology and Public Health*, the Food and Agriculture Organization and the World Health Organization, all of whom have encouraged our work over the last decade or so. It would have been much harder critically to assess the impact of huge societal shifts of food, the supply chain and health without their support and encouragement.

Finally, we want to thank Pirkko Heasman and Liz Castledine for their unstinting support while we wrote this book; Akan Leander, Angela Cairns and the team at Earthscan and James & James for producing it; and especially Jonathan Sinclair Wilson for encouraging us to write it. Any errors and confusions are, of course, our own.

Tim Lang
London UK
March 2004

Michael Heasman
Jokela, Finland
March 2004

List of Acronyms and Abbreviations

AoA	Agreement on Agriculture (of the GATT)
BMI	body mass index
BSE	bovine spongiform encephalopathy
BST	bovine somatotrophin
CAP	Common Agricultural Policy
CEC	Commission of the European Community (also EC)
CHD	coronary heart disease
CI	Consumers International (world body of consumer NGOs)
Codex	Codex Alimentarius Commission (joint WHO/FAO body)
CVD	cardiovascular disease
DALY	disability adjusted life year
EAGGF	European Agricultural Guidance and Guarantee Fund (CAP)
EC	European Commission
EP	European Parliament
EU	European Union
FAO	Food and Agriculture Organization (of the UN)
GATT	General Agreement on Tariffs and Trade
GBD	Global Burden of Disease (a research study)
GM	genetic modification
IARC	International Agency for Research on Cancer
ICN	International Conference on Nutrition (1992)
IFG	International Forum on Globalization
IFPRI	International Food Policy Research Institute
JECFA	Joint Expert Committee on Food Additives
NGO	non-governmental organization
NIDDM	non-insulin-dependent diabetes mellitus
OECD	Organisation for Economic Co-operation and Development
PCB	polychlorinated biphenyl
POPs	persistent organic pollutants
SFS	Surplus Food Scheme
SME	small- or medium-sized enterprise
SPS	Sanitary and Phytosanitary Standards (part of 1994 GATT)
TBT	Technical Barriers to Trade (part of the 1994 GATT)
TNCs	transnational corporations
UN	United Nations
UNCED	United Nations Conference on Environment and Development (1992) (also known as the Earth Summit)
UNEP	United Nations Environment Programme
WFS	World Food Summit (1996)
WHO	World Health Organization (of the United Nations)
WHO-E	World Health Organization Regional Office for Europe
WIPO	World Intellectual Property Organization
WTO	World Trade Organization

'The history of the world, my sweet, is who gets eaten and who gets to eat'

Sweeney Todd

For what can war, but endless war still breed?
Till truth and right from violence be freed,
And public faith clear'd from the shameful brand
Of public fraud. In vain doth Valour bleed,
While Avarice and Rapine share the land.

John Milton, English poet, 1608–1674;
from *To the Lord General Fairfax* (1648)

INTRODUCTION

'Freedom from want of food, therefore, must mean making available for every citizen in every country sufficient of the right kind of food for health. If we are planning food for the people, no lower standard can be accepted.'

Sir John Boyd Orr, first Director-General of the Food and Agriculture Organization of the United Nations (1880–1971)[1]

CORE ARGUMENTS

Food policy is in crisis, in particular over health. Yet health can be the key to the solution to this crisis. For the last half-century, there has been one dominant model of food supply. This is now running out of steam and is being challenged by competing approaches: three major scenarios, each of which is shaping the future of food and health. We argue that, at the heart of any new vision, there has to be a coherent conception of how to link human with ecological health. Humanity has reached a critical juncture in its relationship to food supply and food policy, and both public and corporate policies are failing to grasp the enormity of the challenge. Food policy needs to provide solutions to the worldwide burden of disease, ill health and food-related environmental damage. There is a new era of experimentation underway emerging out of the decades we term the 'Food Wars'. These have been characterized by struggles over how to conceive of the future of food and the shaping of minds, markets and mouths.

Food is an intimate part of our daily lives. It is a biological necessity but it also shapes and is a vehicle for the way we interact with friends, family, work colleagues and ourselves. It is associated with pleasure, seduction, pain, power and caring. As we eat our daily food, bought in the shops that we know, buying brands that we are familiar with, it is hard to imagine that there is such a thing as a global food economy, stretching from the local corner store to the giant food conglomerate, under pressure right from the way food is produced and processed to its impact on our long-term health and well-being.

Our interest here is in food policy: the decision-making that shapes the way the world of food operates and is controlled. We see the world of food policy as formed and fractured by a series of conflicts – the Food Wars – structured around three dominant worldviews or 'paradigms' (a term explained fully in Chapter 1). These offer different conceptions of the relationship between food and health and also offer distinct and sometimes competing choices for public policy, the corporate sector and civil society. We argue that health has often been somewhat marginalized in policy and that the Food Wars are, in part, about a jostling for position by different interest groups seeking to influence the future of food.

Addressing the challenges of health will require better processes for making food policies and reform of the institutions of food governance; they need to be shaped in an integrated way. Unless this is done, we believe that the food supply chain will lose public trust. If it is to achieve popular support and legitimacy, it will need to be infused with what we call 'food democracy', a notion we explore towards the end of the book.

Our focus therefore is on the policy choices that shape how humanity orders its food economy and on urging public policy to play a positive role in promoting the public good. To this end, we explore five key elements of the world of food that we consider to be crucial. These are:

- *health*: the relationships between diet, disease, nutrition and public health;
- *business*: the way food is produced and handled, from farm inputs to consumption;
- *consumer culture*: how, why and where people consume food;
- *the environment*: the use and misuse of land, sea and other natural resources when producing food; and

- *food governance*: how the food economy is regulated and how food policy choices are made and implemented.

These issues are often studied in isolation, and at times deserve such micro-attention. But the scale of the pressures and challenges in the context of the global food supply now suggests that this 'compartmental' approach is no longer a viable way of handling food policy-making. We are calling for a new framework for making food policy choices.

While today's food economy is grounded in a long history of production, experimentation and technological change, the industrial food supply is still relatively young in human history – a little more than 150 to 200 years old. Since World War II the food economy has undergone further remarkable commercial and technological expansion in order to provide food for an unprecedented growth in human population to more than 6 billion in 2003, and can deliver, in theory, enough food to end world hunger. For those with the means and access to purchase them, the modern food system has produced an array of processed, all-year-round, convenient foodstuffs never before available.

Yet at this very pinnacle of success in the way food is produced, the sustainability of food production systems and the quality of foodstuffs in the developed and developing worlds are being challenged as never before. The current food system appears to lurch from crisis to crisis: from new health scares such as BSE to environmental disasters such as over-fishing and the collapse of fish stocks. At the same time, global food supply faces new challenges: a continuing surge in population growth in some parts of the world and an increasingly aged population in others; the introduction of radical new technologies such as genetic modification; a new global scale and scope of corporate control and influence; a breakdown in consumer trust in food governance and institutions; and persistent health problems associated with inadequate diet such as heart disease, obesity and diabetes which, alongside hunger and famine, affect hundreds of millions of people.

It is obvious that something has to be done for the future. *Food Wars* argues that such challenges cannot be met in a piecemeal fashion. There has to be a new vision of public health. Our concern is to make the links across these discrete policy areas and to show continuity in thinking, from the way food is produced to the management of consumption and the healthiness of foodstuffs. We argue that the future viability of the food economy can be framed to deliver effectively to the general public only if there are new and integrated policy choices.

Difficult questions loom. How can population health goals be reconciled with the way people want to live their lives? Can consumers realistically continue to expect ever cheaper food? What sort of intensification in production is best for human and environmental health? How can patterns of food trade benefit more people? What are the acceptable limits to the continuing concentration of market share by giant food companies? To what extent should public money support food production, if at all?

Why Food Wars?

Every day, millions of men, women and children are direct or indirect casualties of failures of food policy to deliver safe, nutritious and life-enhancing diets. People raised in the developed world since World War II may think that the damage is felt only in areas of the world that suffer famine, malnutrition or other deficiencies; but in the rich world too there is a huge toll. While Western societies have increased the caloric content of the diet and boosted the sheer quantity of our food, they have at the same time introduced methods of production, distribution and consumption that threaten the future of the food system that delivers those calories. These methods have at the same time contributed to reducing the quality and nutritional value of many foodstuffs (such as the loss of essential bioactive components like vitamins and minerals). Health has been assumed to follow from sufficiency of supply. While scientific understanding of food is now very complex, too much policy-making has failed to face up to the human and environmental health damage that surrounds it.

All around us, food culture is divided. On the one hand, we have 'celebrity chefs' with top-rating TV shows, cookery and diet books on the best-seller lists, and popular media concerns about food quality, safety and availability. On the other hand, a crisis of food supply still dominates great tracts of the world. Hunger and under-nutrition still dog many lands, as well as premature deaths due to malconsumption and over-consumption. In 2001, for example, the US Surgeon General attributed 300,000 deaths in the US alone to obesity.[2] This book is partly about these dichotomies: over- and under-consumption; over- and under-production; over- and under-availability; intensification versus extensification; and hi-tech solutions versus traditional, culturally based ones.

In the Food Wars, there are numerous conflicts over the quality of food; food safety; nutrition; trade in foodstuffs; corporate control of food supply; food poverty and supply insecurity; the coexistence of the overfed and underfed; the unprecedented environmental damage from food production and the role and purpose of technology. How are organizations, policy-makers, businesses, farmers, non-governmental organizations (NGOs) and even individuals to tackle the enormity of the global and local challenges now confronting the food system? Despite apparent food abundance, the security of the food supply cannot be taken for granted.

We set out to write this book because we were frustrated that the key figures in food policy appeared to be skirting around major problems rather than facing them, or too often dealing with the challenges separately in neat policy boxes rather than holistically. Much of the food industry sees the responsibility for food as lying with the individual consumer, and any 'liberal'-minded intervention in food supply, they argue, is condescending: treating individuals as victims rather than intelligent food consumers. Such an approach, we argue, ignores the realities and the scale of the crisis in food and health which is beyond the scope of either individuals or single companies, and also ignores the power relationships shaping food supply. Much of this book is our attempt to resolve the complex battles over what the 'food and health' problem really is and what to do about it.

Are Radical Options in Food and Health Feasible or Even Possible?

Within the world of food policy there is a creeping recognition that radical solutions are needed. Distinct policy choices are emerging which will frame business and consumer opportunities. In the nutrition sciences, for example, a new 'ecological nutrition' is being developed along evolutionary principles – seeking diets that suit humans' evolutionary legacy. Such thinking has the potential to offer new radical ways of perceiving food and health (in particular the way food is produced) that go beyond the narrow 'technical fix' of many solutions being offered to 'feed the world' or to maintain health and fitness in ageing consumers in the developed world through 'health-enhancing' foods and beverages. Another big factor is the apparent revolution in 'life sciences', based upon an understanding that genes predispose people to diseases and that diet may trigger genetic predispositions. This so-called nutrigenomic understanding could have profound implications for the 'personalization' or 'individualization' of diets.

The word 'radical' is used here in the way that world business expert Gary Hamel uses it, namely that a 'radical idea has the power to change customer expectations. . . to change the basis for competition. . . and the power to change industry economics.'[3] We argue that beneath the apparently calm surface of the food supply chain (which, for all its scandals and monetary crises, has increased output and fed more people than ever before in human history) there are powerful undercurrents. Consumers appear, for example, to be able to change what they demand from the supply chain in fundamental ways; the recent restructuring of the food economy is changing the basis for competition and what is meant by a 'market'; and globalizing influences are changing the industry.

A prominent example of radical thinking entering the food industry is the way in which nutrition science is currently being used for new product development, food marketing and business strategy. In 2003 the Chief Executive Officer of Nestlé, the planet's largest food company, for example, stated that Nestlé aimed to become the world's leading nutrition company within five years.[4] This announced what, in effect, we are starting to recognize: that no food company can remain in business today without creating added value through nutrition and health. Yet where is there a similar vision for food, nutrition and health in public policy? What should such a vision look like? What should it include or exclude? How can it embrace the whole food chain, from growing food to final consumption?

Another radical conflict, in a polarized form, can be seen in the global tussle between a 'GM' (genetically modified) future or an 'organic' future – with both camps making claims of enhanced health benefits as the rationale for their competing ways of producing food. Through examples such as these we try, where possible, to explore the possibility of 'radical' options in food and health policy and whether these are feasible, what their scope might be, or even whether the 'technical fix' approach is the most appropriate way forward.

The foundation on which we build our call for a more radical approach to how food is grown and produced is the following:

- The model of food and agriculture put in place in the mid-20th century has been very successful in raising output but it has put quantity before quality.
- Humanity has moved from an agricultural/rural to a hypermarket/urban food culture in a remarkably short time, a process beginning to roll out fast in the developing world at present.
- While policy attention has traditionally been on agriculture, it is what happens *off* the farm in terms of processing, retail and food service that is in effect changing the food economy, not least for the labour force which can too often suffer low wages and poor conditions.
- Throughout the world, diet is changing in ways that carry huge health implications and challenges. This is in part due to trade liberalization and in part to consumer aspirations; in this respect, there is both a 'push' and a 'pull' in the food system.
- Food, nutrition and health challenges are global. Countries like Brazil, India and China are already in the grip of a double burden of food-related disease: degenerative diseases (heart disease, cancers, diabetes and obesity) take a heavy toll in all countries. At the same time mass hunger persists.
- An individualized medical model of food and nutrition is predominant and is presented as the only source of appropriate solutions to food and health challenges.
- The environmental pressures on food production are reaching crisis scale, from over-fishing (a reality facing Canada and the European Union), through the loss of soil to grow food in, to not enough water for agricultural production and irrigation.
- Without an ecologically integrated perspective, food policy will remain unable to provide long-term consumer confidence or food supply and distribution security.

- There is rising evidence of injustice within the food system. This includes the maldistribution of food, poor access to a good diet, inequities in the labour process and unfair returns for key suppliers along the food chain.

An Outline of the Book

This is a book about ideas of how the future of food is to be shaped and conceived. We address this task by setting up a conceptual framework in Chapter 1 of three paradigms. There we discuss in detail the character of the Food Wars, the assumptions of the paradigms that inform this book and what we mean by a paradigm. We argue that food policy often has a troubled relationship with evidence – sometimes lagging, sometimes leading it. How much evidence is needed to change policy?

Chapters 2 to 7 deal in detail with the evidence in support of our conceptual framework as it relates to health, food policy and the dynamics of the food system. We start by looking at the evidence of how the world's diet is changing and facing the problems of both under-consumption and over-consumption, often within the same countries. There is the mythology that the rich world suffers heart disease while the poor world suffers hunger. Diet-related diseases such as heart disease are becoming rapidly more prevalent in low- and medium-income countries. We show how diet- and health-related problems are growing in scale, not diminishing, as might be assumed with better food supply. Newer concerns such as obesity and diabetes are rising, in addition to the ongoing costs of heart disease.

In Chapter 3 we look at how public policy has responded to evidence about diet and disease. We give a short historical overview of changing conceptions of public health and the importance of nutrition, arguing that nutrition is a battleground between those who see it as framed by social objectives and those who believe that targeting only 'at risk' individuals is a more effective intervention. We review how governments have tended to rely upon health education as the mechanism for improving public health, setting dietary goals and offering guidelines which put responsibility upon individuals for their own health. We question this food policy strategy.

The success or failure of food policy will be dependent on how it relates to the workings of the food economy and affects particular food business interests. In Chapter 4 we present an overview of what is meant by the food system/economy, arguing that, while consolidation and concentration of the power of the food industry is a long-running trend, the scale and pace of this change are new. The food industry is relying on a twin strategy to take it into the future: first, relying on technology and 'technical fixes' to resolve most problems; and second, aligning itself with the interests of consumers.

While most food companies today will describe themselves and their activities as 'consumer-led', we argue that such an epithet is too superficial. We propose that a better grasp of food and consumer culture would help public policy analysts face what is happening in modern food markets. In Chapter 5 we map out what we see as the new consumer culture and landscape. Even at the basic market-led level, the rich-world consumer is developing a very different conception of food: convenience, snacking, ready meals, an eating-out culture and a food lifestyle that meet time constraints, and that recognize the newer role of women in society.

Chapter 6 turns to another war zone: essentially a conflict over food quality. Our case is that the food supply chain is committed to producing a range of foodstuffs in environmentally unsustainable and wasteful ways that militate against human health. Today's food supply chain, while seemingly appropriate for the past, is now shown to damage and threaten the environment. Food, a means for life, is threatening its own continued production. Too many policy-makers still believe that they can merely 'bolt on' an eco-friendly niche market to the crisis of food and the environment. A re-orientation of the entire food supply chain is needed if both human and environmental health are to be delivered.

In the commercial context, there is no respite in the tragedies continually hitting rural and farming communities. While farmers and the land are being squeezed, oligopolies from agribusiness to food processing, retailing and even food service dictate the workings of the food supply chain. We suggest that much public policy response to date has been at best reactive rather than proactive; and in many instances, NGOs and the business and scientific communities, albeit differently, have been more in tune with wider societal trends about food and health than policy-makers and government. But, as we argue in Chapter 7, future food and health choices must ultimately be resolved in public discourse: designing and reworking the institutional 'architecture' of food policy to deliver public goods is a pressing challenge. There are limited public forums for delivering, let alone creating, an integrated food and health policy. There is a crisis of institutions and of governance (that curious English word that refers to the science and practice of government) at all levels – local, national, regional and global. The processes of government are too often trapped in 'boxes' of responsibility, with no one retaining overall responsibility across the different compartments.

Our objective in this book is to contribute to the debate and to suggest that there is already available a wide range of policy options and alternative voices. In Chapter 8 we argue that policy-makers too often assume that they have little choice and consequently discourage the alternatives. But we think that there is a new era of experimentation underway and through our 'paradigms' we show that there are different ways of assessing and making choices.

A new conception of health – linking human and ecological health – has to be at the heart of a new policy vision. To deliver healthy consumption requires a different set of priorities within the food supply chain. We need to generate new areas of knowledge about food and health. Food policy in general needs to develop a range of alternative food scenarios, at the very least as 'insurance policies' against unforeseen crises and to tackle the unacceptable legacy from the last century of disease, ill health and environmental damage.

This book offers a panorama. We argue that throughout the 20th century, food caused problems in public and corporate policy and, vice versa, public and corporate policy caused problems for the world of food. Numerous crises have sparked incremental reforms, most recently over food safety. But still, the framework of public policy on food is too fragmented and restricted. Problems are addressed too often in an ad hoc or interim manner when what is required is a systematic framework for addressing food policy, integrating core drivers such as health, business, environmental impact, consumer experience and policy management. This more coordinated approach may still be in embryonic stages but it already finds itself in an arena of considerable conflict. There is some way to go in the Food Wars before there is Food Peace.

CHAPTER 1

THE FOOD WARS THESIS

'If you know before you look, you cannot see for knowing.'

Sir Terry Frost RA (British artist 1915–2003)

CORE ARGUMENTS

Different visions for the future of food are shaping the potential for how food will be produced and marketed. Inevitably, there will be policy choices – for the state, the corporate sector and civil society. Human and environmental health needs to be at the heart of these choices. Three broad conceptual frameworks or 'paradigms' propose the way forward for food policy, the food economy and health itself. All make claims to raise production and to deliver health benefits through food. The challenge for policy-makers is how to sift through the evidence and to give a fair hearing to a range of choices. This process is sometimes difficult because the relationship between evidence and policy is not what it seems. The world of food is on the cusp of a far-reaching transition.

INTRODUCTION

The world is producing more food than ever to feed more mouths than ever.[1] For the better off there are more food and beverage product choices than it is possible to imagine – globally 25,000 products in the average supermarket and more than 20,000 new packaged foods and beverages in 2002 alone.[2] Yet for many people there is a general feeling of unease and mistrust about the future of our food supply. Food and problems associated with producing and consuming food generate political and policy crises and are regular fodder for media coverage. In addition, along with the food production successes of the past 40 years in reducing famine, hunger continues hand in hand with excess. The optimism of the 20th-century food policy planners that, with good management and science problems associated with food would disappear, has not been fulfilled. Food's capacity to cause problems has not lessened.

As we will show, new relationships are already apparent throughout the entire food supply chain, from the way the food is produced to its consumption. Increasingly, alternatives to the prevailing structures of the food economy are also being widely mooted. No wonder there are such arguments about food. The pace and scale of change engender reactions; forces within the food supply chain are often at odds with each other about their vision for the future; there are competing versions of what the future could be, over which partisan forces argue. Our simple conclusion is that food policy-making matters more now than ever before.

To set the context for the future of food policy over the next two decades, we see the world of food supply currently in the throes of a long-term transition: from a food policy world dominated by farming and agriculture, agribusiness and commodity-style production, to one dominated by consumption: major branded food manufacturers, food retailing and food service. This transition is causing new tensions, challenges, threats and opportunities along the whole food chain, from farm to consumer, which we call the Food Wars: the precursor to what we argue is a fundamental reframing of the assumptions about the way we will come to analyse, research and carry out food production, the Food Wars encompass competing visions and models for the future of food supply driven, in part, by emerging new scientific understanding and accompanying technologies, but also by food politics and shifting demographics in terms of patterns of diet-related disease and illness as well as consumer-lifestyle choices.

In this chapter we set out to capture this complex pattern of change by suggesting a new conceptual model of three competing frameworks or 'paradigms' for food which we term the Productionist paradigm (the dominant and current model), the newly emerging Life Sciences Integrated paradigm and the

Ecologically Integrated paradigm. But first we need to set out some basic assumptions about food policy and the food supply chain that informs this conceptual model.

Food Policy Choices

Throughout this book, we use the terms 'food and health policy', 'food policy' and 'food and farming policy': those policies and the policy-making processes that shape the outcome of the food supply chain, food culture and who eats what, when and how, and with what consequences. Our task here is to unravel the strands of competing interests and policy objectives. There is no one food policy or one food policy-maker: there are policies and policy-makers, all of which contribute to the overall process.

Food policy-making is essentially a social process. The shape of the food supply chain is the outcome of myriad decisions and actions from production to consumption; it can involve people and organizations who may not even call themselves policy-makers. For example, the food industry, when it sets specifications for food products, is in part determining the nutritional intake of consumers; health-care planners, when facing the burgeoning costs of managing the rise of certain diseases (such as diabetes and some cancers) are making decisions that are 'policy', dealing with the results of how food is produced and consumed. Equally, competition authorities or town planners, when making decisions about retail market share or the siting of supermarkets, are determining issues as diverse as prices, access to food shops and local culture. The value of this very broad conception of food policy is that it helps to make sense of what otherwise remains a disparate, inchoate jumble.

Food policy is contested terrain: a battle of interests, knowledge and beliefs. The sort of food economy that exists is the result of a set of conscious policy choices made in the past, including both state and corporate decisions, involving funding for particular types of food production and processing, the setting of research priorities and national and strategic objectives, the provision of education and information, the creation of rules for trade and safe food and law enforcement and sanctions when things go wrong.

Our conception of food policy is that it should embrace decision-making along the whole of what is known as the food supply chain. Figure 1.1 is a simplified version of what is meant by the food system[3] or food supply chain – a term originally promoted by agricultural economists who now use a different term – 'value chain' – to analyse how, from farm to consumer, raw commodities get value added to them. The important point to note is that analysis from a food-chain perspective assumes that change in one part of the chain, intentionally or not, has an impact on other parts. Increasingly, analysis from a food-chain perspective is used to understand trends and the global restructuring of the food supply.

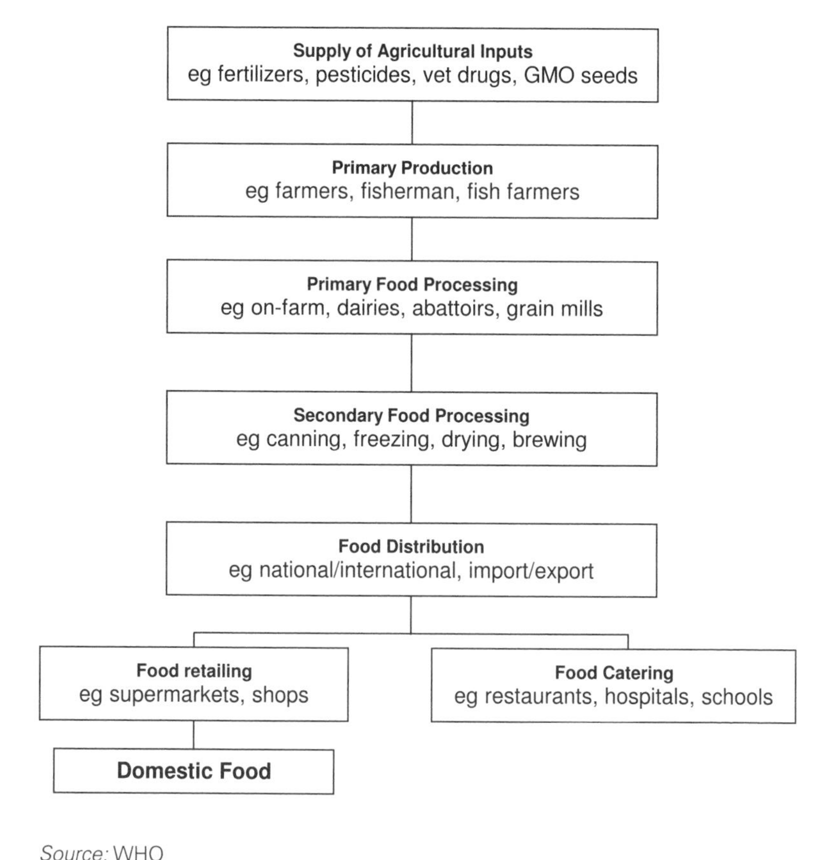

Figure 1.1 *A simple version of the food supply chain*

Key Characteristics of the Food Supply Chain

The model of the food supply chain in Figure 1.1 allows us to note some key characteristics of the modern supply system. We can summarize these under four main arguments:

Pressures 'off the farm' dominate the food system

Traditionally, for the last century, agriculture has dominated food policy thinking and still dominates international budgetary debates (for instance about the rights and wrongs of subsidies). The food supply chain is today driven by forces away from the farm, yet policy still focuses on commodity-producing agriculture. Pressures off the land are more important in framing the food economy than politicians often like to admit. Today, the main drivers of the food supply chain are the powerful forces of processors, traders and retailers, in turn focusing on capturing consumer needs.

Consumption is the key to understanding the food system

Power in the modern food economy is increasingly driven by concerns about the consumption end of the food supply chain. With the rise to dominance of food retailing, the retailer is a broker – between primary producers and consumption – and is a powerful figure in the corridors of power. Yet individual retail consumers are diverse and usually unconscious of their collective influence: they can be badly organized and they carry most of the health costs of current food supply, yet they are made responsible for their own diet-related (ill) health since they are ultimately answerable for what they eat – put another way, food production is being posited as a victim of consumer choice!

Public and corporate interests do not correspond

The pace of development and the structure of the food chain is being increasingly shaped by a small number of powerful food conglomerates. While this has been an evolving process, consolidation in the food industry has now reached a new level of influence in key markets. These corporate interests see food policy-making as part of their business strategy and are often well represented in the food policy arenas. This can be double-edged: on one hand, industry interests are frequently more aware of public objectives and unhappiness than the supposed public guardians themselves, and on the other hand, industry is hardly likely to give due weight to policy that conflicts with its immediate financial and market positioning. This raises a problem for what we call food governance – the role of public democratic control, accountability and public responsibility – an issue raised throughout this book but particularly addressed in Chapter 7.

Health has been marginalized in the food economy

Although the food supply chain model in Figure 1.1 is a simplified description of the food economy, it can imply support for the view that human and environmental health are an outcome of the smooth running of the food supply chain. In fact, health can fall down the gaps between sectors and is not seen as a prime responsibility of any one group. Human and environmental health ought to be the connecting tissue between and within all the economic sectors and be intrinsic to the whole food supply chain. A valuable debate has begun from an environmental health perspective; there now needs to be debate beginning from a human health perspective.

The War of Paradigms: Time for a New Framework?

There are structural tensions between different interests, views and economic investment patterns in food policy which are far-reaching in their consequences. Despite many illusions to the contrary, there is in the real world of food politics much jostling for position and attempts to impose rules on others. Our concern here is that the outcome of these conflicts and the compromises that are hammered out in policy meetings constantly shape the food supply chain. The overriding framework of food policy in relation to health, humanity and society as a whole requires major re-working. It is the whole picture that is our concern.

The choices that we explore in this book can perhaps best be understood in three ways: first, there are a number of key battlegrounds in the Food Wars; second, the outcome of these global conflicts is of immense significance for human health, that is, of individuals, societies and environment, ie ecological health; third, new ways of thinking about the future of food suggest that a paradigm shift is underway.

If a paradigm is a set of assumptions from which new knowledge is generated, a way of seeing the world which shapes intellectual beliefs and actions, then science is a process, not an endgame of neutral fact-finding; it expresses values even in its facts. We use the term 'food paradigm' to indicate a set of shared understandings, common rules and ways of conceiving problems and solutions about food. A paradigm for us is an underlying, fundamental set of framing assumptions that shape the way a body of knowledge is thought of.

The term 'paradigm' is associated with the work of Thomas Kuhn, the philosopher of science who first popularized the term.[4] In fact, he merely built on a concept spelled out by Ludwig Wittgenstein (1889–1951), the Austrian philosopher. Kuhn took Wittgenstein's concept of paradigms and applied it to science as a process of making ideas: a set of 'universally recognised scientific achievements that for a time provide model problems and solutions to a community of practitioners'. Kuhn was interested in how scientific understanding went through momentous crisis points and what determined why one accepted framework of thinking fell by the wayside while another triumphed in its place. (For example, the work of Isaac Newton transformed how humans thought of the physical world; nearly three centuries later the new physics of relativity which Albert Einstein and others introduced created another paradigm shift which replaced the Newtonian worldview, transforming and superseding its tenets.) Kuhn himself was said to have used the term with at least 21 different shades of meaning,[5] and academics today use the term 'paradigm' more fluidly or metaphorically than even Kuhn originally intended.

The food system that developed rapidly after World War II exemplified a way of thinking that we call the Productionist paradigm: it remains the dominant worldview, but one which is now contested in respect of the future of food by newer 'models', chief amongst which are two emerging frameworks, which we call the Life Sciences Integrated paradigm and the Ecologically Integrated paradigm. Both are grounded in the science of biology, but each interprets biological and societal systems in ways that offer differing choices for our food future: how food is produced, who produces it and how it is sold; questions of social justice, where the food is produced (global versus local sourcing) and the place of food in human health. Figure 1.2 illustrates how we see the situation in the era of the Food Wars.

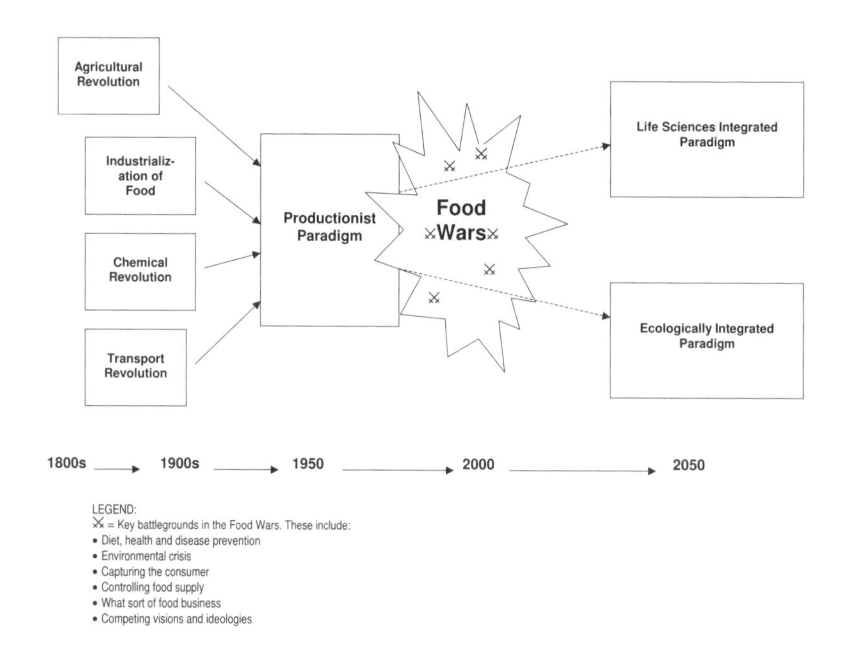

Figure 1.2 *The era of the Food Wars*

The Productionist Paradigm

Very powerful forces are lining up behind these new paradigms for food – but it should be realized that the existing Productionist paradigm still has influence – as is evidenced, for example, by the continued failure to fundamentally reform the European Union's Common Agricultural Policy. We argue that the status quo is no longer a viable option because the very methods it uses, such as animal husbandry, chemical inputs and its patterns of trade, are making it a poor policy option, no longer serving the public interest. We see the assumptions of this paradigm being challenged on many fronts and failing in credibility in significant areas.

The origins of the Productionist paradigm lie in the industrialization of food over the last 200 years and its concomitant advances in chemical, transport and agricultural technologies. Over this period food supply in many parts of the world has moved from often local, small-scale production to concentrated production and mass distribution of foodstuffs. Such a shift is a defining characteristic of the Productionist paradigm (even though it should be noted that much global food production is still local or regionally based).

With the arrival of industrialization and the explosion of urban populations in the last two centuries, social division of food became even more politically sensitive. Reliance on trade in food commodities, such as spices and sugar,[6] already considerable for some foods, grew; pressures to intensify production accelerated, increasing rural poverty: increased output from the land reduced the actual labour required on the land. The features of this agricultural revolution include: the increased use of inputs and of plant and animal breeding, the growth of fewer but larger farms, mechanization and a reliance on fossil fuels.[7, 8]

For us the Productionist paradigm goes far beyond the farm: it typified the whole 20th-century outlook – in particular from the 1930s onwards – in which the food supply chain became production-led in order to increase the quantity of food over other priorities. It developed a science base to further the goals of increasing output. Universities, colleges of agriculture, extension services and a whole panoply of support were gradually incorporated into this paradigm, which came to dominate food policy after the shortages and failures of the pre-World War II period. The production-driven model was built not just upon the agricultural revolution of the 18th century onwards (and of the chemical and transport revolutions too), but on the capacity of food processors to preserve, store and distribute food en masse.[9]

The triumph of the Productionist paradigm was cemented in the experience of mid-20th century starvation, food shortages, and maldistribution in many countries.[10] Throughout the world, governments created new national and international policies designed to increase production by applying large-scale industrial techniques that applied modern chemical, transportation, processing and farming technologies. The overarching goal of this paradigm was to increase output and efficiencies of labour and capital for increasingly urbanized populations. It is an irony that, while historically one of its policy goals was to increase national self-sufficiency and production, its surpluses are now being used to weaken the self-sufficiency policies of many developing countries who are being urged to open their local markets to global trade.

Now, half a century on from the consolidation of the Productionist paradigm, it is under strain and showing up major limitations. Although Productionism has been successful in raising production in line with an unprecedented rise in world population, 1.9 to 2.2 billion people in the world are estimated to remain directly or indirectly untouched by modern agricultural technology.[11] Health and environmental strains are threatening its survival: matters such as oil shortages, climate change, labour 'efficiencies', water depletion, pollution and public concern about animal welfare and the nature of plant and animal breeding. There is also a serious battle over who owns the food supply, not just in terms of companies (which we discuss in Chapter 4) but also of the intellectual property, even the genetic basis, of foods.[12]

To achieve its objectives, industrialized production historically focused on monocultures (single crops in a field) rather than diversity, but this created a reliance on artificial inputs (pesticides and fertilizers) and energy-intensive engineering both on and off the farm. The sustainability and profitability of the Productionist paradigm is now far from certain, with agribusiness and politics, as well as markets and consumers, now questioning how our food is produced.

Two New Paradigms of Food Supply?

The limitations of the Productionist paradigm manifest themselves in the increasing number of Food Wars over such issues as the health implications of chemicals used in production and the treatment of animals, through environmental damage and pollution and global trading practices to corporate control and power and the mistrust of consumers. But there are two strong alternative paradigms emerging, and we gather these under two conceptual frameworks: both offer human and environmental health benefits and both are grounded in new understanding of the science of biology.

Out of the Food Wars, therefore, we see emerging two possible science-informed visions for the future. They are competing paradigms for the future of food, both seeking to transform the Productionist paradigm. One is what we call the Life Sciences Integrated paradigm and the other the Ecologically Integrated paradigm. (Figure 1.2 on page 18 situates both paradigms historically in the context of the Food Wars.) Both derive from a common root: the argument that the 21st century will be the century of biology.[13] 'Bio' is now the language of innovation and represents a fundamental shift in the understanding of life: if the 20th century was characterized by the emergence of post-industrialization and 'information', then the 21st century promises to be the age of biological science. It is already giving rise to new controversies, for example over genetically modified foods and cloning. Languages in many tongues are being forced to inject new 'bio'-words into their lexicon: there is now bioprocessing, bioprospecting, bioprivacy, bioextinction, biodiversty, bioscience, bioinformatics, biovigilance, biosafety, bioterrrorism and, of course, biotechnologies.

In short, the future of the food economy will rely more upon the biological rather than the chemical sciences to deliver its vision for production, even though the chemical sciences will continue to play a prominent role in the medium term. As most of us know, a critical battle has been waged over the application of biotechnology to plants designed to resist specific chemical weed killers. The industrial nature of the Productionist paradigm is being softened and reshaped by new biological thinking.

The Life Sciences Integrated Paradigm

The Life Sciences Integrated paradigm describes the rapidly emerging scientific framework that is heralding the application of new biological technologies to food production. We propose this paradigm as a way of capturing a body of thought that has as its core a mechanistic and fairly medicalized interpretation of human and environmental health. In this, food is perceived as almost like a drug, a solution to diseased conditions, part of a planned, controllable and systemic manipulation of the determinants of health and ill health. This highly sophisticated thinking about food and health is at the heart of the application of biotechnology to food production, and its application on an industrial scale is at the core of the Life Sciences Integrated paradigm. A distinguishing characteristic is its reliance on biological rather than chemical sciences to deliver its vision for production. Techniques in biotechnology are already delivering many advances in food production methods, food handling and consumer products.

We should stress here that this paradigm means more than genetic modification (GM) alone, and includes the whole spectrum of biotechnology: that is, the use and manipulation of living materials in the manufacture and processing of foodstuffs. Enzymes, for example, are a key processing aid within biotechnology that receive only negligible publicity or adverse publicity.[14] But it is GM that has become the central defining characteristic of the Life Sciences Integrated paradigm and the focus of media, consumer, and policy attention. GM seeds and the chemical inputs they require are reshaping the biological base of agricultural production at a speed that is unprecedented in human food production. Despite the relatively crude state of the technology, GM is being introduced into world food systems at a rate that some see as irreversible. The long-term implications for agricultural environments, and for the structure and power relationships in the food chain, are unknown.

The novelty of the science – taking the genes from unrelated species and inserting them into another to forge a new plant or animal that would not be possible in nature (known technically as 'recombinant DNA biotechnology') – represents a revolutionary technological shift set to change the economics of whole industries. One of the attractions of the paradigm is that in many respects it relies on a simple re-interpretation of the existing Productionist paradigm but claims to remedy a number of its limitations: from lessening environmental impacts, through improving human health from greater food production, to creating new products with enhanced, yet often contested, health benefits.

From an agricultural point of view, the commercialization of crops through GM has so far been spectacular and is being heralded as the new Green Revolution, even though there are reservations about the technology.[15] As a result, the Life Sciences Integrated paradigm is well placed to become the dominant paradigm of the early 21st century. Plantings of GM crops have risen from zero in the mid-1990s to more than 50 million hectares planted worldwide by 2001, 68 per cent of all plantings taking place in the United States. In terms of global crop production, 40 per cent of soy, 7 per cent of maize, 20 per cent of cotton and 11 per cent of oilseed rape were given over to GM production as early as 2001.[16] For the GM seed companies, the technology is proving something of a bonanza, generating sales of US$3.67 billion in 2001 and forecast to grow more than 50 per cent to $5.57 billion by 2005.

Nutrigenomics is another line of research being pursued within the emerging Life Sciences Integrated paradigm and, to some extent, it typifies what that paradigm offers. Nutrigenomics seeks to understand how nutrition and particular dietary intakes interact with the structure and expression of genes and with genetic pre-potential. Why can one person eat a diet high in fats and not get cancer or heart disease, when another cannot?[17] If it were possible to unravel the interaction of genes, diet, ingredients and lifestyle, the promise of delivering an individualized or personalized approach to food and health might be realized.

Nutrigenomics is the application of the new understanding of genes and how they operate in plants, animals and microorganisms to help deliver that goal. Researchers try to unravel the mechanisms involved – which foods and which ingredients have an impact on which genes and which diseases. This would have been inconceivable without the completion of the mapping of the human genome by US and European geneticists.

Nutrigenomics promises a targeted fix to the diet and health policy problem, based on an acceptance that both micronutrients and macronutrients alter the metabolic programming of cells, and on an understanding of how diet is a key factor in disease.[18] The commercial as well as academic search is on for bio-active ingredients which could be exploited for health.[19,20] For instance, Guy Miller, head of Galileo Laboratories Inc, a US biotech company working in this area, has stated that:

by being able to elucidate genetic profiles of individuals, diets will be formulated from crop to fork to confer prevention or retard disease progression. As basic science advances converge with e-commerce, new opportunities will emerge to deliver to consumers, whose genetic susceptibility to specific diets and diseases are known, products tailored to individual dietary needs.[21]

Although researchers are attracting funds to this work, realizing health gains is probably some time off. Even if nutrigenomics does yield more precise understanding of the diet–gene–health connection, many observers consider that existing population dietary advice still stands. Even if some people are more likely to trigger degenerative diseases from eating a particular balance of nutrients, the population as a whole would benefit from attaining already known dietary goals such as restricting consumption of saturated and total fats and increasing intake of vitamins and trace elements from fruit and vegetables.

Nutrigenomics, argue the sceptics, may offer commercial wealth by selling to the 'worried well' and rich consumers, but it is probably of little relevance to global public health. Already concerns have been raised. One concern is that big business logic is outstripping the public's ability to make informed choices. Other ethical dilemmas relate to the problem of privacy and cost, where a nutrigenomic test costs US$400.[22] Nevertheless, nutrigenomics suits the more individualized policy approach of looking after one's own health. It says little about the need to alter the environment that reduces the chance of whole populations taking exercise or consuming a wholesome diet.[23]

The Life Sciences Integrated paradigm is already informing some key trends that are transforming food supply. It continues to work within monocultural production and its commercial structure is characterized by the concentration of large-scale production processes and agribusiness companies operating on a global scale. It is an unusual Food War in relation to health in that large companies not usually associated with humanitarian activism are now advocating that its technologies, particularly genetic modification, be rapidly implemented in order to feed the world.

GM biotechnology seems a near-perfect solution to the many agribusiness problems of the Productionist paradigm, and many experts in this area argue that GM plantings and scientific developments are the beginnings of a total revolution in production. The former chairman and chief executive officer of the leading biotechnology company Monsanto, Bob Shapiro, articulated his confidence in GM technology in a letter to shareholders in 1998: '. . . the previously separate domains of agriculture, nutrition and health should now be managed as an interconnected system. We use the term "life sciences" to describe that system.'[24] But the troubles that Monsanto faced from protestors in the late 1990s is a reminder of how demanding the battle for the hearts and minds of consumers can be: Monsanto was taken over by Pharmacia and Shapiro was retired as CEO. Although the science and technology that is potentially shaping the Life Sciences Integrated paradigm is at the cutting edge, equally powerful scientific arguments are being proffered by its critics who protest that it is little more than a modernization of the Productionist paradigm, with the same weaknesses and potential for damage. These concerns are what brought to greater prominence a politically much weaker and, until recently, highly fragmented body of scientific knowledge which we call the Ecologically Integrated paradigm.

The Life Sciences Integrated paradigm is being supported by considerable investment, mostly private but also public, and mostly in the US. Biotech research expenditure in 2002 in the US was €11.4 billion and €5 billion in the EU.[25] A European survey in 1999 reported that eight countries represented 83 per cent of the total funds spent on biotechnology. This state funding from the European Union and its members was €712 million for plant biotechnology and €674 million for animal biotechnology.[26] The corporate giants are tantalized at the prospect of biotechnology becoming the defining science for the 21st century, while a sizeable fraction of food capital – especially those companies which are closer to consumers – remains hesitant about investing in GM and is increasingly interested in the competing, but so far marginal, Ecologically Integrated paradigm. A key concern, not just for potential big-business supporters of either paradigm is that the Life Sciences Integrated paradigm, far from freeing the world from the agricultural treadmill and the commercial dependencies of the Productionist paradigm, might chain us to them. This is a battle that will ultimately be fought out in the stock exchange, in company boardrooms and at the supermarket shelves where consumers meet the food supply chain.

The Ecologically Integrated Paradigm

Since the 1930s, there has been persistent criticism of industrial agribusiness which argues that agriculture, nutrition and health are indeed interconnected and that it is workings of the Productionist paradigm that threatens that connection.[27] The moral high ground Monsanto was thinking it had discovered with its new 'life sciences' vision was already occupied by activists, policy-makers, nutritionists, environmentalists and ecological and biological scientists, not to mention tens of millions of consumers whose reflex response is to be suspicious of anyone who tampers with their food.

Ecological thinking on food is not new but remains mostly on the outside track of mainstream policy-making. For example, in his 1979 book *The Sane Alternative*, James Robertson outlined five scenarios for the future which could be applied to food and public policy.[28] These were: business as usual; disaster; totalitarian conservatism; hyper-expansionism; and sane/human/ecology. Trying to apply Robertson's five scenarios to food, Professor Joan Gussow argued that no meaningful proponent of food in public policy could argue a case for the promotion of 'disaster'. Further, since the 'business as usual' scenario would ultimately lead to ecological disasters, and since the track record of 'totalitarian conservatism' engenders mass resistance (often yielding some form of democracy), within any democratic public food policy there are only two meaningful scenarios to consider: namely, 'hyper-expansionism' (with the world of food being dominated by big business, biotechnology and factory farming, among others), or 'sane/human/ecology (which is the world of, among other things, small-scale production and low inputs).[29]

In the sense we use here, the Ecologically Integrated paradigm is also grounded firmly in the science of biology, but it takes a more integrative and less engineering approach to nature. Its core assumption recognizes mutual dependencies, symbiotic relationships and more subtle forms of manipulation, and it aims to preserve ecological diversity. It takes a more holistic view of health and society than the more 'medicalized' one of the Life Sciences paradigm.

In its thinking about agriculture, the Ecologically Integrated paradigm framework corresponds closely to the body of thinking described as agroecology.[30] Agroecology is gaining support among experts working with farmers in the developing world, but it also offers a new vision of food for the developed world. The world's poor farmers and citizens facing food crisis can rely only on self-reliance and small-scale farming; for them agroecological technologies offer one of the few viable alternatives. Cuba, for example, has become the global model of a successful case study of sustainable agriculture using agroecology technologies. Poor farmers in the developing world are synonymous with traditional, sustainable agriculture, but often poorly served by the top-down transfer-of-technology approach with its bias in favour of modern scientific knowledge, and its application of large-scale production methods. Agroecological methods, however, are re-discovering local skills and traditional knowledge, but applied with modern understanding to meet the challenges of food production. This is because a guiding principle of the Ecologically Integrated paradigm is that diverse natural communities are productive and should be supported.[31] A hurdle to overcome in this respect is the specificity of regional ecosystems and the need for specialist local knowledge. This paradigm therefore contrasts with the homogeneous technological packages characteristic of both the Productionist paradigm and the Life Sciences Integrated paradigm, relying upon bio-pesticides technologies to combat insect pests and develop resistant plant varieties and crop rotations; on microbial antagonists to combat plant pathogens and produce better rotations; and on cover cropping to suppress weeds, replacing synthetic fertilizers with bio-fertilizers. There is an increasing emphasis on skills and knowledge management in contrast to the single technician managing thousands of hectares on a 'recipe' basis; it would re-link the people with the land, encourage small-scale management units and return alienated farm workers to the land.

Agroecology is emerging as the discipline that provides the basic ecological principles of the study, design and management of agroecosytems with a view to productivity conserving natural resources, and to systems that are culturally sensitive, socially just and economically viable.[32] Such technologies include organic matter accumulation, nutrient cycling, soil biological activity, natural control mechanisms (disease suppression, biocontrol of insects and weeds), resource conservation and regeneration (to include soil, water and germplasm), general enhancement of agrobiodiversity and synergisms between components. The agroecology model, however, has yet to demonstrate its widespread applicability, not only in the developing, but also in the developed world. More work is needed clearly to demonstrate how the Ecologically Integrated, like the Life Sciences Integrated paradigm and the Productionist paradigm will offer viable alternatives for the future of food supply.

A key distinction is that within the Ecologically Integrated paradigm, thinking ecologically about health requires a consideration of the circumstances, experiences and dynamics of groups and populations.[33] Individual and population health depends upon the stocks of natural resources, the functioning of ecosystems and cohesive social relations. The Ecologically Integrated paradigm is integrative, not disintegrative: whereas the Life Sciences Integrated paradigm's vision for agriculture relies heavily on the laboratory, the Ecologically Integrated paradigm asks: why develop seeds modified to resist pesticides so that the company that owns the intellectual property to both can prosper? It claims that new regimes of mixed planting of crops can provide successful alternatives with fewer risks, either to the environment, to consumer acceptance or to human health.[34]

The Three Paradigms Summarized

The three paradigms, although explored throughout this book, are so central to our thinking and arguments that we now provide a summary of each: Table 1.1 of the Productionist paradigm (page 29), Table 1.2 of the Life Sciences paradigm (page 31), and Table 1.3 of the Ecologically Integrated paradigm (page 32).

All three paradigms share some features. Although we characterize the Productionist paradigm, for example, as more focused on quantity than quality, in fact all paradigms take a position on quality; Productionism, for instance, takes a cosmetic approach. Similarly, all paradigms assume some kind of market economy, and have concerns, for example about competition practices and control of markets. Throughout the food supply chain, adherents of each of the paradigms know the importance of macro-economic frameworks such as the rules for trade and the need to deliver food safety; the Productionist paradigm, for instance, was in part born out of evidence from the 1930s about the waste of food due to problems of storage and losses due to contamination, notably in the developing world.

And even today, while both the Ecologically Integrated and Life Sciences Integrated paradigms promote an environmental dimension, they differ on how they deliver it: how, for instance, to optimize use of resources such as land, energy, chemicals and water. Both paradigms see biology as central, but, whereas the latter looks to biology to control the relationship between food and health, the Ecologically Integrated paradigm views this position as biological reductionism; the Life Sciences Integrated Paradigm posits that food is to be (re)made applying the advances in biological science and the unity of thinking between chemistry,

Table 1.1 *Features of the Productionist paradigm*

Drivers	commitment to raise output; immediate gains sought through intensification
Key food sector	commodity markets; high-input agriculture; mass processing for mass markets
Industry approach	homogeneous products; pursuit of quantity and productivity (throughput) over quality
Scientific focus	chemistry + pharmaceuticals
Policy framework	largely set by agriculture ministries; reliance on subsidies
Consumer focus	cheapness; appearance of food; homogeneous products; convenience for women; assumes safety of foods
Market focus	national markets; emergence of consumer choice; shift to branding
Environmental assumptions	cheap energy for inputs and transport; limitless natural resources; monoculture; externalization of waste/pollution
Political support	historically strong but declining, as reflected in policy battle over subsidies
Role of knowledge	agroeconomists as important as scientists
Health approach	marginal interest; assumes that health gains follow from sufficiency of supply

biology, engineering and management control, whereas the Ecologically Integrated paradigm argues that biology has to be approached from a less controlling perspective: working *with* nature, rather than *on* it, is its ethos. Thus, a key distinction between the paradigms is not just what they propose in science but what forms of control and ownership they represent. For example, the paradigms differ in how they view commercial control over intellectual property and whether they appeal to the more controlling or the more democratic elements in society.

It should also be noted that the paradigms can be interpreted differently in political terms: early proponents of the Ecologically Integrated paradigm, for instance, were linked to both far right and more democratic 'left' political movements, the former interpreting 'nature' in authoritarian terms and arguing that hierarchies and top-down rules are essential to allow intrinsic values to be asserted. The left position, in contrast, argued that ecology needed champions to protect it from the depredations of advanced capitalism.[35] Although Tables 1.1, 1.2 and 1.3 (pages 29–32) highlight differences, it should be noted that all the paradigms have a strong health orientation; the role of the food supply chain is to deliver health. But as the rest of this book explores, what is meant by health, and what the determinants of health are, are matters of conjecture.

WHICH WILL DOMINATE?

There are many obstacles to the widespread adoption of health alternatives, the greatest presented by political–corporate power and vested interests. Yet at times there is a psychological barrier to believing that the alternatives can work. The Ecologically Integrated paradigm is currently the underdog in the contest for paradigmatic dominance but its chances have been significantly improved by the growth of environmental and consumer resistance to GM foods. By contrast, the reputations of the life sciences companies are being tarnished by their association with the economic, 'neo-liberal' rhetoric which argues that consumers should be given a choice; yet when consumers demand labelling of foods that contain GM ingredients, they are sometimes accused of being anti-science or Luddites. In practice, as has occurred with soya, the uptake of GM seed by farmers in the US has meant

Table 1.2 *Features of the Life Sciences Integrated paradigm*

Drivers	science-led integration of food supply chain; tight managerial control
Key food sector	capital-intensive use of Life Sciences (agrofood); food retailers dominate supply chain; reliance on intensive agriculture for economies of scale
Industry approach	aims for industrial-scale application of biotechnology primarily in agriculture but increasingly in manufacturing (enzymes not just GM); uses a mixture of chemical and biological inputs
Scientific focus	links genetics, biology, engineering, nutrition; control from laboratory to field and factory; science presented as neutral but tailored by industry-led/ oriented funding
Policy framework	top-down, expert-led; backed by trade and finance ministries; challenges regulatory, industry, policy and public boundaries
Consumer focus	production of 'champion' products (eg functional foods to appeal to individual choice); structured choice; food features can be designed to appeal to market-derived characteristics
Market focus	global ambitions; large companies dominate; 'Life Science' fix is the only mainstream business model
Environmental assumptions	intensive use of biological inputs; claims to deliver environmental health benefits
Political support	fast-developing; divisions among both rich and poor countries about how to interpret Life Sciences paradigm
Role of knowledge	top-down; expert-led; hi-tech skills; laboratory science base
Health approach	relies on novel but unproven impact; argues that health can be fixed technically by new combination of screening on an individualized basis; seeks to improve beneficial traits of crops for human health

that there is little choice: with ingredients derived from GM soya already in use by food processors, it is hard for even the most dedicated activist actually to avoid consuming GM-derived products.

Table 1.3 *Features of the Ecologically Integrated paradigm*

Drivers	environmental; energy/waste reduction; diversity 'ground upwards'; reduction of certain inputs; aims for diversity on and off the field; risk minimization by building diversity
Key food sector	integration of all; but emphasis on whole-farm systems approach (land and watersheds); biodiversity enhancement to stabilize and maximize yields over the long term
Industry approach	aims to move organic foods from marginal to mainstream; nervous about increasing the scale of production and capacity of quality controls; select use of biotechnology (fermentation, not GM)
Scientific focus	biology; ecology; multidisciplinary; agroecological technology instead of chemicals
Policy framework	partnership of ministries; collaborative institutional structures needed; promotes advantages of decentralization and team-work
Consumer focus	citizens not consumers; improved links between the land and consumption; greater transparency
Market focus	regional and local focus – 'bio-regionalism'; nervous about export-led agriculture; favours smaller companies but increasingly adopted by larger ones
Environmental assumptions	resources are finite; need to move away from extensive monoculture and reliance on fossil fuels; need to integrate environmental, nature and conservation policy with industrial and social policy
Political support	weak, but low base strengthening in many countries; some merging of fragmented 'movements' claiming high ground
Role of knowledge	knowledge-intensive, rather than input-intensive; skills needed across whole supply chain; knowledge as empowerment
Health approach	presents itself as 'healthy' alternative but as yet on a weak evidence base; promotes diet diversity

It would be wrong to dub the Ecologically Integrated paradigm as reactionary, anti-science or anti-big business: it offers a particular view of science, business and consumption, aiming to prove it can be 'big' in the sense of ensuring a viable food economy in developed world markets. In contrast, one of the strengths of the Life Sciences Integrated paradigm is that it has immense influence in the corridors of power and also among the decision-makers in many large food companies. It also builds on the structures of the Productionist paradigm which has proven remarkably successful in food output. Meanwhile, the Ecologically Integrated paradigm still remains marginal in mainstream food business and is often portrayed as quirky or backward-looking. But while this image may be changing due in part to its popularity with consumers and the media, its scientific basis is growing in stature and evidence in both developed and developing countries. This is the hard business logic for supporting ecological production and we predict increasing polarization between those companies tending towards the values and culture of the Ecologically Integrated paradigm and those favouring the Life Sciences Integrated paradigm. A key battleground is already the regulatory arena – which products can get approved, with what scientific evidence and credibility? – but there is also a battle ahead for access to public monies and political credibility to support the development of different paradigms.

Thus, the existing Productionist paradigm will be the inevitable opponent in protracted Food Wars because its methods have been demonstrated to be unsustainable and harmful to human and environmental health (see Chapters 2 and 6). Only two new contenders will battle it out for future dominance; both are science-based; both make claims to environmental and human health benefits. It is important to remember that Professor Thomas Kuhn, who coined the notion of paradigm shifts in science, acknowledged that two paradigms can coexist. This may be the case for food, in which case an ever-present danger is that there will be a polarized food supply shaped by competing paradigms. The tragedy would be if this supply polarized around choices available only for the food-rich at the expense of the food-poor.

One scenario is not a period of mutual tolerance between the paradigms but an era of serious conflict, with proponents seeing little middle ground. If the Life Sciences Integrated paradigm becomes well ensconced in the corridors of power, the Ecologically Integrated paradigm may have greater public acceptance (Chapters 4 and 5 explore the drivers of actual behaviour more closely), and vice versa.

The Place of Food and Health in the 'Paradigm' Framework

In the Productionist paradigm (Figure 1.3), health is portrayed as being enhanced, above all, by increasing production, which required investment in both monetary and scientific terms. Agriculture, the prophets of Productionism argued, deserved massive support if it was to move away from 'peasant', low-yield systems.[36] (This, incidentally, was their rationale for the now much-derided subsidy system throughout the West.) As long as food could be adequately and equitably distributed, health benefits would result. This Productionist view of health saw the main problems as under-consumption, under-production and poor distribution. The health goal of public policy, therefore, should be to increase production of key health-enhancing ingredients such as milk, meat, wheat, and other 'big' agricultural commodities.[37] Figure 1.3 shows how this policy relationship might connect inputs and outputs in health.

The health assumptions on which the Productionist paradigm was built were based on what today would be regarded as a very narrow understanding of nutrition and health. For example, the observation in the 1800s that animal protein aided human growth led to massive resources in countries such as the US and Europe being invested in the development of the dairy and meat industries.[38] The agricultural and agribusiness focus of the Productionist paradigm has also been weakened by the shift of power and finance down the food supply chain to the retailing, trade and consumer industries such as food service, where most of the money from food is now made (a feature spelled out in Chapter 4). In the US, for example, about half of all food expenditure is on consumption outside the home.

This change in consumption patterns suggests just how out of touch the Productionist paradigm is with health needs. Public policy responses from within the Productionist paradigm look

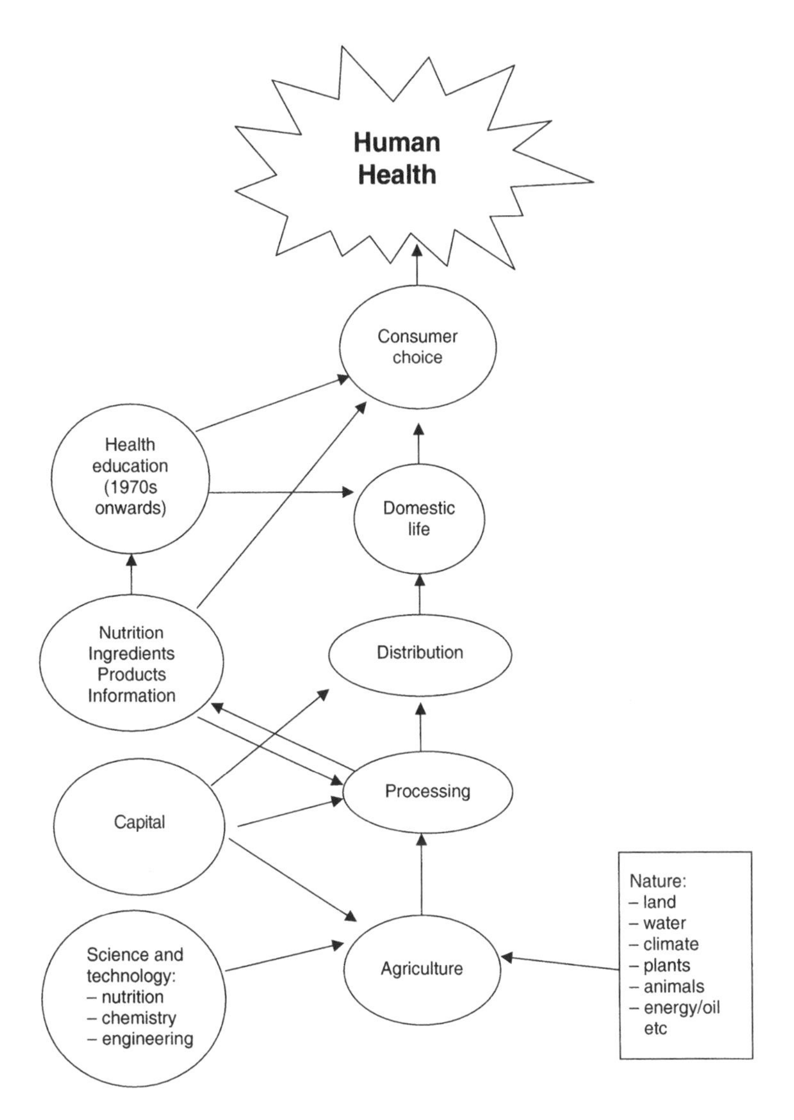

Figure 1.3 *Productionist approach to health (1950s to present, with 'health education' included post 1970s)*

increasingly like rearguard actions, when just half a century ago it promoted a proactive approach – policy intervention, new initiatives and new ways of farming. Today, the paradigm is increasingly reliant on drugs, special foods with purported health benefits and fringe crops to try to gain a foothold in 'health consumer' markets. No longer does it drive and structure the food supply chain in a way that makes sense for health. (We expand on this theory in Chapter 2.) Even in developing countries, the paradigm has problems. For example, the Green Revolution, while delivering more energy-rich macronutrients, has at the same time exacerbated rates of maternal anaemia and childhood deficiencies in iron, zinc and betacarotene because the higher-yielding strains of wheat and rice contain relatively fewer micronutrients.[39]

In addition, with support from World Bank and International Monetary Fund policies, the Productionists encouraged the production of cash crops in order to increase income. 'Health education' was bolted on in the face of growing evidence about the impact of diet on cardiovascular diseases.[40] Rather than rethink the paradigm, governments and corporate interests decided simply to put dietary warnings in place. In other words, the policy was to tell consumers to eat more sensibly and to look after themselves. They were bombarded with leaflets and public education programmes, with only mixed results. Other areas of health education, such as nutritional labelling, became hotly contested areas within food policy, rather than being a channel of information to alter the relationship between supply and demand. The paradigm's policy solutions thus took highly individualized approaches, and nutrition became an increasingly politically charged issue (this is discussed further in Chapter 3).[41, 42]

The Productionist paradigm encouraged a reductionist view of the relationship between food and health: that food's contribution to health comes from the 'right' ingestion of the 'correct' balance of ingredients. There is no good or bad food, according to this ethos, only good or bad diets. The onus of responsibility for diet-related health is thereby put on consumers – mainly through what they were expected to read and interpret from food labels. Health was defined as the absence of disease. If consumers wanted to improve their life-expectancy chances, they should eat healthy foods. The old Roman statement of consumer responsibility – *caveat emptor*: 'buyer beware' – prevailed.

We take issue with this 'production-led' position. As we show in Chapter 2, much of diet-related health, like health itself, is socially determined.[43] Yet although we have many reservations about the Productionist model – and note too that the failure to ensure equitable distribution of food is not entirely its fault – it has been spectacularly successful in its own key terms. Outputs and total yields have increased, albeit more in some developed areas of the world due to land ownership and macro-economic policies,[44] a welcome fact, given the world reality of too many people chasing too little food.[45] The world's population has doubled since the 1950s, yet the rate of hunger has dropped, notwithstanding that 800 million people still face daily hunger, but millions of new mouths have been fed by rising outputs and this has to be judged a major success.

But has the Productionist paradigm run its course? The current era of the Food Wars (see Figure 1.2 on p18) suggests global doubts about sustainability. Evidence is mounting as to its financial costs to the taxpayer; there are also costs in human and environmental health and, just as importantly, doubts that the current food system can deliver the requisite amount, range and quality of foods needed for 9 billion mouths predicted for the world by the mid-21st century.

Within our paradigm-based analysis, it is important to note that neither emerging paradigm replicates the elitist bias of the Productionist paradigm which favours wealthy individuals in developed nations over their poor counterparts in other parts of the world. A policy priority must be the capabilities of the working classes and the poor to feed their families.

The Life Sciences and Ecologically Integrated Paradigms' Approaches to Health

Increasing the food supply is still a critical policy concern, and one espoused in particular by the Life Sciences paradigm's approach to health, with its focus on individualized health (Figure 1.4). It offers an almost industrial model of health in that it promises the capacity to understand the constituent parts of disease and the human's capacity to fall prey to particular diseases, and then offers long-term personalized dietary solutions, such as nutrigenomics, implying a highly sophisticated understanding of the minutiae of the biological and genetic 'cogs' in the human 'machine'. The Life Sciences approach disaggregates the

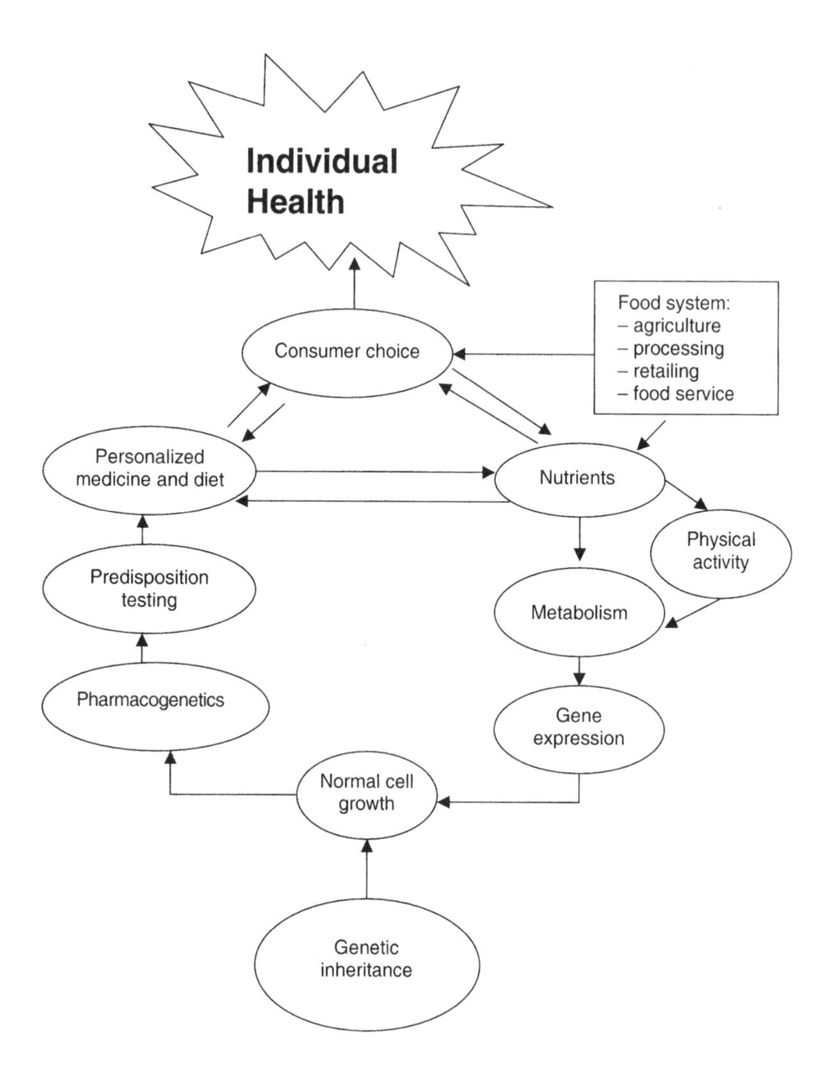

Figure 1.4 *Life Sciences Integrated approach to health*

complexities of the food–disease–health nexus into discrete parts and offers food or food-derived ingredients as potential aids: health is delivered by science.

A fundamental difference between the Productionist and the Life Sciences paradigms on one hand, and the Ecologically Integrated paradigm (Figure 1.5) on the other hand, is that the first two paradigms conceive of health as an outcome of a long process

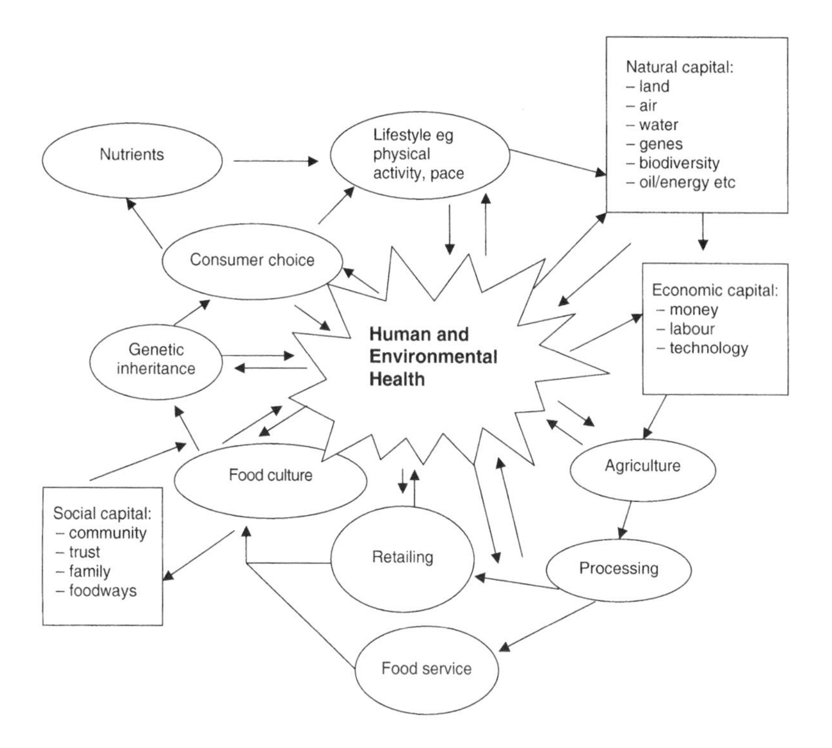

Figure 1.5 *Ecologically Integrated approach to health*

(the food chain or bio-food chain), while the latter conceives of health as something that is intrinsic to each stage of the growing and distribution process. Both emerging paradigms, we repeat, build on the new understanding of the centrality of biology to life.

The health approach of the Ecologically Integrated paradigm, as the name implies, is centred on ecology: understanding the working of systems and cycles that are characteristic of biological systems in nature. The emphasis is on process – notably feedback loops, cycles, symbiosis and interconnections. It proposes that the goal of food policy should be to understand these processes and to work with them, rather than to engineer, constrict or fragment them. For example, an Ecological approach sees monoculture, whether in the field or in diet, as anathema, whereas the other two paradigms see monoculture as a matter of business reality and efficiency – specialism being the route to enterprise.

Both alternative paradigms claim to deliver environmental and health benefits that they say are not being delivered by the Productionist paradigm: reduced chemical inputs, foods with better nutrient profiles, and food security. Both believe that the Productionist approach will be unable to deliver enough food for burgeoning world populations, or not without unbearable dislocation.

Where the paradigms differ, however, is on how to deliver that food increase and its assumed benefits to the general health. The Life Sciences paradigm is espoused by the agrochemical and pharmaceutical end of the food system while the Ecologically Integrated paradigm looks to learn from and modernize more traditional farming knowledge. Both embrace concepts of intellectual property, the Life Sciences through the patenting of genetic materials, the Ecologically Integrated paradigm by building on and modernizing knowledge built up over many generations. The Life Sciences paradigm argues that its technologies will deliver human and environmental health.

Ending the Food Wars through Policy and Evidence

In Figure 1.2 (page 18), the depiction of the era of the Food Wars features the symbol of the crossed swords used by historians and map-makers to designate the site of a battlefield.

The war analogy is apt: unnecessary millions of lives are being lost by the way that the food economy currently operates. The wars might go on sometimes behind closed doors in boardrooms as well as in parliaments and intergovernmental discourse, but the battles for supremacy are very real: lives hang on their outcome; people get hurt in the clashes. The Food Wars must come to an end, but when and with what result?

In wars, propaganda prevails. In the last decade, UK citizens, for instance, have become only too aware of how, when the sordid tale of bovine spongiform encephalopathy (BSE) or 'mad cow disease' unfolded, very powerful vested interests in the food chain – exporters, farmers and other interests – fought to present the risks as either being under control or not serious.[46, 47] As we write (in spring 2004), the death toll from the human variant of

BSE, variant Creutzfeldt-Jakob Disease (vCJD), is 139 in the UK,[48] with a handful of cases in other countries. Although this number has had huge political attention – and led to the restructuring of European food control systems and the creation of food agencies to attempt to recapture public confidence in the ability of governments to protect consumers – the huge death rate due to diet-related ill health – such as the 300,000 annual deaths in the US related to obesity – is somehow accepted as unavoidable.[49] Both forms of premature death and distress should be unacceptable.

Capturing the Consumer

A healthy diet – one low in saturated fats (eg dairy and meats) and sugars, and high in fruits and vegetables and a diversity of foods – also has benefits on environmental grounds. The trend towards the Westernized diet with its high content of meat and dairy, and a reliance on a narrow range of crops is warping and distorting ecology. The rhetoric of economic neo-liberalism has long accorded consumers primacy; they are said to be the keystone of why market economies deliver efficiencies. Yet in practice consumers are under-informed, sometimes patronized and heavily targeted by marketing and sponsorship. High calorific foods such as burgers and soft drinks are relatively cheap; labelling does not compensate for price signals. For example, nutritional labelling is mandatory in the European Union only once a health claim has been made. Choices – particularly by children – are easily influenced by heavy advertising and cause-related marketing.[50] Image is everything.

The focus on the consumption end of the chain food chain has led to a reinvention of industries dedicated to the marketing and branding of foodstuffs. Major food brands mostly emanating from the West have consolidated their position and are being promoted worldwide, entering Eastern and Southern markets and changing dietary patterns.[51, 52] However, in developed countries, health has become a marketing battleground with increasing awareness of the links between diet and disease in a rapidly ageing population. Consumers are demanding products with health benefits, a demand which the food industry now is scrabbling to satisfy (see Chapter 5).

The key issue here for policy is whether a proliferation of particular products with presumed health benefits is the right solution to diet-related ill health, as is often claimed by the food producers. After decades of being highly resistant to the evidence that our food has an impact on ill health,[53, 54] large sectors of the food industry now, ironically, see health as a potential growth area. At first sight, this is welcome but there is a real danger that this will either be a short-lived technical fix or merely create a new niche – 'health' – in already saturated markets.

Another fundamental concern is what sort of food culture is emerging. Many food traditions and common eating behaviours are altering rapidly in the face of new products, marketing and lifestyles. There is rampant 'burgerization', for one thing – the domination of US-style fast food, often washed down by soft drinks across the whole developed world. (Americans themselves now spend more per annum on fast food than on higher education, personal computers, computer software or new cars.)[55] In addition, marketing departments of food companies often celebrate their role in offering consumers an astonishing range of choice. But this choice comes at considerable environmental cost – long-distance trucking, excessive energy use, monocropping on intensive farms – as well as the health costs of obesity, diabetes and other degenerative diseases (see Chapter 6).

Evidence-based Policy?

Why is policy not being changed and what is stopping such change? Why does the Productionist paradigm have such a grip in the face of increasing weight of 'evidence' as to its shortcomings? Indeed, food policies and processes tend to be developed, not by evidence, but by political expediency and much more. If policy were based on evidence we would see, for example, immediate action utilizing all available policy levers to deliver a reduction in the incidence of heart disease and diet-related cancers.[56, 57] Instead, there have been decades of delay and obfuscation. The diet-related epidemics of heart disease, cancers, obesity and diabetes are now spreading to the South. Of deaths from all chronic diseases, 72 per cent occur in low and lower-middle income countries.[58] Approximately two-thirds of all cardiovascular disease deaths before the age of 70 years are now occurring in developing countries.[59] As noted, in the US alone 300,000 annual deaths are associated with obesity.[60] Worldwide, according to the WHO, there are more than 1 billion adults overweight and 300 million clinically obese,[61] leading to 3 million deaths annually from overweight and obesity.[62]

In addition to mortality, there is also morbidity – the loss of capacity due to suffering and days lost at work, representing considerable external costs. Yet for the last decade or so politicians and policy-makers have claimed that food safety is the main issue in food policy. In the European Union, for instance, 1.5 million premature deaths occur annually due to heart disease,[63] yet political attention within Member States and in the European Commission (the civil service) has been dominated by food safety, which in fact accounts for relatively few deaths. Obesity rates, by contrast, are rising alarmingly worldwide and the WHO now describes obesity as an epidemic.[64] The associated health costs have long been documented,[65, 66] yet leading politicians and world policy-makers hold to the notion that exercise and diet are solely issues of individual choice, as though the proliferation of foods based on saturated or hydrogenated fats, sugars and refined carbohydrates, and the sophisticated marketing techniques used to promote them, are not critical components of the health crisis.

The *evidence* about environmental damage is similarly overwhelming.[67] This includes falling water tables (especially in key areas of agricultural production), the deterioration of range lands which supply most of the world's animal protein; soil erosion and loss of top soil to grow crops, and the continuing destruction of, and damage to, croplands.[68] Above all, we will later point to the collapse of global fisheries and climate change.

Already, it should be clear that this book explores the problematic and often highly contentious relationship between evidence, policy and practice. It is often argued in modern medicine that practice should be based only on solid evidence. This so-called Cochrane 'gold standard' approach argues that health care should be based on an excellent quality of evidence. The Cochrane Collaboration is a network of 50 review groups around the world, made up of health specialists committed to systematic reviews of the effects of health-care intervention. Such reviews should be rigorous, and based on peer-reviewed journals (rather than the 'grey' literature that has not gone through the anonymous process of peer review), and take note of first-rate evidence, ideally from placebo-controlled double blind studies, and so on.[69] There are good grounds for arguing that a Cochrane-type approach could be applied to nutrition interventions.[70] The ideal relationship between policy and evidence requires an increase in the availability of evidence to which policy-makers have to listen.

In practice, the relationship between food policy and evidence falls into a number of discrete possibilities which include the following:

- Policy with evidence.
- Policy without evidence.
- Policy despite evidence.
- Policy burying evidence.
- Policy claiming evidence.
- Evidence searching for a policy.
- Evidence but no policy.
- Evidence despite or in the face of policy.
- Evidence in line with policy.

In our view there is enough evidence to know that the current post-World War II Productionist paradigm of food and farming is no longer credible. Its impact on health is sufficiently counterproductive for the judgement that its time is up to be safely delivered to policy-makers. Therein lies the credibility of the Life Sciences Integrated and Ecologically Integrated paradigms – neither would be so credible if it was not for the gaping human and environmental failures in the Productionist paradigm.

What is missing is a more innovative and organized health lobby to pursue and ensure a change in policy-making in relation to food and health. Figure 1.6 illustrates the breadth of thinking and policy input that is now required in the decision-making process and in the management and delivery of food policy. The ministries which affect food supply include health, trade, environment, agriculture and food, fisheries, consumer protection, development, foreign affairs and industry, but rarely is there a constructive dialogue across these ministries.

Why is there no reasoned, coordinated overview with rationally derived concerns to drive food and health policy? Is government controlling the food supply chain or is it the other way

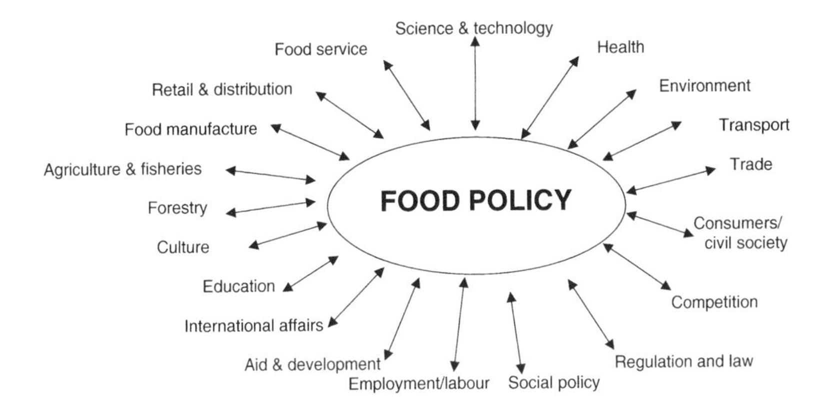

Figure 1.6 *The food policy web*

round? Is commerce framing public policy? Too often public policy-making is ad hoc or gets distracted by special pleas and particularities when what is needed is some long-term strategic thinking. Ironically, commercial interests do tend to take much longer-term strategic views. Governments come and go, and work to short-term horizons. The mismatch between public and commercial policy is particularly worrying. It is now time for reform. Long-term health must be the key driver of any local, national, regional and global food supply chain. The benchmark must be policy's impact upon positive human and environmental health outcomes. This will also require institutional reform (an issue which is taken up in Chapters 7 and 8).

One thing is already certain. Food policy is moving from a prime concern with raising productivity at any cost to addressing, at the very minimum, sustainable production and a more wary consumer response and food consciousness. There is a real danger that the Food Wars may themselves become the dominant ethos of the age, rather than the pursuit of resolution – a new peace. Food is highly sensitive and can be a weapon of control. Food's impact on health requires that it be produced, distributed and maintained in ways that do not threaten the ecosphere.

The public health movement – all those professions, institutions and organizations dedicated in principle to promoting better public health – has been marginalized in the debate about the future of food and on the direction that food policy could go, when it has much to contribute to the debate about the shape and purpose of the food chain. If public health affects everyone and everything, it ought to be central to the debate about food policy. There are some welcome signs that the proponents of the public health are stirring. New arguments and a new preparedness to stand up against the easy ideological route of individual choice and market forces is seen on occasions. The evidence mounts.[71] This is one good result from the sudden policy interest in obesity, after years in which it was ignored or downplayed.

Key questions for the future of food addressed in this book include:

- Who eats and what?
- Who controls access to food?
- How is food grown and processed?
- How is it traded and distributed?
- How is food supply regulated?
- Who and what shapes food policies?
- How is its impact on society and the environment addressed?

A new conception of health has to be at the heart of the reshaping of the food supply and this will require new imagination from all drivers of food policy.

Chapter 2

Diet and Health: Diseases and Food

'Let Reason rule in man, and he dares not trespass against his fellow-creature, but will do as he would be done unto. For Reason tells him, is thy neighbour hungry and naked today, do thou feed him and clothe him, it may be thy case tomorrow, and then he will be ready to help thee.'

Gerrard Winstanley, English Leveller, 1609–1676[1]

Core Arguments

The Productionist paradigm is critically flawed in respect of human health. Half a century ago it responded to issues then seen as critical but which now require radical revision. While successfully raising the caloric value of the world food supply, it has failed to address the issue of quality, and as a result, there is now a worldwide legacy of externalized ill-health costs. The world's human health profile is now very mixed. Within the same populations, in both developed and developing countries, there exists diet-related disease due both to under- and over-consumption. The pattern of diet that 30 years ago was associated with the affluent West is increasingly appearing in the developing countries, in a phenomenon known as the 'nutrition transition': while the incidence of certain diet-related diseases has decreased, such as heart disease in the West, others are increasing, particularly diabetes and obesity worldwide, and heart disease in the developing world. Massive global inequities in income and expectations contribute to this double burden of disease, and current policies are failing to address it.

Introduction

One of the key Food Wars is over the impact of the modern diet on human health. In the last quarter of the 20th century, nutrition moved from the sidelines of public health to being central to the marketing of foodstuffs, and major public health campaigns urged consumers to improve their diets.

This human health dimension is central to our critique of the Productionist paradigm in two respects. First, even though global food production has increased to meet caloric needs, its nutritional content may be less than desirable. Second, food distribution remains deficient: nearly a billion people remain malnourished. In this chapter, we explore the relationship between diet and the range of disease and illnesses that are associated with food choices. We discuss, too, the existence of gross inequalities within and between countries in the form of food poverty amidst food abundance and wealth.

In late 2002 and 2003, a wave of new public health reports reminded the world that diet is a major factor in the causes of death and morbidity. Although deeply unpalatable to some sections of the food industry, these reports were sober reminders of the enormity and scale of the public health crisis. The joint WHO and FAO's 2003 report on diet, nutrition and the prevention of chronic diseases drew attention to high prevalence of diseases which could be prevented by better nutrition, including:[2]

- obesity;
- diabetes;
- cardiovascular diseases;
- cancers;
- osteoporosis and bone fractures;
- dental disease.

Of course, these diseases are not solely exacerbated by poor diet but also by lack of physical activity. In truth, this report was only reiterating the story of nutrition's impact on public health that had been rehearsed for many years, and the evidence for which was judged to be remarkably sound, but as Dr Gro-Harlem Brundtland, then the Director-General of the WHO, stated in the report: 'What is new is that we are laying down the foundation for a global policy response.' To this end, the WHO set up an international consultation dialogue to prepare its global strategy on diet, physical activity and health, scheduled to be launched in 2004. By international agency standards, this relatively speedy shift from evidence to policy-making indicates the real urgency of the problem. The draft strategy was launched ahead of schedule in December 2003.[3]

Already by 2002, the WHO had produced a major review of the national burdens that such diseases cause. Of the top ten risk factors associated with non-communicable diseases, food and drink contribute to eight (with the two remaining – tobacco and unsafe sex – are not associated with diet and food intake):[4]

- blood pressure;
- cholesterol;
- underweight;
- fruit and vegetable intake;
- high body mass index;
- physical inactivity;
- alcohol;
- unsafe water, sanitation and hygiene.

In the 2003 World Cancer Report, the most comprehensive global examination of the disease to date, the WHO stated that cancer rates could further increase by 50 per cent to 15 million new cases in 2020.[5] To stem the rise of this toll, the WHO and the International Agency for Research on Cancer (the IARC) argued that three issues in particular need to be tackled:

- tobacco consumption (still the most important immediate avoidable risk to health);
- healthy lifestyle and diet, in particular the frequent consumption of fruit and vegetables and the taking of physical activity;
- early detection and screening of diseases to allow prevention and cure.

In addition to these UN reports, the International Association for the Study of Obesity (the IASO) revised its figures of the global obesity pandemic: it estimates that 1.7 billion people are overweight or obese, a 50 per cent increase on previous estimates. The IASO's International Obesity Task Force stated that the revised figures meant that most governments were simply ignoring one of the biggest risks to world population health.[6]

These reports testify to an extensive body of research and evidence from diverse sources around the world of the link between food availability, consumption styles and specific patterns of disease and illness. Table 2.1 confirms some of the diet-related causes of death throughout the world. Good health and longevity were intended to result from ensured sufficiency of supply; at the beginning of the 21st century, far from diet-related ill health being banished from the policy agenda, it appears to be experiencing a renewed crisis.

Under the old Productionist paradigm, the main focus was under-nutrition. Yet at the end of the 20th century, with diseases such as heart disease, cancers, diabetes and obesity rampant worldwide, not just in the affluent West, a new focus must be placed on diet and inappropriate eating. In this chapter, we begin to explore wider societal changes which impose progress in this regard through demographic shifts, maldistribution of and poor access to food, and spiralling health-care costs. These factors add weight to our argument that the Productionist paradigm is beyond its own sell-by date.

Policy-making is failing to address the causes of these food-related health problems and too often resorts to only palliative measures. This is partly because the Productionist paradigm's approach to health narrows the framework for considering alternative solutions: by being centred on striving to increase output, it has taken only a medicalized, rather than a socially determined, view of health.

Could the proponents of Productionism have anticipated the scale of these most recent health concerns? To some extent, they could not. Even excessive intake of fats as causing ill health might have been something of a shock for the Productionist paradigm, as it was almost heretical to argue that too much of a nutrient could be harmful to health.[9] Part of the problem here was the essential paternalism of the paradigm which assumed total knowledge of all variables needed to make good food policy; governments and companies could be trusted to look after the public health; the consumers' role was to select products to create their own balanced diets. Recent history, however, has shown that governments and the food supply chain failed to adapt to new scientific knowledge in relation to food and health. Nationally and internationally, the influence of health scientists on public policy has been minimal. Consumerism triumphed.

Table 2.1 *Some major diet-related diseases*

Problem	Extent/comment
Low birthweights	30 million infants born in developing countries each year with low birthweight: by 2000, 11.9 per cent of all newborns in developing countries (11.7 million infants)
Child under-nutrition	150 million underweight pre-school children: in 2000, 32.5 per cent of children under 5 years in developing countries stunted, amounting to 182 million pre-school children. Problem linked to mental impairment. Vitamin A deficiency affects 140–250 million schoolchildren; in 1995 11.6 million deaths among children under 5 years old in developing countries.
Anaemia	Prevalent in schoolchildren; maternal anaemia pandemic in some countries.
Adult chronic diseases	These include adult-onset diabetes, heart disease and hypertension, all accentuated by early childhood under-nutrition.
Obesity	A risk factor for some chronic diseases (see above), especially adult-onset diabetes.[7] Overweight rising rapidly in all regions of the world.
Underweight	In 2000, an estimated 26.7 per cent of pre-school children in developing countries.
Infectious diseases	Still the world's major killers but incidence worsened by poor nutrition; particularly affects developing countries.
Vitamin A deficiency	Severe vitamin A deficiency on the decline in all regions, but sub-clinical vitamin A deficiency still affects between 140 and 250 million pre-school children in developing countries, and is associated with high rates of morbidity and mortality.

Source: adapted from ACC/SCN 2000[8]

To some extent, too, the public health world has colluded in its own marginalization from 'live' policy-making by its fixation with deficiency diseases: for example, on programmes of food fortification or on protein shortages which could be made up by increased meat and dairy production. Despite a successful worldwide campaign to increase intake of folic acid following the discovery of its connection with spina bifida (neural tube defect syndrome), the overall impact of nutritional science in policy-making has been negligible. Its response to the current epidemic of heart disease has been 'health education' – advice, leaflets and exhortations to change behaviour – explaining it as caused by modern lifestyles, rather than by preventable dietary deficiencies.

Almost as soon as the Productionist paradigm was put in place worldwide in the last half of the 20th century, global campaigns were needed to address the increase in degenerative diseases. However, the necessary policy instruments were not in place to tackle the health impact of long-term shifts in diet. The UN bodies which noted the evidence of new patterns of ill health were merely intergovernmental bodies who lacked any administrative power and influence to act on the global and national level. Commercial interests, on the other hand, had no such limits and could pursue their global ambitions, selling foods and a lifestyle around the world without regard to their consequences, and being able to defend their actions as being in the public interest.

Instead, the developed world now must confront one of the most challenging food and health disasters ever to face humankind: an epidemic of obesity with little prospect of an end in sight and the prospect of a new wave of diet-related diseases in its wake. It has little in its armoury with which to combat the causes of obesity, now affecting significant numbers of children and with even graver implications for future population health. Health education is ineffective; consumerism is part of the problem, but politically it is nearly sacrosanct.

Meanwhile, hunger and insufficiency continue, ironically, to prevail. As a 1995 FAO review of the global picture starkly put it: '[H]unger . . . persists in developing countries at a time when global food production has evolved to a stage when sufficient food is produced to meet the needs of every person on the planet.'[10] Over-consumption and under-consumption coexist. There is gross inequality of global distribution and availability of food energy. The same review asserted that Western Europe, for example, has in theory 3500 kcalories available per person per day and North America has 3600, while sub-Saharan Africa has 2100 and India has 2200. By 2015 the FAO calculates that 6 per cent of the world's population (462 million people) will be living in countries with under 2200 kcalories available per person. And by 2030, in the most optimistic scenario, in sub-Saharan Africa 15 per cent of the population will be under-nourished. Numbers of the under-nourished look set only to decline much more slowly than suggested by targets, for example those of the World Food Summit of 1996.[11]

The transnational nature of these patterns of diet-related disease demands public policy attention. The enormity of this human health problem cannot be over-emphasized. Diseases associated with deficient diet account for 60 per cent of years of life lost in the established market economies.[12]

Scientists categorize diseases into two broad groups: communicable (carried from person to person or via some intermediary factor; these include diseases such as malaria, food poisoning, SARS); non-communicable (acquired by lifestyle or other mismatch between humans and their environment, such as cardiovascular disease and cancers). Figure 2.1 indicates that in the developed world, deaths through infectious and parasitic diseases are very low compared to developing countries, while diet-related non-communicable diseases like coronary heart disease (CHD) and cancers are high in both developed and developing worlds. Degenerative disease rates are already high in the developing world. Figure 2.2 gives the leading causes of mortality by age to give another view of the global disease patterns.

The WHO and the FAO reports stress that world health in general is in transition with non-communicable diseases now taking a higher toll than communicable diseases. Figure 2.3 shows the WHO prognosis of how the rates of non-communicable disease are expected to rise. Factors in this health transition include diet, demographic change (such as an ageing population) and cultural factors related to globalization.

The Nutrition Transition

In a series of papers, Professor Barry Popkin and his colleagues have argued that there is what they term a 'nutrition transition' occurring in the developing world, associated primarily with rising wealth.[13,14] The thesis, which has been extensively sup-

	Africa	Western Pacific	Europe	The Americas	Eastern Mediterranean	South-East Asia	Total world-wide
Infectious & parasitic diseases	5787	794	212	394	959	2968	11,114
Cardiovascular diseases	1136	3817	4857	1927	1080	3911	16,728
Cancers	410	2315	1822	1115	272	1160	7094
Respiratory infections*	1071	511	273	228	365	1393	3841
Perinatal and maternal causes	585	371	69	192	371	1183	2771
Injuries	747	1231	803	540	391	1267	4979

* This does NOT include respiratory diseases; includes upper and lower respiratory infections and otitis media

Source: WHO, *Shaping the Future*, World Health Report, Geneva, 2003, calculated from Annex Table 2

Figure 2.1 *Number of deaths by WHO regions, estimates for 2002 (thousands)*

ported by country and regional studies,[15] argues simply that diet-related ill health previously associated with the affluent West is now becoming increasingly manifest in developing countries.[16,17] The 'nutrition transition' suggests shifts in diet from one pattern to another: for example, from a restricted diet to one that is high in saturated fat, sugar and refined foods, and low in fibre. This transition is associated with two other historic processes of change: the demographic and epidemiological transitions. Demographically, world populations have shifted from patterns of high fertility and high mortality to patterns of low fertility and low mortality. In the epidemiological transition, there is a shift from a pattern of disease characterized by infections, malnutrition and episodic famine to a pattern of disease with a high rate of the chronic and degenerative diseases. This change of disease pattern is associated with a shift from rural to urban and industrial lifestyle.

WHO researchers have noted that changes in dietary patterns can be driven not just by rising income and affluence but also by the immiseration that accompanies others' rising wealth;[18] low-

Leading causes of mortality by age

Adults, 2002

15–59		60 and over	
2279	HIV/AIDS	5823	Ischaemic heart disease
1331	Ischaemic heart disease	4692	Cerebrovascular disease
1037	Tuberculosis	2399	Chronic obstruc pulmonary disease
811	Road traffic accidents	1398	Lower respiratory infections
783	Cerebrovascular disease	929	Trachea, bronchus, lung cancers
672	Self-inflicted injuries	754	Diabetes mellitus
475	Violence	735	Hypertensive heart disease
382	Cirrhosis of the liver	606	Stomach cancer
352	Lower respiratory infections	496	Tuberculosis
343	Chronic obstruc pulmonary disease	478	Colon and rectal cancers

Source: WHO, *World Health Report 2003*

Figure 2.2 *Leading causes of mortality, by age, 2002*

income countries are experiencing the effects of the transition but cannot afford to deal with them.[19] Popkin argues that, while the nutrition transition brings greater variety of foods to people who previously had narrow diets, the resulting health problems from the shift in diet should not be traded off against the culinary and experiential gains. Consumers might enjoy the new variety of foods that greater wealth offers but they are often unaware of the risk of disease that can follow. The implications of the nutrition transition now ought to exercise the minds of global as well national policy-makers: certainly health policy specialists are concerned at the rise of degenerative diseases in low- and middle-income countries.[20, 21]

Nutrition may have recently become a key notion in modern dietary thinking but it only echoes the insights of an earlier generation of researchers which included nutrition and public health pioneers such as Professors Trowell and Burkitt, whose observations from the 1950s to the 1980s led them to question 'whether Western influence in Africa, Asia, Central and South America and the Far East is unnecessarily imposing our diseases on other populations who are presently relatively free of them'.[22] Trowell and Burkitt, both with long medical experience in Africa,

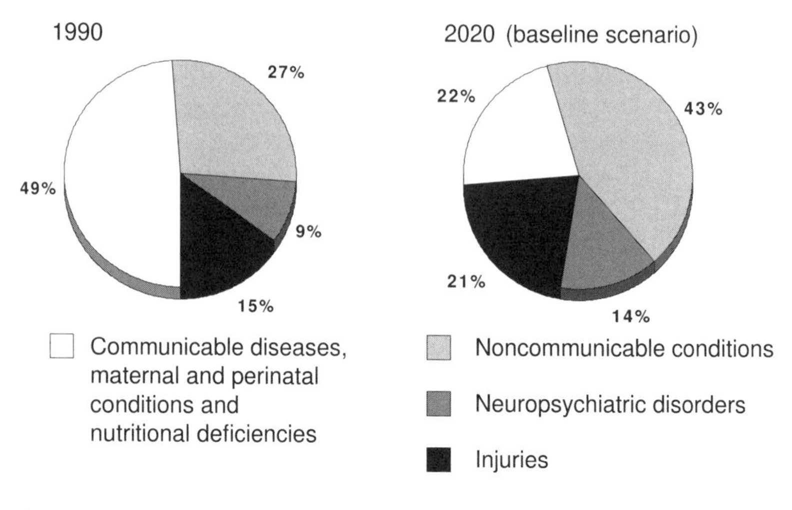

Source: WHO, *Evidence, Information and Policy,* 2000

Figure 2.3 *Anticipated shift in global burden of disease 1990–2020, by disease group in developing countries (WHO)*

could easily explain the variation in infectious diseases, but not the variation in rates of non-infectious diseases such as heart disease between countries at different economic levels of wealth and development. In Africa in the post-World War II period, they witnessed the rise of key indicators for diseases such as heart disease and high blood pressure in peoples who had previously had little experience of them.[23] The dietary transition is swift. An FAO study of very under-nourished Chinese people (living on 1480 kcalories per day) shows that they derive three-quarters of their energy intake from starchy staples such as rice, while better-fed Chinese (living on 2500 kcal per day) are able to reduce their energy intake from such staples and to diversify their food sources (see Figures 2.4 and 2.5 which compare the diets of under-nourished and well-nourished people in China).

Popkin has shown how this same process occurs in both urban and rural populations in developing countries with rising incomes. Figures 2.6 and 2.7 show the relationship between per capita income and what predominantly rural and predominantly urban populations eat as both get wealthier:[24] both eat more meats and fats, and reduce carbohydrate, as a proportion of their

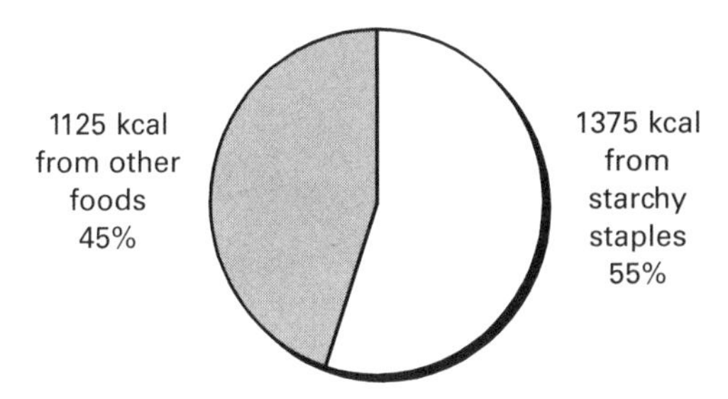

Source: National Survey of Income and Expenditure of Urban Households, Government of China, 1990; FAO, *State of Food Insecurity*, 2000, http://www.fao.org DOCREP/X8200E/x8200e03.htm#P0_0

Figure 2.4 *Diet of a well-nourished Chinese adult (2500 kcal/person/day)*

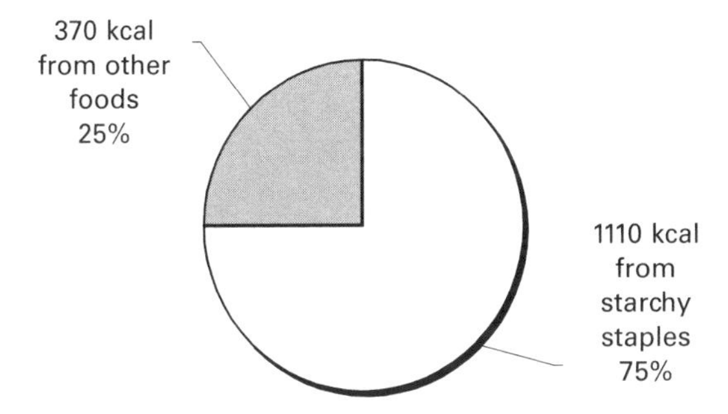

Source: National Survey of Income and Expenditure of Urban Households, Government of China, 1990; http://www.fao.org/DOCREP/X8200E/X8200e03.htm#PO_O

Figure 2.5 *Diet of an under-nourished Chinese adult (1480 kcal/person/day)*

overall diet. But there still remain differences between urban and rural populations, probably due to their different levels of activity, access to dietary ingredients and cultural mores.[25] The more urban population also consumes more added sugars as it gets wealthier, whereas the rural population consumes less. Popkin and his colleagues' point is that changing economic circumstances markedly shape the mix of nutrients in the diet and that lifestyle factors – such as the degree of urbanization[26] and changing labour patterns – have a major effect on health.

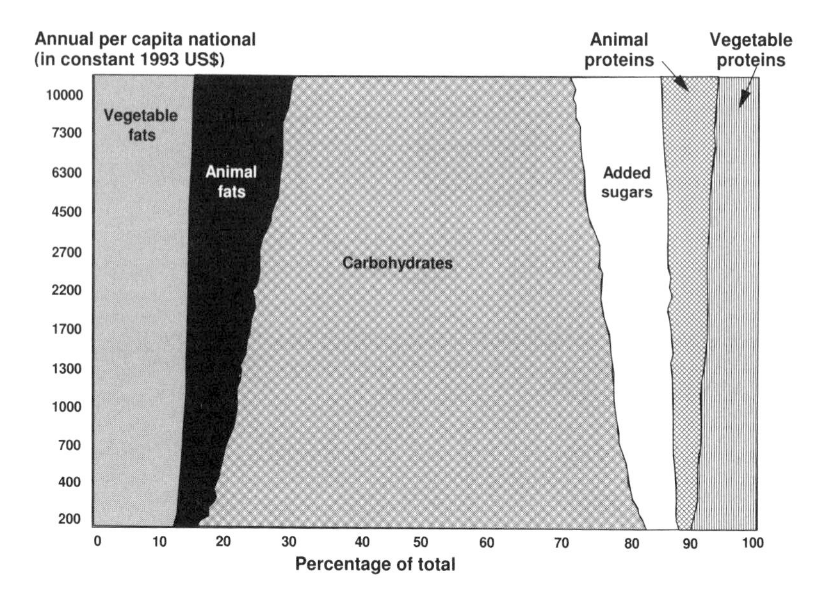

Source: FAO/World Bank/Popkin, B (1998) 'The Nutrition Transition and its health implications on lower income countries', *Public Health Nutrition*, 1, 5–21

Figure 2.6 *Relationship between the proportion of energy from each food source and GNP per capita, with the proportion of the urban population at 25 per cent, 1990*

The transition is occurring in areas that usually receive little food policy attention. A study by the WHO has reported that in the Middle East changing diets and lifestyles are now resulting in changing patterns of both mortality and morbidity there too.[27] Dietary and health changes can be rapid. In Saudi Arabia, for instance, meat consumption doubled and fat consumption tripled between the mid-1970s and the early 1990s; in Jordan, there has, in the same timescale, been a sharp rise in deaths from cardiovascular disease. These problems compound older Middle-Eastern health problems such as protein-energy malnutrition, especially among children. In China, the national health profile began to follow a more Western pattern of diet-related disease as the population gradually urbanized,[28] coinciding with an increase in degenerative diseases. Consumption of legumes such as soyabean was replaced by animal protein in the form of meat. One expert nutritional review of this problem concluded that

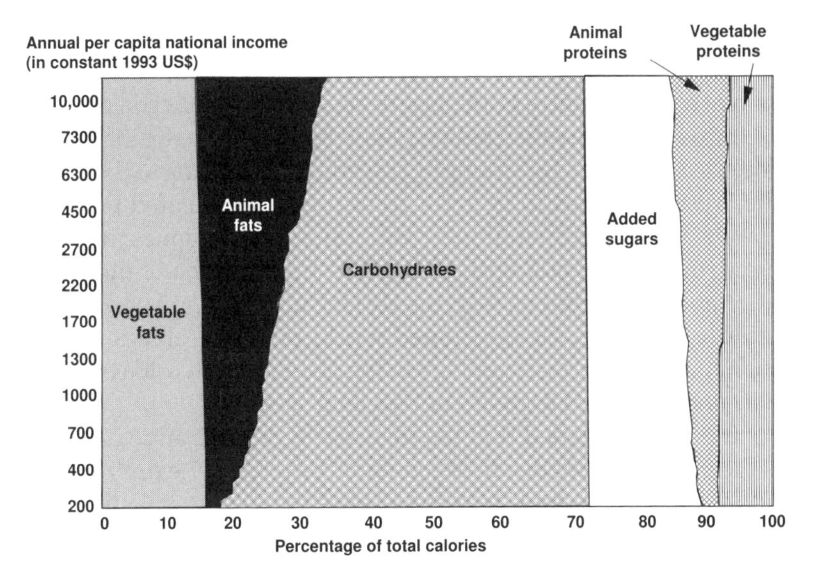

Source: FAO/World Bank/Popkin, B (1998) 'The Nutrition Transition and its health implications on lower income countries', *Public Health Nutrition*, 1, 5–21

Figure 2.7 *Relationship between the proportion of energy from each food source and GNP per capita, with the proportion of the urban population at 75 per cent, 1990*

exhorting the Chinese people to consume more soy when they were voting with their purses to eat more meat would be ineffective 'in the context of an increasingly free and global market'.[29] Such studies can suggest that the battle to prevent Western diseases in the developing world appears already to have been lost. If the nutrition transition is weakening health in China, the world's most populous and fastest economically growing nation, which has 22 per cent of the world's population but only 7 per cent of its land, what chance is there for diet-related health improvements throughout the developing world?

As populations become richer, they substitute cereal foods for higher-value protein foods such as milk, dairy products and meat, increased consumption of which is associated with Westernization of ill health. Relatively better-off populations also consume a greater number of non-staple foods and have a more varied, if not healthier, diet.[30] Thus we have the modern nutritional paradox: in the same low-income country there may be ill health caused by both malnutrition and over-nutrition; in the same rural area of a poor country both obesity and underweight can coexist.

In policy terms the challenge is whether India, China, Latin America or Africa, for example, can afford the technical fixes that the West can resort to in order to improve diet-related health:[31] coronary by-pass operations; continuous drug regimes; expensive drugs and foods with presumed health-enhancing benefits;[32] and subscriptions to gyms and leisure centres. The affluent middle classes in the developed world might be able to afford such fixes but the vast numbers in the developing world certainly will not. Technical fixes are not societal solutions.

It could be argued that the increase in degenerative diseases is the inevitable downside of economic progress. The problem for policy-making is how to differentiate between protecting the already protective elements of traditional, indigenous diets such as legumes, fruit and vegetables, and opening up more varied food markets, which is deemed to be good economic policy. In practice, too few policy-makers in the developing world have been prepared to fight to keep 'good' elements of national and local diets or to constrain the flow of Western-style foods and drinks into their countries lest they infringe support for trade liberalization. Thus, in stark terms, trade and economic policies have triumphed over health interests. US-style fast foods – the 'burgerization' of food cultures – have been hailed as modernity. We must now expose the production, marketing and prices of fast food,[33, 34] their nutritional value and their impact on health.[35]

Three Categories of Malnutrition: Underfed, Overfed and Badly Fed

More than 2 billion people in the world today have their lives blighted by nutritional inadequacy. On one hand, half of this number do not have enough to eat; on the other hand, a growing army of people exhibit the symptoms of overfeeding and obesity. In both cases, the international communities are floundering for solutions, and malnutrition results, as indicated by the following table.

Table 2.2 *Types and effects of malnutrition*

Type of malnutrition	Nutritional effect	No. of people affected globally (× billion)
Hunger	deficiency of calories and protein	at least 1.2
Micronutrient deficiency	deficiency of vitamins and minerals	2.0–3.5
Over-consumption	excess of calories, often accompanied by deficiency of vitamins and minerals	at least 1.2–1.7

Source: Gardner and Halweil (2000), based on WHO, IFPRI, ACC/SCN data[36]

Figure 2.8 highlights the role of the mother in infant health. Even before conception, the mother's own nutrition is vital.[37] It is now understood that children who are born with a low birthweight are at increased risk of developing heart disease and that good nourishment of the foetus is key. That nutrition affects disease patterns and life expectancy is now well documented.[38]

One of the particularly tragic consequences of under-nourishment is its impact on the world's children. UNICEF calculates that 800 million children worldwide suffer malnutrition at any given time (Figure 2.9 gives the FAO's estimated locations of these millions. Table 2.3 then gives the sobering projections for 2015 and 2030.) High proportions of Asian and African mothers are under-nourished, largely due to seasonal food shortages, especially in Africa. About 243 million adults in developing countries are deemed to be severely under-nourished.[39] This type of adult under-nutrition can impair work capacity and lower resistance to infection.

Against a rapid growth in world population, well-informed observers agree that greater food production is needed for the future.[40, 41, 42, 43] One estimate suggests that by 2020 there will be 1 billion young people growing up with impaired mental development due to poor nutrition. At a conservative estimate, this means there will 40 million young people added to the total each year.[44]

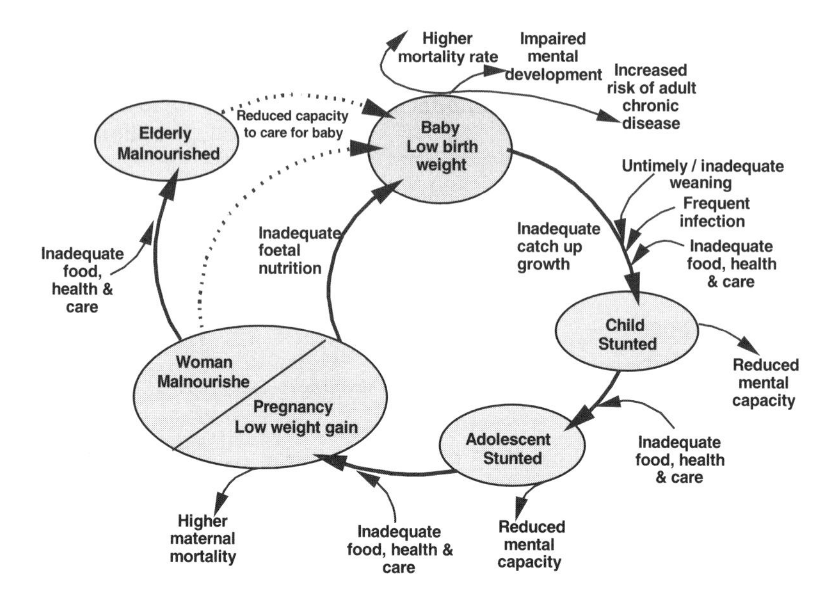

Source: ACC/SCN (2000) *Nutrition through the Life Cycle: 4th Report on the World Nutrition Situation*, New York: UN Administrative Committee on Co-ordination, Sub-committee on Nutrition, p8

Figure 2.8 *Life cycle – the proposed causal links*

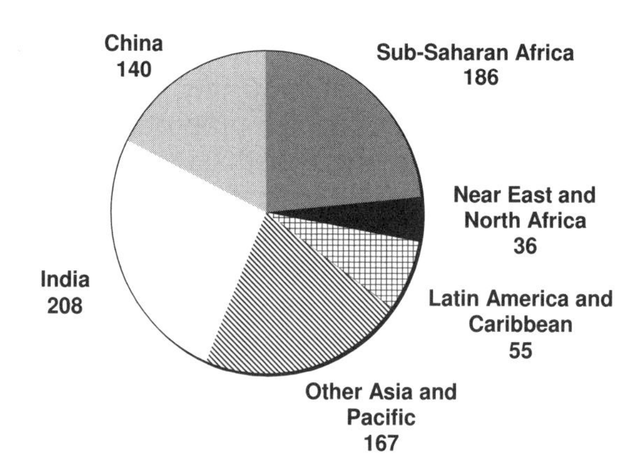

Source: State of Food Insecurity in the World 2000, http://www.fao.org/DOCREP/X8200E/x8200e03.htm#P0_0

Figure 2.9 *Number of under-nourished by region, 1996–1998, millions*

Table 2.3 *Projected trends in under-nourishment by region, 1996–2030*

	1996–98	2015	2030	1996–98	2015	2030
	Per cent of population			Millions of people		
Sub-Saharan Africa	34	22	15	186	184	165
Near East/North Africa	10	8	6	36	38	35
Latin America and the Caribbean	11	7	5	55	45	32
China and India	16	7	3	348	195	98
Other Asia	19	10	5	166	114	70
Developing countries	18	10	6	791	576	400

Source: FAO (2000) *Agriculture: Towards 2015/30*, Technical Interim Report, April, Rome, FAO, www.fao.org

The Obesity Epidemic

As early as 1948, there were medical international groups researching the incidence of obesity in various countries.[45] There were official reports at country level by the early 1980s,[46] and there has also been a commercial and consumer response to obesity for even longer.[47] But the grip of international obesity was in fact confirmed by the WHO's Task Force on Obesity in 1998. Today, overweight and obesity are key risk factors for chronic and non-communicable diseases.[48] In developing countries obesity is more common amongst people of higher socio-economic status and in those living in urban communities. In more affluent countries, it is associated with lower socio-economic status, especially amongst women and rural communities.[49] Historically and biologically, weight gain and fat storage have been indicators of health and prosperity. Only the rich could afford to get fat. By 2000, the WHO was expressing alarm that more than 300 million people were defined as obese, with 750 million overweight, ie pre-obese: over a billion people deemed overweight or obese globally.[50] But by 2003, this figure had been radically revised upwards when the International Association for the Study of Obesity (the IASO) calculated that up to 1.7 billion people were now overweight or obese. The new figures were in part due to more accurate statistics but also to the recalculation of obesity benchmarks, which acknowledged rising obesity in Asia.[51]

Particularly worrying is that extreme degrees of obesity are rising even faster than the overall epidemic: in 2003, 6.3 per cent of US women, that is one in sixteen, were morbidly obese, with a body mass index of 40 or more.

Obesity is defined as an excessively high amount of body fat or adipose tissue in relation to lean body mass. Standards can be determined in several ways, notably by calculating population averages or by a mathematical formula known as 'body mass index' (BMI),[52] a simple index of weight-for-height: a person's weight (in kilos) divided by the square of the height in metres (kg/m^2). BMI provides, in the words of the WHO, 'the most useful, albeit crude, population-level measure of obesity'. A personal BMI of between 25 and 29.9 is considered overweight; 'obesity' means a BMI of 30 and above; a personal BMI of less than 17 is considered underweight. There is some argument about whether the definition of overweight (a BMI within the 25–29.9 range) should be lowered from 25 to 23, in which case tens of millions more people would be considered overweight, and such an unofficial re-classification has led to the disparity between current world obesity figures.

BMI levels are a useful predictor of risk from degenerative diseases. Unutilized food energy is stored as fat. Currently, the US National Institutes of Health consider that all adults (aged 18 years or older) who have a BMI of 25 or more are at risk from premature death and disability as a consequence of overweight and obesity.[53] Men are at risk who have a waist measurement greater than 40 inches (102 cm); women are at risk who have a waist measurement greater than 35 inches (88 cm). Whilst height is obviously also taken into consideration, we should regard these measurements as key health benchmarks.

Figure 2.10 shows how, in a remarkably short time, the rate of obesity within countries is rising. In the UK, for instance, between 1980 and 2000, obesity trebled from 7 per cent of the population to 21 per cent.[54] Particularly alarming is that the 'North Americanization' of obesity is spreading down Latin America.[55] Figure 2.11 illustrates the level of obesity in comparison to underweight in developed and developing countries. In many countries, levels of obesity are double what they were 15 years ago.[56] In Peru, Tunisia, Colombia, Brazil, Costa Rica, Cuba, Chile, Mexico and Ghana, for example, overweight adults outnumber those who are thin. Even Ethiopia and India, tradition-

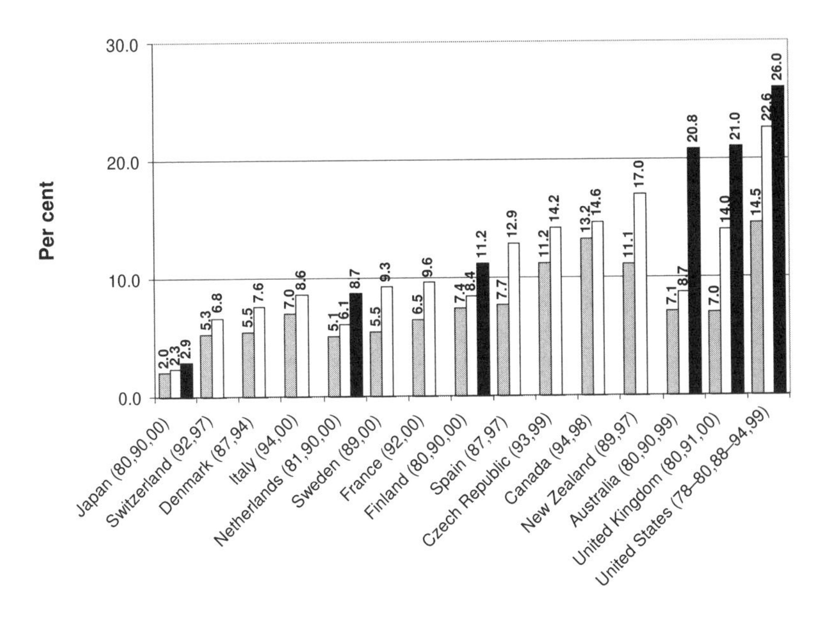

Source: OECD Health Data 2002, http://www.oecd.org/pdf/M00031000/M00031130.pdf

Figure 2.10 *Obesity in adult population across OECD countries*

ally beset by under-nutrition[57] and starvation now have the added burden of an emerging obesity problem. The trend to obesity is occurring in countries with different economic profiles, from the Asian 'Tiger' economies to the oil-rich Middle East.[58]

Rising obesity rates among children are particularly troubling to health professionals, as this trend suggests massive problems of degenerative disease for the future. In Jamaica and Chile, for instance, one in ten children is obese; in Japan, a country historically with a very low incidence of fat in its diet and with a low incidence of obesity, the frequency of obesity in school children has increased from 5 per cent to 9 per cent for girls and 10 per cent for boys in 1996.[59] (Table 2.4 summarizes the rapid rise in obesity as measured by comparing initial surveys with follow-up worldwide studies. The final column of the table shows how obesity is becoming out of control in developed and developing countries alike.) Even in Australia, obesity rose 3.4-fold for boys and 4.6-fold for girls between 1985 and 1995; in Egypt, 3.9-fold between 1978 and 1996; in Morocco, 2.5-fold in just five years, from 1987

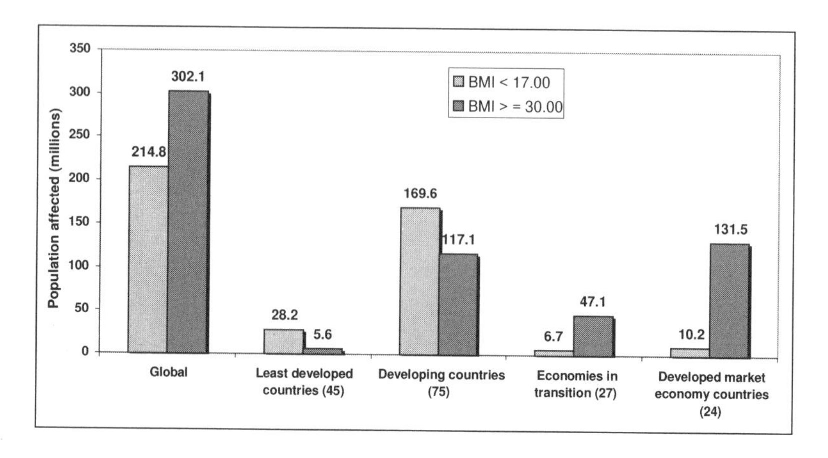

Source: WHO, *Nutrition for Health and Development: A Global Agenda for Combating Malnutrition*, 2000, http://www.who.int/nut/db_bmi.htm

Figure 2.11 *Global population affected by underweight and obesity in adults, by level of development, 2000*

to 1992; in Scotland by 2.3-fold for boys and 1.8-fold for girls between 1984 and 1994. A child's weight can be thrown off balance by a daily consumption of only one sugar-sweetened soft drink of 120 kcals; over ten years, this intake would turn into 50kg of excess growth. Although their review also fully acknowledged the role of genetics, the authors pointed to pressures on children's diets from advertisements to help explain the rapidity of consumption and obesity changes.[60]

Health education seems to be powerless before this rising tide of obesity. On the island of Mauritius, for instance, a study which examined adults over a period of five years found that, despite a national programme promoting healthy eating and increased physical activity, obesity levels had increased dramatically:[61] men with a BMI above 25 increased from 26.1 per cent to 35.7 per cent and for women the figure grew from 37.9 per cent to 47.7 per cent during the five-year study. The government of Mauritius concluded that a National Nutrition Policy and National Plan of Action on Nutrition was needed.[62] Even in the US, the homeland of fast food, President George W Bush was so alarmed by the obesity crisis that in 2002 he launched a national debate. He has long had good reason for concern,[63] as even as far back as 1986,

Table 2.4 *Global increases in the prevalence of childhood obesity*[1]

Country	Index of measurement	Age of children	Date of first study (% obesity)	Date of second study (% obesity)	Growth of obesity incidence from first to second study
USA[2]	BMI=95th	6–11	1971/74 (4%)	1999 (13%)	Up 3.3-fold
	percentile	12–19	1971/74 (6%)	1999 (14%)	Up 2.3-fold
England[3]	Age-adjusted BMI cut-off linked to adult value of 30 kg/m^2	4–11	1984 (0.6% boys; 1.3% girls)	1994 (1.7% boys; 2.6% girls)	Up 2.8-fold (boys) Up 2.0-fold (girls)
Scotland[3]	Age-adjusted BMI cut-off linked to adult value of 30 kg/m^2	4–11	1984 (0.9% boys; 1.8% girls)	1994 (2.1% boys; 3.2% girls)	Up 2.3-fold (boys) Up 1.8-fold (girls)
China[4]	Age-adjusted BMI cut-off	6–9	1991 (10.5%)	1997 (11.3%)	Up 1.1-fold
	linked to adult value of 25 kg/m^2	10–18	1991 (4.5%)	1997 (6.2%)	Up 1.4-fold
Japan[5]	≥120% of standard weight	10	1970 (<4 boys; ≈4% girls)	1996 (≈10% boys; ≈9% girls)	Up 2.5-fold (boys) Up 2.3-fold (girls)
Egypt[6]	Weight-for-height >2 SD from median	0–5	1978 (2.2%)	1996 (8.6%)	Up 3.9-fold
Australia[7]	Age-adjusted BMI cut-off linked to adult value of 30 kg/m^2	7–15	1985 (1.4% boys; 1.2% girls)	1995 (4.7% boys; 5.5% girls)	Up 3.4-fold (boys) Up 4.6-fold (girls)
Ghana[6]	Weight-for-height >2 SD from median	0–3	1988 (0.5%)	1993/94 (1.9%)	Up 3.8-fold
Morocco[6]	Weight-for-height >2 SD from median	0–5	1987 (2.7%)	1992 (6.8%)	Up 2.5-fold
Brazil[4]	Age-adjusted BMI cut-off	6–9	1974 (4.9%)	1997 (17.4%)	Up 3.6-fold

Table 2.4 *(Continued)*

Country	Index of measurement	Age of children	Date of first study (% obesity)	Date of second study (% obesity)	Growth of obesity incidence from first to second study
	linked to adult value of 25 kg/m^2	10–18	1974 (3.7%)	1997 (2.6%)	Up 3.4-fold
Chile[8]	Weight-for-height >2 SD from median	0–6	1985 (4.6%)	1995 (7.2%)	Up 1.6-fold
Costa Rica[6]	Weight-for-height >2 SD from median	0–6 (1982) 1–7 (1996)	1982 (2.3%)	1996 (6.2%)	Up 2.7-fold
Haiti[6]	Weight-for-height >2 SD from median	0–5	1978 (0.8%)	1994/95 (2.8%)	Up 3.5-fold

Sources: 1 Ebbeling, CB, Pawlak, DB and Ludwig, DS (2002) 'Childhood obesity: public-health crisis, common sense cure', *The Lancet*, 360, 10 August, 473–482; 2 National Center for Health Statistics (1999) Prevalence of overweight among children and adolescents: United States, 1999–2000, available at www.cdc.gov/nchs/products/pubs/pubd/hestats/overweight99.htm (accessed 29 January 2002); 3 Chinn, S and Rona, RJ (2001) 'Prevalence and trends in overweight and obesity in three cross-sectional studies of British children, 1974–94', *BMJ*, 322, 24–26; 4 Wang, Y, Monteiro, C and Popkin, BM (2002) 'Trends of obesity and underweight in older children and adolescents in the US, Brazil, China, and Russia', *Am J Clin Nutr*, 75, 971–977; 5 Murata, M (2000) 'Secular trends in growth and changes in eating patterns of Japanese children', *Am J Clin Nutr*, 72 (suppl), 1379S–1383S; 6 deOnis, M and Blossner, M (2000) 'Prevalence and trends of overweight among preschool children in developing countries', *Am J Clin Nutr*, 72, 1032–1039; 7 Magarey, Am, Daniels, LA and Boulton, TJC (2001) 'Prevalence of overweight and obesity in Australian children and adolescents: reassessment of 1985 and 1995 data against new standard international definitions', *Med J Aust*, 174, 561–564; 8 Filozof, C, Gonzalez, C, Sereday, M, Mazza, C and Braguinsky, J (2001) 'Obesity prevalence and trends in Latin American countries', *Obes Rev*, 2, 99–106.

the economic costs of illness associated with overweight in the US were estimated to be $39 billion; today the estimated cost of obesity and overweight is about $117 billion.[64] The rise in US obesity is dramatic: between 1991 and 2001, adult obesity increased by 74 per cent. The percentage of US children and adolescents who are defined as overweight has more than doubled since the early 1970s, and about 13 per cent of children and adolescents are now seriously overweight.[65] These general US figures disguise marked differences between ethnic groups and income levels: according to the Centers for Disease Control and Prevention, 27 per cent of black and about 21 per cent of Hispanics of all ages are considered obese – that is, a third overweight – compared with a still worrying but lower 17 per cent among whites.[66] The poor are more obese than the more affluent within the US. The price of food is a key driver of obesity: saturated fats from dairy and meat and hydrogenated (trans) fats are relatively cheap.[67]

The connection between overweight and health risk is alarmingly highlighted by the following list of the physical ailments that an overweight population (with a BMI higher than 25) is at risk of:[68]

- high blood pressure, hypertension;
- high blood cholesterol, dyslipidemia;
- type-2 (non-insulin-dependent) diabetes;
- insulin resistance, glucose intolerance;
- hyperinsulinemia;
- coronary heart disease;
- angina pectoris;
- congestive heart failure;
- stroke;
- gallstones;
- cholescystitis and cholelithiasis;
- gout;
- osteoarthritis;
- obstructive sleep apnea and respiratory problems;
- some types of cancer (such as endometrial, breast, prostate and colon);
- complications of pregnancy;
- poor female reproductive health (such as menstrual irregularities, infertility and irregular ovulation);
- bladder control problems (such as stress incontinence);
- uric acid nephrolithiasis;
- psychological disorders (such as depression, eating disorders, distorted body image, and low self esteem).

There is a vocal position – particularly articulated in the US – arguing that the critique of obesity is an infringement of personal liberty and 'size-ist', making cultural value statements. If someone wants to be fat and is content and loved by others, goes this argument, what does it matter? The list of health problems given above is surely an answer to this position. The costs of what is presented as an 'individual' problem are, in fact, society wide. The ill-health that results is paid for either in direct costs or in a societal drag – lost opportunities, inequalities and lost efficiencies. This is why policy-makers have to get to grips with obesity and the world's weight problem.

Both obesity and overweight are preventable. At present the debate about obesity is divided about which of three broad strategies of action is the best to address. One strand argues that it is a problem caused by over-consumption (diet and the types of food) and over-supply; another that it is lack of physical activity; and the third that there might be a matter of genetic predisposition. Certainly, the emphasis has to be on changing the environmental determinants that allow obesity to happen. A pioneering analysis by Australian researchers in the mid-1990s proposed that the obesity pandemic could only be explained in 'ecological' terms: Professors Garry Egger and Boyd Swinburn set out environmental determinants such as transport, pricing and supply; they claimed that environmental factors were so powerful in upsetting energy balances that obesity could be viewed as 'a normal response to an abnormal environment'.[69] So finely balanced are caloric intake and physical activity than even slight alterations in their levels can lead to weight gain. Swinburn and Egger assert that no amount of individual exhortation will reduce worldwide obesity;[70,71] transport, neighbourhood layout, home environments, fiscal policies and other alterations of supply chains must be tackled instead.

Calculating the Burden of Diet-related Disease

During the 1990s, world attention was given to calculating the costs of what has been called the 'burden of disease'. Five of the ten leading causes of death in the world's most economically advanced country, the US, were, by the 1980s, diet-related: coronary heart disease, some types of cancer, stroke, diabetes mellitus and atherosclerosis. Another three – cirrhosis of the liver, accidents and suicides – were associated with excessive alcohol intake.[72] Together these diseases were accounting for nearly 1.5 million of the 2.1 million annual deaths in the US. Only two categories in the top ten – chronic obstructive lung disease and pneumonia and influenza – had no food connection.

In a 1990s study published by the World Bank, 'The Global Burden of Disease',[73] the authors Murray and Lopez gave a detailed review of causes of mortality in eight regions of the world. Ischaemic heart disease accounted for 6.26 million deaths. Of these, 2.7 million were in established market economies and formerly socialist economies of Europe; 3.6 million were in developing countries (out of 50.5 million deaths from all causes in 1990). Stroke was the next most common cause of death (4.38 million deaths, almost 3 million in developing countries), closely followed by acute respiratory infections (4.3 million, 3.9 million in developing countries). Other leading causes of death include diarrhoeal disease (almost totally occurring in developing countries), chronic obstructive pulmonary disease, tuberculosis, measles, low birthweight, road-traffic accidents and lung cancer, with only diarrhoea and low birthweight having a diet-related aetiology. They also calculated that cancers caused about 6 million deaths in 1990. About 2.4 million cancer deaths occurred in established market economies and former socialist economies of Europe. By 1990, therefore, there were already 50 per cent more cancer deaths in less developed countries than in developed countries.

For their analysis, Murray and Lopez created a new index they called the DALY, standing for the 'disability adjusted life year'. A DALY is the sum of life years lost owing to premature death, and years lived with disability (adjusted for severity). It is thus a measure of both death and disability (both mortality and morbidity). The top ten DALYs in all developing regions combined already included ischaemic heart disease and cerebovascular disease. Murray and Lopez's report concluded: 'Clearly, the focus of research and debate about health policy in developing regions should address the current challenges presented by the epidemiological transition now, rather than several decades hence.' Table 2.5 gives their original breakdown for the world of the DALYs by main disease, present and anticipated.

Table 2.5 *DALYs lost by cause, for the developed and developing countries, 1990 and 2020*

Cause	Developed 1990 (%)	Developed 2020 (%)	Developing 1990 (%)	Developing 2020 (%)
Infectious diseases	7.8	4.3	48.7	22.2
Cardiovascular disease	20.4	22.0	8.3	13.8
Coronary heart disease	9.9	11.2	2.5	5.2
Stroke	5.9	6.2	2.4	4.2
Diabetes	1.9	1.5	0.7	0.7
Cancer	13.7	16.8	4.0	9.0
Neuropsychiatric disorders	22.0	21.8	9.0	13.7
Injuries	14.5	13.0	15.2	21.1

Source: Murray, CJL and Lopez, AD (1996) *The Global Burden of Disease: A Comprehensive Assessment of Mortality and Disability from Diseases, Injuries and Risk Factors in 1990 and Projected to 2020*, Cambridge, MA: Harvard University Press on behalf of the World Bank and WHO

The authors anticipated that the greatest increase in cardiovascular disease-related DALYs would occur in developing countries, up 8.3 per cent in 1990 to 13.8 per cent in 2020 – a rising burden of disease for those countries which could least afford it. The corresponding increase in developed countries' DALYs associated with non-communicable diseases was calculated to be only relatively slight, rising from 20.4 per cent to 22.0 per cent. (The developed world already had a high base rate of DALYs from diet-related disease). (Interestingly, there is hardly any movement in diabetes figures for developing countries and a fall for developed countries, yet it should be noted that diabetes figures are in fact rising rapidly worldwide. The newness of this diet-related epidemic might have been too late for Murray and Lopez's 1990 data.)

One purpose of the DALY method is to enable policy-makers to estimate the relative risk of major factors for health. Table 2.6 gives the Swedish National Institute of Public Health's summary of the calculated impacts of smoking, alcohol, diet and physical activity for key DALYs in the EU and Australia. Again, the diet-related disease toll is very high. Smoking, as was noted at the start of this chapter, is a major contributory factor in heart disease but the dietary factors, when separated, were almost as great.

Table 2.6 *DALYs lost by selected causes, for the EU and Australia, around 1995*

Cause	EU %	Australia %
Smoking	9.0	9.5
Alcohol consumption	8.4	2.1
Diet and physical activity	8.3	16.4
Overweight	3.7	2.4
Low fruit and vegetable intake	3.5	2.7
High saturated fat intake	1.1	2.6
Physical inactivity	1.4	6.8

Sources: National Institute of Public Health, Stockholm (1997)[74, 75]

The DALY approach was extended in the 'World Health Report 2002' which was based on a series of massive multi-country studies designed to test and refine the methodology. Figure 2.12 details risk factors by level of development. The results, however, merely deepened the insights from the earlier study. If anything, the burden of diet-related disease and of lack of physical activity received even higher profile. Special studies on the impact of lack of fruit and vegetables in the diet showed great impact. The WHO–FAO 2003 report underlined how a variety of diseases, from heart disease to diabetes, were all associated with the same dietary pattern: over-consumption, excess fat, under-consumption of fruit and vegetables and excess added sugar and salt.[76]

The financial costs

In 2001, the Commission on Macroeconomics and Health, created by the WHO, argued that there were mutual benefits to be had from improved health and for the economy, particularly for those in low-income countries.[77] Table 2.7 shows how general health care costs are rising rapidly in many developed economies; in the developing world, the costs of health care for degenerative diseases are now also looming as a serious concern. The growth of health expenditure is sometimes higher than the growth of gross domestic product (GDP). Table 2.8 gives a breakdown of the direct and indirect costs for a number of key diet-related diseases in the US; these costs are immense, even for such a rich society.

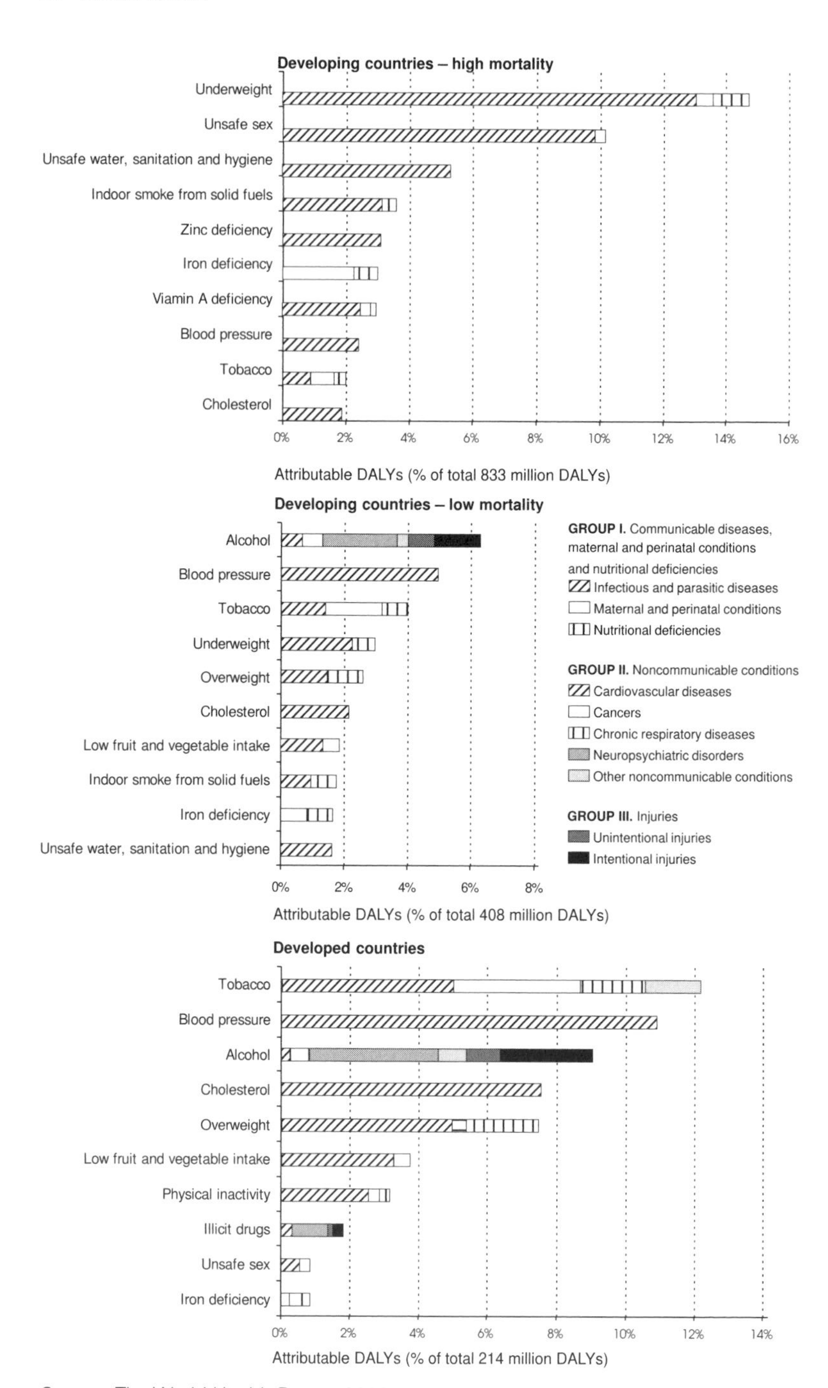

Source: The World Health Report 2002

Figure 2.12 *Burden of disease attributable to ten selected leading risk factors, by level of development and type of affected outcome*

Health ministries, it appears, are locked in a model which tends to be curative rather than preventative. The UK health care system, for instance, costs £68 billion for around 60 million people, costs that are anticipated to rise to between £154 billion ($231 billion) and £184 billion ($276 billion) by 2022–2023 in 2002 prices.[79] In other words, at constant prices, UK health care costs are doubling.

In the context of diet-related disease, the direct and indirect financial tolls of ill health could offer opportunities for positive policy intervention through a health-enhancing food supply chain. An estimate for the UK by the Oxford University British Heart Foundation Health Promotion Research Group has calculated that coronary heart disease (CHD) – constituting about half of all cases of cardiovascular disease – costs the UK £10 billion per annum. These costs are made up of £1.6 billion in direct costs (primarily to the taxpayer through the costs of treatment by the NHS) and £8.4 billion in indirect costs to industry and to society as a whole, though loss of productivity due to death and disability.[80] (This is probably an underestimate of the direct costs to the UK's National Health Service as these costs do not include the cancer treatment costs.)

A report chaired in 2002 by Derek Wanless, a former head of the NatWest Bank, for the Chancellor of the Exchequer produced not dissimilar calculations.[81] It estimated that costs for the health service will rise alarmingly if targets are not met to reduce CHD and cancers. CHD treatment costs (drugs like statins and surgical techniques like re-vascularisation) would add an additional £2.4 billion per annum by 2010–11, doubling CHD expenditure. Such calculations remind us of the multi-headed nature of ill health. Smoking, diet, physical activity, genetics, environment and socio-economic background all have direct health outcomes. Wanless and his team were convinced by US scientific work that high cholesterol – 'which is mainly due to diet' – accounts for 43 per cent of CHD incidence, compared to 20 per cent for smoking. This sort of evidence shows that the poor diet has such far-reaching financial implications that it warrants higher political attention. This case was confirmed by a second Wanless study arguing for the economic value of facing the public health costs of poor diet, lifestyle and education.[82] However, for the last quarter of a century policy attention has been directed to cutting costs, not by altering the food supply chain, but by such policies as contracting

Table 2.7 *Growth of expenditure on health, 1990–2000*

	Real per capita growth rates, 1990–2000 (%)		Health spending as per cent of GDP		
	Health spending	GDP	1990	1998	2000
Australia	3.1	2.4	7.8	8.5	8.3
Austria	3.1	1.8	7.1	8.0	8.0
Belgium	3.5	1.8	7.4	8.5	8.7
Canada	1.8	1.7	9.0	9.1	9.1
Czech Republic	3.9	0.1	5.0	7.1	7.2
Denmark	1.7	1.9	8.5	8.4	8.3
Finland	0.1	1.8	7.9	6.9	6.6
France	2.3	1.4	8.6	9.3	9.5
Germany	2.2	0.2	8.7	10.6	10.6
Greece	2.8	1.9	7.5	8.7	8.3
Hungary [(a)]	2.0	2.7	7.1	6.9	6.8
Iceland	2.9	1.6	7.9	8.3	8.9
Ireland	6.6	6.4	6.6	6.8	6.7
Italy	1.4	1.4	8.0	7.7	8.1
Japan	3.9	1.1	5.9	7.1	7.8
Korea	7.4	5.1	4.8	5.1	5.9
Luxembourg [(b)]	3.7	4.5	6.1	5.8	6.0
Mexico	3.7	1.6	4.4	5.3	5.4
Netherlands	2.4	2.3	8.0	8.1	8.1
New Zealand	2.9	1.5	6.9	7.9	8.0
Norway	3.5	2.8	7.8	8.5	7.5
Poland [(b)]	4.8	3.5	5.3	6.4	6.2
Portugal	5.3	2.4	6.2	8.3	8.2
Slovak Republic	..	4.0	..	5.9	5.9
Spain	3.9	2.4	6.6	7.6	7.7
Switzerland	2.5	0.2	8.5	10.6	10.7
United Kingdom	3.8	1.9	6.0	6.8	7.3
United States	3.2	2.3	11.9	12.9	13.0
OECD Average [(c,d)]	**3.3**	**2.2**	**7.2**	**8.0**	**8.0**
EU Average	**3.1**	**2.3**	**7.4**	**8.0**	**8.0**

Source: OECD (2002) Health Data 2002, www.oecd.org/pdf/M00031000/M00031130.pdf (p1). (a) Hungary: 1991–2000. (b) Luxembourg and Poland: 1990–1999. (c) OECD averages exclude the Slovak Republic because of missing 1990 estimates. (d) Unweighted averages.

Note: No recent estimates are available for Sweden and Turkey.

Table 2.8 *Economic costs of diet- and exercise-related health problems, US*

Disease	Direct costs US$ billion (medical expenditures)	Indirect costs US$ billion (productivity losses)	Total costs US$ billion
Heart disease	97.9	77.4	**175.3**
Stroke	28.3	15.0	**43.3**
Arthritis	20.9	62.9	**83.8**
Osteoporosis	n.a.	14.9	**14.9**
Breast cancer	8.3	7.8	**16.1**
Colon cancer	8.1	n.a.	**8.1**
Prostate cancer	5.9	n.a.	**5.9**
Gall bladder disease	6.7	0.6	**7.3**
Diabetes	45.0	55.0	**100.0**
Obesity	55.7	51.4	**107.1**
		Total =	**561.8**

Sources: National Institutes of Health (1998) and Wolf and Colditz (1998)[78]
Note: Costs are expressed in constant 1998 dollars, using the Consumer Price Index.

out services and by privatization. In the UK, less than £5 million a year is spent on food-related health education. Meanwhile, drug companies and surgeons only offer expensive but highly sophisticated solutions when the patient is already sick.

Indeed, drug treatments can be hugely expensive. A trial on over 20,000 UK people with high risks for heart disease showed that giving patients a type of drug known as statins reduced the risk of a first coronary attack by 25 per cent but would cost £1 ($1.5 or €1.5) per patient per day.[83] Currently, 1.8 million people are prescribed statins, costing UK£750 million a year. Taking statins for three years can reduce the risk of a heart attack by up to a third.

Coronary heart disease (CHD)

Since 1999, the WHO has attributed 30 per cent of all annual global deaths – that is, of 15 million people – to cardiovascular disease.[84, 85] The majority of those deaths are in low- and middle-income countries. In 1998, 86 per cent of DALYs were lost to cardiovascular disease worldwide.

The main risk factors for heart disease are high blood pressure, smoking and lipid concentrations (cholesterol levels). Others include age, sex, family history and the presence of diabetes. WHO recommendations for reducing CVD include:[86]

- regular physical activity
- linoleic acid
- fish and fish oils
- vegetables and fruits, including berries
- potassium
- low to moderate alcohol intake.

The WHO judges that there is convincing evidence for the increasing risks from:

- myristic and palmitic acids
- trans-fatty acids
- high sodium intake
- overweight
- high alcohol intake.

In regard to CHD, public health policy has tended to focus on two things: health education as prevention, and improved medical treatment through drug, hospital and surgical care. It has also urged behavioural change, in particular a reduction of total fat intake and especially of saturated fats (mainly from animal meat and dairy fats). This health promotion policy has had an effect: rates of heart disease are declining in most affluent Western countries, after years of steady increase since the immediate post-World War II period (see Tables 2.9 and 2.10).[87]

The global picture is more complex, however.[88] For example, the steep rise in CHD in the newly independent countries of Eastern Europe (such as Belarus, Azerbaijan and Hungary) is worrying. Leaving the strictures of the Soviet era means only that already high rates of CHD have risen further. Even in countries considered to have a healthy diet, like Greece and Japan, social change is being accompanied by changing patterns of diet-related disease: Greece's CHD and obesity rates are rising as it changes to a more Northern European diet high in animal fats, following entry to the European Union and increased tourism.

Table 2.9 *Age-standardized deaths per 100,000 population from CHD selected countries, 1968–1996: men*

Men	1968	1978	1988	1998
Finland	718	664	477	340
UK	517	546	434	297
Austria	327	349	262	226
US	694	504	292	224
Australia	674	409	315	202
Canada	543	457	296	200
Italy*	230	249	172	150
Belgium*	345	313	184	147
Spain	99	165	146	125
France	152	154	118	92
Japan	92	74	52	58

Note: *latest statistics for 1994

Table 2.10 *Age-standardized deaths per 100,000 population from CHD selected countries, 1968–1996: women*

Women	1968	1978	1988	1996
UK	175	182	156	107
Finland	204	177	141	93
US	273	185	119	92
Austria	120	119	84	81
Australia	268	186	117	73
Canada	198	155	100	72
Belgium*	111	100	61	46
Italy*	87	82	51	43
Spain	33	46	39	34
France	49	44	30	22
Japan	45	99	21	21

Note: *latest statistics for 1994
Source: British Heart Foundation from WHO country statistics

Death rates from CHD may have dropped in the US and Finland, but it should be remembered that their morbidity and costs are still high, as was shown by the Global Burden of Disease studies.

This complexity keeps epidemiologists busy around the world, but the rapidity of change should bring little surprise. In

1981 Trowell summarized the emergence of CHD amongst East Africans: in the 1930s, he reported, autopsies had shown zero CHD in East Africa, and only one case among 2994 autopsies conducted in Makere University Medical School over the period 1931–1946. However, by the 1960s CHD in this region was emerging as a major rather than peripheral health problem.[89]

In China, between 1991 and 1995,[90] CHD accounted for 15 per cent of all deaths. Cholesterol levels here, compared to those found in Western populations, were low but were increasing rapidly among the urban populations where a more affluent lifestyle was being adopted. Daily intake of meat, eggs and cooking oil had increased while intake of legumes and cereals had decreased. A reduction in the consumption of Western fast foods was also recommended as were increasing physical activity levels, an urging which could be applied to many urbanizing developing countries.

Food-related cancers

Since the 1980s, dietary factors have been thought to account for around 30 per cent of cancers in Western countries, making diet second only to tobacco as a preventable cause of cancer;[91] in developing countries diet accounted for around 20 per cent.[92] Table 2.11 gives the 1997 review of food–cancer research by the World Cancer Research Fund. An updated report is due out in 2006.

The annual WHO World Health Report has shown that cancers are increasing worldwide,[93] and the 2003 World Cancer Report suggested that, like obesity, rising cancer rates are preventable. By virtue of steadily ageing populations, cancer could further increase by 50 per cent to 15 million new cases a year by 2020. In 2000, 6.2 million people died of cancer worldwide (12.5 per cent of all deaths), but 22.4 million were living with cancer. In the South, cancers of the oesophagus, liver and cervix are more common, while in the North, there is a predominance of cancers of the lung, colon, pancreas and breast.

The most significant cause of death among men is lung cancer and among women, breast cancer, but certain lifestyle changes, such as to diet or smoking habits, would alter these patterns. Some cancers are closely associated with diets centred on well-cooked red meats, animal proteins and saturated fats in large quantities, with a daily routine that takes in little physical activity.[94] Indeed, many cancers could be prevented by modifying dietary habits to include more fruits, vegetables, high-fibre cereals, fats and oils derived from vegetables, nuts, seeds and fish, and by limiting the intake of animal fats derived from meat, milk and dairy products.[95,96] A number of published studies show that an increase in antioxidant nutrients such as beta-carotene, vitamins C and E, zinc and selenium could also decrease the risk of certain cancers and there seems to be strong evidence that eating a diet rich in fresh fruit and vegetables will reduce the risk of stomach cancer.[97] Yet the nutrition transition is being driven in a different direction – towards a diet actually higher in processed foods and animal fats, key food industries within the Productionist paradigm.

Diabetes

The incidence of Type 2 diabetes is, alarmingly, on the increase. This form of diabetes was formerly known as non-insulin-dependent diabetes mellitus (NIDDM), occurring when the body is unable to respond to the insulin produced by the pancreas; it accounts for around 90 per cent of cases worldwide. In Type 1 diabetes (formerly known as insulin-dependent), the pancreas fails to produce the insulin which is essential for survival; this form develops most frequently in children and adolescents, but is now being increasingly noted later in life.[98] It is anticipated that cases of Type 2 diabetes will rise coming years (see Table 2.12): the WHO anticipates a doubling in the number of cases from 150 million in 1997 to 300 million in 2025, with the greatest number of new cases being in China and India.[99]

Diabetes is the fourth main cause of death in most developed countries. Research demonstrates the association between excessive weight gain, central adiposity (fat around the waist) and the development of Type 2 diabetes. Diabetics are two to four times more likely to develop cardiovascular diseases than others, and a stroke is twice as common in people with diabetes and high blood pressure as it is for those with high blood pressure alone.

In 2000, India recorded 32.7 million diabetics, China 22.6 million and the US 15.3 million, while Brazil recorded only 3.3

Table 2.11 *Cancers preventable by dietary means*

	Global Ranking (incidence)	Global Incidence (1000s)	Dietary Factors (convincing or probable)	Non-dietary Risk Factors (established)	Preventable by Diet			
					Low estimate (%)	High estimate (%)	Low estimate (1000s)	High estimate (1000s)
mouth and pharynx	5	575	↓ vegetables & fruits[a] ↑ alcohol[a]	? smoking[a] ↑ betel[a]	33	50	190	288
nasopharynx			↑ salted fish[b]	↑ EBV[b]				
larynx	14	190	↓ vegetables & fruits ↑ alcohol	↑ smoking	33	50	63	95
oesophagus	8	480	↓ vegetables & fruits ↑ deficient diets ↑ alcohol	↑ smoking ↑ Barrett's oesophagus	50	75	240	360
lung	1	1320	↓ vegetables & fruits	↑ smoking ↑ occupation	20	33	254	436
stomach	2	1015	↓ vegetables & fruits ↓ refrigeration ↑ salt ↑ salted foods	↑ *H. pylori*	66	75	670	761
pancreas	13	200	↓ vegetables & fruits ↑ meat, animal fat	↑ smoking	33	50	66	100
gallbladder	–	c	–	–	–	–	–	–
liver	6	540	↑ alcohol ↑ contaminated food	↑ MBV and HCV	33	66	178	356
colon, rectum	4	875	↓ vegetables ↓ physical activity ↑ meat ↑ alcohol	↑ smoking ↑ genes ↑ ulcerative colitis ↑ *S. sinensis* ↓ NSAIDs	66	75	578	656
breast	3	910	↓ vegetables	↓ reproductive	33	50	300	455

			↑ rapid early growth ↑ early menarche ↑ obesity ↑ alcohol	↑ genes ↑ radiation				
ovary	15	190	–	↑ genes ↓ reproductive	10	20	19	38
endometrium	16	170	↑ obesity	↑ OCs ↑ oestrogens ↓ reproductive	25	50	43	85
cervix	7	525	↓ vegetables & fruits	↑ HPV ↑ smoking	10	20	53	105
prostate	9	400	↑ meat or meat fat or dairy fat		10	20	40	80
thyroid	–	100[d]	↑ iodine deficiency	↑ radiation	10	20	10	20
kidney	17	165	↑ obesity	↑ smoking ↑ phenacetin	25	33	41	54
bladder	11	310		↑ smoking ↑ occupation ↑ *S. haematobium*	10	20	31	62
other		2,355	–	–	10	10	236	236
Total (1996)		**10,320**					**3022** **29.3%**	**4187** **40.6%**

Notes: Included as 'dietary factors' in this table are various foods, nutrients, alcoholic drinks, body weight and physical activity. The panel has estimated the extent to which specific cancers or cancer in general are preventable by the dietary and associated factors described in this report. The figures suggested are ranges consistent with current scientific knowledge as reviewed and assessed in Chapters 4–7, and take established non-dietary risk factors, notably the use of tobacco, specific infections and occupational exposures to carcinogens, into account. The arrows represent either decreasing risk (↓) or increasing risk (↑).

Figures on global ranking and incidence: Parkin et al (1993); WHO (1997)

[a]: mouth and pharynx; also chewing tobacco

[b]: nasopharynx

[c]: reliable worldwide data are not collected by IARC for this site

[d]: conservative estimate based on the IARC (1993)

Source: Table 9.1.2 in World Cancer Research Fund (1997), *Food, Nutrition and the Prevention of Cancer: A Global Perspective*. Washington, DC: World Cancer Research Fund/American Institute for Cancer Research. Reproduced by permission.

Table 2.12 *Prevalence of diabetes worldwide*

	2000	2030	Projected growth (%)
Africa	7,020,553	18,244,638	160
Mediterranean	15,189,760	43,483,842	186
Americas	33,014,823	66,828,417	102
European	33,380,754	48,411,977	45
SE Asia	45,810,544	122,023,693	166
Western Pacific	36,138,079	71,685,158	98
Total	171,000,000	366,000,000	114

Source: WHO (2004) *Diabetes Action Programme*, Geneva, WHO; http://www.who.int/diabetes/facts/world_figures/en/, accessed 2 June 2004

million and Italy 3.1 million. In 2000, the five countries with the highest diabetes prevalence in the adult population only were Papua New Guinea (15.5 per cent), Mauritius (15 per cent), Bahrain (14.8 per cent), Mexico (14.2 per cent) and Trinidad & Tobago (14.1 per cent).[100, 101] Such disparate statistics reflect a transition from traditional diet and from an activity-based lifestyle to a more sedentary one. By 2025, the prevalence of diabetes is anticipated to triple in Africa, the Eastern Mediterranean, the Middle East and South Asia. It is expected to double in the Americas and the Western Pacific and to almost double in Europe. In India, incidence is much higher in urban than rural populations:[102] in urban Chennai (Madras), for example, cases of diabetes rose by 40 per cent in 1988–1994. Incidence is rising among male urban dwellers of South India compared to the rural male population. In addition to *Diabetes mellitus*, the prevalence of non-insulin-dependent diabetes (NIDDM) increased dramatically within the urban populations of India within just a decade.[103] In Thailand, also, NIDDM is more pronounced amongst females in the urban population than it is in the rural population,[104] whilst in the rural environment, incidence of NIDDM amongst males is higher.

In the UK, Professor David Barker and colleagues have shown that adult diabetes is associated with low birthweight,[105] while studies in India suggest that poor interuterine growth, combined with obesity later in life is associated with insulin resistance, diabetes and increased cardiovascular risk.[106] Once again, a single disease seems attributable to a pattern of poor nutrition related to the lifecycle, and is one whose costs are externalized onto society as a whole and health care in particular. Devastating complications of diabetes, such as blindness, kidney failure and heart disease, are imposing a huge financial burden: in some countries 5–10 per cent of national health budgets.

Food Safety and Food-borne Diseases

Whilst attention to such non-communicable diseases is of vital importance, food safety, food-borne diseases and other communicable diseases remain uppermost within food and public health policy, partly due to consumer campaigns about risks and to heightened media awareness of poor food processing standards. Food safety problems include risks from:[107]

- veterinary drug and pesticide residues;
- food additives;
- pathogens (ie illness-causing bacteria, viruses, parasites, fungi and their toxins);
- environmental toxins such as heavy metals (eg lead and mercury);
- persistent organic pollutants such as dioxins;
- unconventional agents such as prions associated with BSE.

In particular, companies have had to respond to new public awareness about food safety issues, and new regimes of traceability have been implemented to enable companies to track food ingredients in order to eliminate subsequent legal or insurance liability consequences. In this respect, food companies are anxious to present themselves as guardians of the public health.[108] The attention food safety receives is predictably higher in affluent countries when, on the evidence, the burden of ill health is far greater in the developing world, due to lack of investment and infrastructure, including drains, housing, water supplies and food control systems. The World Health Report 2002 pointed out that, in developing countries, water supply and general sanitation remain the fourth highest health-risk factor, after underweight, unsafe sex and blood pressure.[109] In developing countries which are building their food export markets, there is too often a bipolar structure, with higher standards for foods for export to affluent countries than for domestic markets. There ought to be a cascading down into internal markets of these higher standards.[110]

Environmental risks to health are a significant problem on the global scale and, in Western countries in the 1990s, new strains of deadly bacteria such as *E. coli 0157* captured policy attention, an estimated 30 per cent of people having suffered a bout of food-borne disease annually. The US, for instance, reports an annual 76 million cases, resulting in 325,000 hospitalizations and 5000 deaths.[111] The WHO estimates that 2.1 million children die every year from the diarrhoeal diseases caused by contaminated water and food,[112, 113] asserting that each year worldwide there are 'thousands of millions' of cases of food-borne disease.[114]

In early industrializing countries, a grand era of engineering made dramatic health improvements in public health (a story we return to in Chapter 3). Part of that investment included the introduction of effective monitoring and hygiene practice systems, such as the establishment of local authority laboratories and training, the packaging of foods and processes such as milk pasteurization. Today, public health proponents are actively trying to promote a 'second wave' of food safety intervention but this time using a risk-reduction management system known as Hazards Analysis Critical Control Point (HACCP), an approach designed to build safety awareness and control of potential points of hygiene breakdown into food handling and management systems. HACCP also encourages the creation of a 'paper' trail to enable tracking along the production process, essential in order to obviate errors and enable learning. Breakdowns in food safety have in the past led to major political and business crises, with governments under attack and new bodies responsible for food safety being set up in many countries. As food supply chains become more complex and as the scale of production, distribution and mass catering increases, so the chances for problems associated with food contamination rise; mass production breakdowns in food safety spread contamination and pathogens widely. An outbreak of *Salmonellosis* in the US in 1994, for example, affected an estimated 224,000 people.[115] *Listeria monocytogenes* has a fatality rate of 30 per cent, a fact that seriously dented UK public confidence in the 'cook–chill' and 'oven-ready' foods of the late 1980s.

Cross-border trade in agricultural and food products, as well as international pacts have brought food safety to the fore.[116] The Director-General of the WHO, in a speech on food safety to the UN Codex Alimentarius Commission, said: 'globalisation of the world's food supply also means globalisation of public health concerns.'[117] Crises over BSE, *Salmonellosis* and *E. coli*, for example, had had a significant political impact throughout both the UK and EU, for instance,[118] and many countries have experienced a fast rise in incidences of *Salmonellosis* and *Campylobacter* infections since the 1980s, both bacteria being associated with meat and meat products. Despite countries such as Denmark and Sweden having strict policies governing the extermination of flocks and herds found to be carrying *Salmonella*, the incidence continues through the contamination of feedstuffs, and in Denmark in 1998 the percentage of positive flocks with *Campylobacter* was 47.1 per cent.

Thus, in many developed countries with good monitoring systems, the incidence of food-borne disease has in fact risen during the era of the Productionist paradigm: in West Germany cases of infectious *S. enteritis* rose from 11 per 100,000 head of population in 1963 to 193 per 100,000 in 1990;[119] in England and Wales formal notifications of the same disease rose from 14,253 cases in 1982 to 86,528 in 2000. These cases resulted in millions of days lost from work but, fortunately, relatively few deaths.

Bacteria fill gaps left by nature, evolving new strains; but they are constantly evolving even as science combats existing strains. The new food processes and systems of distribution ushered in by the food technology revolution of the second half of the 20th century provided many opportunities for bacteria to develop and colonize new niches. The incidence of *Salmonella* in the UK, for example, first rose, and then, following good monitoring, hygiene intervention and political pressure, fell right back – in two decades.

Table 2.13 gives a list from the WHO of some of the pathogenic organisms that are associated with food and food hygiene: viruses, bacteria, trematodes (flukeworms), cestodes (tapeworms) and nematodes (roundworms), the last three all small worms that can be found either in soil, fish or meats. The first two are concerns in global food trade particularly. In the case of bacteria such as *Listeria monocytogenes*, only 657 cases were reported throughout the European Union in 1998;[120] in the same period, deaths

Table 2.13 *Some pathogenic organisms associated with public health, which may be transmitted through food*

Bacteria	**Protozoa**
Bacillus cereus	*Cryptosporidium spp*
Brucella spp	*Entamoeba histolytica*
Campylobacter jejuni and coli	*Giardia lamblia*
Clostridium botulinum	*Toxoplasma gondii*
Clostridium perfringens	
Escherichia coli	
(pathogenic strains)	**Trematodes** (flukeworms)
Listeria monocytogenes	*Fasciola hepatica*
Mycobacterium bovis	*Opistorchis felineus*
Salmonella typhi and paratyphi	
Salmonella (non-typhi) spp	**Cestodes** (tapeworms)
Shigella spp	*Diphyllobotrium latum*
Staphylococcus aureus	*Echinococcus spp*
Vibrio cholerae	*Taenia solium and saginata*
Vibrio parahaemolyticus	
Vibrio fulnificus	**Nematodes** (roundworms)
Yersinia enterocolitica	*Anisakis spp*
	Ascaris lumbricoides
Viruses	*Trichinella spiralis*
Hepatitis A	*Trichuris trichiura*
Norwalk agents	
Poliovirus	
Rotavirus	

Source: WHO European Centre for Health and Environment, Rome, 2000

from cardiovascular disease in the EU totalled 1.5 million, 42 per cent of all deaths,[121] while, in 1990, diarrhoeal diseases accounted for 11,000 years of life (DALYs) lost out of a total of 22.7 million in Europe; in the same year, cardiovascular disease accounted for 7 million diabetes for 371,000 and cancer of the colon and rectum for 593,000,[122] and five times as many years of life were lost due to drug addiction than to diarrhoeal diseases.

Despite a low health burden in the developed world, the financial costs of food poisoning can be significant. Estimates in the US suggest that the diseases caused by major pathogens cost up to $35 billion each year in medical costs and lost productivity.[123] Policy-makers must be concerned about both food-borne illness and degenerative diseases, the latter of which do not as yet receive sufficient political attention.

INEQUALITIES AND FOOD POVERTY

After all the corporate hyperbole in the 1990s about globalization unleashing wealth for all, a series of UN and other reports reminded the world that, whatever the wealth accrual, its distribution would remain unequal. As one commentator on globalization put it: '. . . half a century of accelerated globalisation has clearly not eliminated poverty from the face of the earth. On the contrary, although the abject poor have decreased since 1960 as a proportion of the world's population, their absolute number has grown.'[124] A new global class structure has emerged from the crisis in health. Diet-related ill health is greatest among poorer socio-economic groups who are locked into a cycle of either hunger and premature death, or of malnutrition, and obesity and degenerative diseases. An equitable food supply would help break this cycle. Even in India where there are extremes of pressure – social, economic, environmental and climatic – lack of adequate diet for a whole-population health is socially determined[125] by caste, politics and economic policy.

In absolute terms, more people live in poverty today than 20 years ago. About a fifth of the world's population, 1.3 billion people, live on a daily income of less than US$1; by 2015 the number of people subsisting below this international poverty line is on line to reach 1.9 billion.[126] Although most of the world's poor live in South and East Asia, sub-Saharan Africa has the fastest growing proportion of people who live in poverty.[127] But poverty is not confined to developing countries and women, children, and older people are at greatest risk. The health of the poor is, concomitantly, at risk from environmental hazards such as unsafe food and water, and from urban hazards such as air, and water pollution and accidents. In 1900 around 5 per cent of the world's people lived in cities with populations exceeding 100,000. Today, an estimated 45 per cent – more than 2.5 billion people worldwide – live in large urban centres, and this is expected to rise to 61 per cent by the year 2025.[128]

According to the tenth annual UN Development Programme Human Development Report of 1999, the richest 20 per cent of the world's population now account for 86 per cent of world gross domestic product (GDP), while the poorest 20 per cent have just 1 per cent.[129] Two hundred of the world's richest people doubled their net worth in the last four years of the 20th century.

The richest three people in the world have assets greater than the combined gross national product of all the least developed countries in the world, accounting for 600 million people. The net worth of the 358 richest people equals the combined income of the poorest 45 per cent of the world's population – about 2.3 billion people.[130]

Income differentials are also increasing. In 1960, 20 per cent of the world's population living in the richest countries earned 30 times the income of the poorest 20 per cent; by 1997, the richest 20 per cent earned 74 times the income of the poorest 20 per cent. The UNDP 1999 report called for tougher rules on global governance, including principles of performance for multinationals on labour standards,[131] fair trade and environmental protection; by 2003, the tone being taken by UNDP was harsher: it reported the 1990s as a period when inequalities widened rapidly, with 50 countries suffering falling living standards in the 1990s. The richest 1 per cent of the world's population, around 60 million, now receives as much income as the poorest 57 per cent, while the income of the richest 25 million Americans is equivalent to that of almost 2 billion of the world's poorest people.[132]

Attempting to face this crisis, its findings helped to create what are known as the UN's Millennium Development goals all to be met by 2015.[133] Food features in six out of the eight goals:

- to eradicate extreme poverty and hunger;
- to achieve universal primary education;
- to promote gender equality and empower women;
- to reduce child mortality;
- to improve maternal health;
- to combat HIV/AIDS, malaria and other diseases;
- to ensure environmental sustainability;
- to develop a global partnership for development.

Some critics argue, worthy though such goals are, they are unlikely to yield any narrowing of inequalities if unaccompanied by firm policies to redistribute wealth from the rich to the poor, an economic anathema in dominant policy circles.[134] The widening disparity between social classes means that the rich, even in poor societies, have access to healthier dietary choices; they are also most likely to be tempted by imported, processed items containing higher levels of fat, sugar and salt. The poor living in rural areas may have a more restricted diet based on a staple food usually grown on their own land, but the urban poor eat only what they can afford to buy, often suffering vitamin and mineral deficiencies as a result.

In the late 1990s, UNICEF was providing evidence of over 6 million annual deaths of children under five attributable to malnutrition.[135] More than 25 per cent of all children under five years old are underweight.[136] In South America and sub-Saharan Africa, the child malnutrition rates have increased. The risk of mortality rises swiftly if a child is even mildly malnourished: dehydration from diarrhoea kills 2.2 million children every year; children who are malnourished develop lifetime disabilities and weakened immune systems and are therefore more susceptible to infectious diseases, as highlighted in the lifecycle model presented (Figure 2.8 on page 62). In addition, under-nourished children suffer impairment of their cognitive skills through lack of nutrients and calories.

UNICEF calculates that South Asia has the highest numbers of stunted children, followed next by sub-Saharan Africa, whereas Latin America has the lowest prevalence of under-fives suffering from underweight, wasting and stunting. In South Asian countries such as Sri Lanka over 25 per cent of children under five years suffer from moderate or severe stunting and over 33 per cent suffer from moderate and severe underweight. In addition, UNICEF claims that iodine deficiency is the biggest cause of preventable brain damage in foetuses, estimating that, in the late 1990s, there were 11 million adults worldwide suffering from cretinism and 760 million with goitres. A global campaign to iodize all salt would reduce this number; indeed, by 2002, the percentage of countries selling only iodized salt had increased from 20 per cent to 72 per cent. But there were still 32 countries where less than half the households consumed iodized salt, and health advice is now to restrict salt intake, as it is a risk factor for hypertension.[137] Generally, though, iodization is a rare policy success; perhaps because it did not threaten food industry interests.

The Changing Meanings of Food Security

In public policy, the term 'food security' is often invoked in respect of a new food system to reduce worldwide figures on the under-nourished.[138] The 1996 World Food Summit defined food security as the situation in which at the individual, household, national, regional and global levels 'all peoples, at all times, have physical and economic access to sufficient, safe and nutritious food to meet their dietary needs and food preferences for an active and healthy life'.[139] In the 1940s, when the WHO and the FAO were set up under the new United Nations, the policy priority was to increase food supply in every continent and nation; this macro-focus was still the dominant paradigm at the time of the 1974 World Food Conference, after which research on food security has mushroomed, and by the 1980s attention had shifted more towards household and individual access to food and towards improving what might be called 'micro-food' security. Four core foci emerged:

- **sufficiency** of food for an active healthy life;
- **access** to food and entitlement to produce, purchase or exchange food;
- **security** in the sense of the balance between vulnerability, risk and insurance;
- **time** and the variability in experiencing chronic, transitory and cyclical food insecurity.

Summing up thinking in this micro-focus period, one research team remarked that 'flexibility, adaptability, diversification and resilience are key words. Perceptions matter. Intra-household issues are central . . . Food security must be seen as a multi-objective phenomenon, where the identification and weighting of objectives can only be decided by the food insecure themselves.'[140] Maxwell, who coordinated the research after working on food security in the Sudan in the late 1980s, argued:

> *A country and people are food secure when their food system operates in such a way as to remove the fear that there will not be enough to eat. In particular, food security will be achieved when the poor and*

> *vulnerable, particularly women and children and those living in marginal areas, have secure access to the food they want. Food security will be achieved when equitable growth ensures that these people have sustainable livelihoods. In the meantime and in addition, however, food security requires the efficient and equitable operation of the food system.*[141]

NGOs tend to take a more value-led approach. The Canadian World Food Day Association, for example, sees food security with the following principles:

- that the ways and means in which food is produced and distributed are respectful of the natural processes of the earth and are thus sustainable;
- that both the production and consumption of food are grounded in and governed by social values that are just and equitable as well as moral and ethical;
- that the ability to acquire food is assured;
- that the food itself is nutritionally adequate and personally and culturally acceptable;
- that the food is obtained in a manner that upholds human dignity.[142]

By the 1990s, whether in the North or South, there was a commonality of ideas about what was meant by food security: the issue of appropriateness of food supply rather than just sufficiency was now accepted to be central.[143] An ecological tinge was infusing what had first been a more conventional notion of nutritional adequacy. Over time, the notion of food security had taken on a concreteness and value within global food governance. This happened mainly because some key professions adopted and promoted the term. International anxieties, mass-media coverage of famines and demographic and supply studies had all focused on the need for enough food to feed the growing world population. Attempts to reduce international debt, featuring highly in G8 meetings, built on concerns about large-scale food production initiatives such as the Green Revolution. Development economists had measured food consumption, through household budget surveys, in terms of calories purchased, as a key indicator of general poverty. The World Bank used food security analysis as a way of mapping the poverty of a country.

Nutritionists and famine relief agencies took measures to aid in the prediction of world food emergencies and to prevent their worst effects. Finally, an increasingly active body of development workers, articulating the view that maldistribution of power was what lay behind food insecurity, developed strategies to help empower local communities to address their own food needs. For all of those people, food security remains a goal, not just a tool of analysis.

This notion of food security has long historical roots. The first director of the FAO, John Boyd Orr, mapped out what he thought the post-War vision to ensure food security should be. This was a vision largely framed by the experience of rich countries in the 'Hungry Thirties'.[144] The vision focus was on availability and increase of the food supply:

1. Countries should set targets within a new global system and foster intergovernmental cooperation to help each other over good times and bad, to ease out booms and slumps in production.
2. Targets should be based on science, above all on nutrition and agricultural science.
3. Targets should be set to achieve health. Premature death from under-nutrition is inexcusable; investment in better food will yield health and economic gains and savings.
4. Agriculture should be financially and politically supported to produce more.
5. Industry should be geared to produce tools to enable agricultural productivity to rise: new buildings, tractors, equipment.
6. Trade should be encouraged to meet the new markets and to ease the over-productive capacity of some areas and match them with under-consumption in other areas.
7. International cooperation will have to follow the (proposed) UN Conference on Food and Agriculture.
8. New organizations will have to be created such as a new International Food and Agricultural Commission, National Food Boards to monitor supplies, Agricultural Marketing Boards, Commodity Boards.[145]

FOOD POVERTY IN THE WESTERN WORLD

Most public health concern about food poverty rightly centres on the developing world, but it is also important to recognize that the impact of food poverty is significant in the developed world. The new era of globalization has unleashed a reconfiguration of social divisions both between and within countries; these social divisions are particularly marked in societies such as the UK and the US which have pursued neo-liberal economic policies. Indeed, one review of EU food and health policies estimated that food poverty was far higher in the UK than any other EU country,[146] where inequalities in income and health widened under the Conservative government of 1979–1997. The proportion of people earning less than half the average income grew[147] and the bottom tenth of society experienced a real, not just relative, decline in income and an increase in social health distinctions. This was a the converse of the post-World War II years of Keynesian social democratic policies during which inequalities narrowed: lower UK socio-economic groups now experience a greater incidence of premature and low birthweight babies, and of heart disease, stroke and some cancers in adults. Risk factors such as bottle-feeding, smoking, physical inactivity, obesity, hypertension, and poor diet were clustered in the lower socio-economic groups[148] whose diet traditionally derives from cheap energy forms such as meat products, full-cream milk, fats, sugars, preserves, potatoes and cereals with little reliance on vegetables, fruit, and wholemeal bread. Essential nutrients such as calcium, iron, magnesium, folate and vitamin C are more likely to be ingested by the higher socio-economic groups:[149, 150] their greater purchasing power creates a market for healthier foods such as skimmed milk, wholemeal bread, fruit and other low-fat options.

Similarly, in the US, hunger has been a persistent cause of concern for decades and rising during the 1990s when the Census Bureau calculated that 11 million Americans lived in households which were 'food insecure' with a further 23 million living in households which were 'food insecure without hunger' (in other words at risk of hunger).[151] Other US surveys of the time estimated that at least 4 million children aged under 12 were hungry and an additional 9.6 million were at risk of hunger during at least one month of the year. Despite political criticisms of these surveys, further research suggested that even self-reported hunger, at least by adults, is a valid indication of low intakes of required nutrients. It should be noted that, ironically, the US, spent over $25 billion on federal and state programmes to provide extra food for its 25 million citizens in need of nutritional support.[152]

Implications for Policy

This chapter has sketched the bare bones of a highly complex global picture of diet-related health. Over the last half-century, epidemiologists have generated many facts, figures and arguments about the role of food in the creation and prevention of ill health, linking what humans eat with their patterns of disease. They raise a number of important questions: how much of a risk does poor diet pose? What proportion of the known incidence of key diseases like cancer, heart disease, diabetes and microbiological poisoning can be attributed to the food supply? What levels of certainty can be applied to the many studies that have been produced? Is diet a bigger factor than, say, tobacco or genetics? For policy-makers, the uncomfortable fact is that the pattern of diet-related diseases summarized in this chapter appears to be closely associated with the Productionist paradigm. Whilst the paradigm had as its objective the need to produce enough to feed people, its harvest of ill health was mainly sown in the name of economic development. Yet the public health message is clear: if diet is inappropriate or inadequate, population ill health will follow. Diet is one of the most alterable factors in human health, but despite strong evidence for intervention, public policy has only implemented lesser measures such as labelling and health education while the supply chain remains legitimized to produce the ingredients of heart disease, cancer, obesity and their diet-related degenerative diseases.

In making these tough assertions, we are aware that to piece together all food research evidence is immensely complex: more research is always needed; scientific understanding inevitably advances and is refined along the way. But surely, there is enough evidence for action. Certainly there is no shortage of reports and studies with which to inform policy. Calling for more research ought not to be an excuse for policy inaction. Policy procrastination is merely poor political prioritization.

Policy attention needs to shift from the overwhelming focus, enshrined in the Productionist paradigm, on under-consumption and under-supply to a new focus on the relationship between the over-supply of certain foodstuffs, excessive marketing and mal-consumption, and do so simultaneously within and between countries. Historically, there has been too much focus on public education as the main driver of health delivery; the diet and health messages, while welcome, have not always had the widespread or long-lasting effect that current data suggests is needed. While there have been reductions, for example, in coronary heart disease mortality rates in affluent societies, this is not universally true, and health education as framed in the West may not be universally appropriate. The food supply chain itself must be re-framed and must target wider, more health-appropriate goals.

Even rich countries are struggling to provide and fund equitable solutions to problems caused by diet: drugs and surgery, designer health foods, scientific research and public health education. But for developing countries, the majority of humanity, who have even fewer resources and weaker health care infrastructure, the picture is even more desperate. At the heart of the food policy challenge is the need to reinforce the notion of entitlement to food. While the 1948 Universal Declaration of Human Rights asserted the right to food for health for all, even into the new millennium the call is still not being adequately met, and, for humanity's sake, it must now be pursued with more vigour.

CHAPTER 3

POLICY RESPONSES TO DIET AND DISEASE

'War is probably the single most powerful instrument of dietary change in human experience. In time of war, both civilians and soldiers are regimented – in modern times, more even than before. There can occur at the same time terrible disorganization and (some would say) terrible organization. Food resources are mobilized, along with other sorts of resources. Large numbers of persons are assembled to do things together – ultimately, to kill together. While learning how, they must eat together. Armies travel on their stomachs; generals – and now economists and nutritionists – decide what to put in them. They must do so while depending upon the national economy and those who run it to supply them with what they prescribe or, rather, they prescribe what they are told they can rely upon having.'

Sidney Mintz, anthropologist of food, USA, b 1922[1]

CORE ARGUMENTS

There has long been a struggle to inject nutrition into state food policy. In the 20th century, understanding of food's role in meeting public health objectives fluctuated considerably with the mixes of scientific advance and social upheaval, notably war and domestic change. Nutrition has now split into two strands: one focused on social objectives such as poverty reduction, the other on biochemical mechanisms. For the Productionist paradigm, dietary guidelines have been the main battleground in nutrition and food policy. There is now a renewed interest in creating an integrated approach to food, diet and health. This could sit comfortably with the Ecologically Integrated paradigm and, in part, with the Life Sciences Integrated paradigm, too. A recurring concern in the discourse about nutrition and health is whether public and corporate policy should be focused on individuals or on populations.

Introduction

The scientific evidence for diet-related disease has not gone unnoticed within governments. In fact, governments' health ministries have helped create the evidence through statistical surveys and funding academic studies, but ensuing policy actions have in many instances been blunted or recommendations left to gather dust on forgotten shelves. The state apparatus of the Productionist paradigm has been controlled by the ministries most associated with production: usually agriculture and not health. Until recently, any connection between health ministries and the mass of evidence in relation to diet, disease and food supply has been subverted or resisted. This is despite more than 100 authoritative scientific reports between 1961 and 1991 recommending dietary change in relation to disease and health being published throughout the world.[2]

Despite this mounting body of evidence, health-focused state intervention in food supply, has been rare. Yet, in theory, there has been international public and nutrition policy commitment to address disease and health. But this has not happened by chance. In this chapter we describe what has been a 100-year food battle to bring nutrition policy to the forefront in state thinking on diet and health. We look at supply chain policy later.

The key sticking point in public health and nutrition policy is this: is it the 'individual' or the 'population-based' approach to food policy which is better? It is our view that health is not simply a personal choice, but that it reflects processes at work in wider society that require a full public response in order to set the framework within which individuals can make health-enhancing choices. However, to date much policy action has been directed solely at individuals, usually exhorting them to greater self-control and dietary restraint and balance. But what should a public policy response for the future be? We see this discussion as crucial since it will demonstrate the deep roots of the tensions between the Life Sciences Integrated paradigm and the Ecologically Integrated paradigm and the urgent need for an integrated ecological approach to nutrition and health policy.

CHANGING CONCEPTIONS OF HEALTH

Current public health has its roots in 19th-century Western reactions to industrialization. The downside of the newly rich and mechanized nation states was disease and poverty on an unprecedented scale,[3] and a new generation of social reformers began to argue that ill health both penalized its victims and threatened the fabric of society. In Victorian Britain a new approach to public health gradually emerged from this realization, pushed into legislation not by medical men but by civil servants such as Edwin Chadwick who framed the Public Health Act of 1848,[4, 5] and its implementation had to be fought for every step of the way.

One can get an idea of the nature of that battle by comparing today's unplanned urban sprawl in poorer countries with the paved orderliness of British cities where this reform took place over a century ago. Despite the railway age enabling the middle classes to live away from the squalor of the urban sources of their wealth, it was clear that ill health could not be escaped: diseases and pollution still prevailed. Containment, investment and prevention policies were reluctantly adopted. Unmade roads and open sewers were replaced by pavements, tarmac and drains. Slowly, over decades, better standards of housing and water and food provision were achieved, but not without contradictions. Pollution (industrial and human), for example, was flushed away from areas of habitation and work into the sea and the land in a manner that today would be unacceptable.

This huge investment in public health engineering – drains, roads, better housing – was based in part on the realization by the state that this same investment would deliver widespread health gains. This engineering solution is the classical notion of public health, but if offered material solutions in a 'moral' package for both the affluent and the poor. Today, this classical notion of public health as social engineering has been marginalized; in its place is an individualized conception in which we become 'consumers' of health and choose our own options for health – or not.[6] This was rooted in and legitimized by the triumph of neo-liberal economics in the 1970s and 1980s in Western democracies. Their ostensibly new notion of health was new in appearance only, since its roots lay in 19th-century liberal economics: in the academic world and, more importantly, in key global financial institutions such as the World Bank and the International Monetary Fund. A new orthodoxy proposed that state involvement in health should be reduced, that pricing and market mechanisms could manage services more efficiently and that privately run insurance was preferable to public schemes.[7]

Margaret Thatcher, the UK Prime Minister at the time, famously asserted that there was no such thing as society; social goods (such as investment in public health) were no more than multiple individual transactions which would be better left to the private sector and private transactions between the consumer and the service provider. The core notion was consumer choice rather than citizens' rights.[8] The challenge which she and others posed to defenders of the classical notion of public health was to fight out whether public health remained relevant and whether or not health investment was a societal responsibility. Despite this ideological marginalization in which food has been a key battleground, the case for a new public health has become ever more powerful.

Changing Conceptions of Public Health

A former Chief Medical Officer of England defined the practice of public health as 'the science and art of preventing disease, prolonging life and promoting health through the organised efforts of society'.[9] At the core of many definitions of public health is the notion that health is not an individual but a social phenomenon and that the social and natural environment frames the chances of people getting or preventing disease. We make a distinction between public health and a 'new' public health for the purposes of emphasizing significantly different approaches to health, and the methods of solving public health problems.[10,11]

Despite, in fact, having been used as early as 1911, the term 'new public health' is associated with a modern analysis of the ways in which lifestyles and living conditions determine health status. It also recognizes the need to mobilize resources and make sound investments in policies, programmes and services which create, maintain and protect health by supporting healthy lifestyles and creating supportive environments for health.

A concept of 'ecological public health' has also emerged in response to the changing nature of health issues and their interface with emerging global environmental problems,[12, 13] such as the destruction of the ozone layer, air and water pollution and global warming, all of which have a substantial impact on health and often elude simple models of causality and intervention.[14, 15] It emphasizes the common ground between achieving health and sustainable development, and it focuses on the economic and environmental determinants of health and on guiding investment towards optimal public health and sustainable use of resources.[16]

In 1998 the WHO proposed a 'new public health' focused upon 'lifestyles and living conditions [which] determine health status', and whose challenge is to 'mobilise resources and make sound investments in policies, programmes and services which create, maintain and protect health by supporting healthy lifestyles and creating supportive environments for health'.[17] A more recent definition called for:

> *an approach which brings together environmental change and person preventative measures with appropriate therapeutic interventions, especially for the elderly and disabled. [H]owever, the New Public Health goes beyond an understanding of human biology and recognises the importance of those social aspects of health problems which are caused by lifestyles.*
>
> *In this way it seeks to avoid the trap of blaming the victim. Many contemporary health problems are therefore seen as being social rather than solely individual problems; underlying them are concrete issues of local and national public policy, and what are needed to address these problems are 'Healthy Public Policies' – policies in many fields which support the promotion of health. In the New Public Health, the environment is social and psychological as well as physical.*[18]

This interpretation, though forward thinking at the time, still emphasizes 'those social aspects of health problems which are caused by lifestyles' whereas our conception of ecological public health emphasizes that unhealthy lifestyles are influenced by social and environmental factors. Such a perception of health is

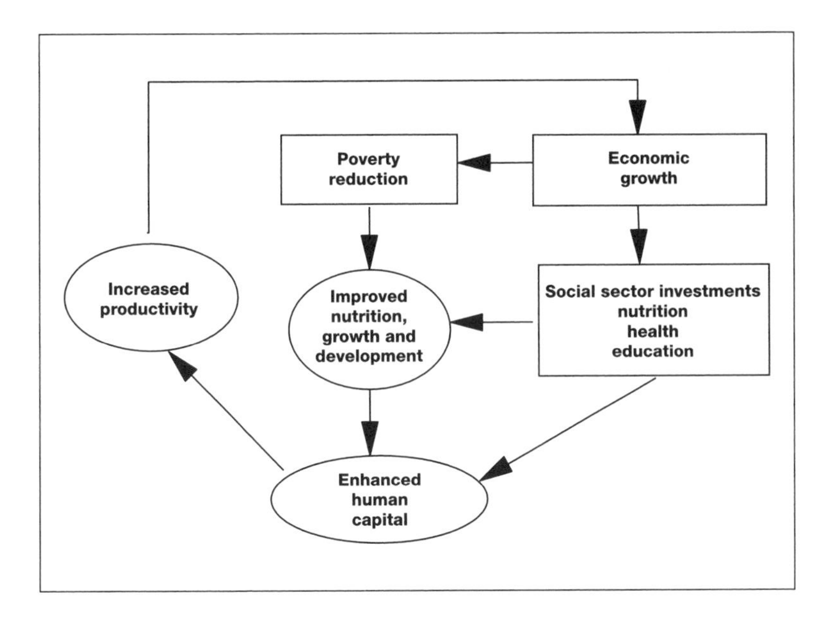

Source: R Martorell (1996) 'The role of nutrition in economic development', *Nutrition Reviews*, 54, 4, S66–S71

Figure 3.1 *Nutrition, health and economic growth*

illustrated by a model developed by Martorell who integrates economic and social factors into nutrition and health (see Figure 3.1), and places importance on social investment not just in health and education but also in nutrition.

This discussion of public health is a reminder that public health can mean different things. There is now a consensus that a public health perspective has to focus on the causes of ill health and on the factors which promote good health, rather than on the symptoms of ill health. The ethos of the new ecological public health is prevention rather than cure on an environmental, not just societal, basis. This has considerable implications for the food supply chain.

The Nutrition Pioneers: A 100-years War

Nutrition policy is a distinct part of public health policy which, like public health, has fought hard to gain a place at the political table. Dr James Lind,[20] although not the first to note the connection between diet and ill health, is often credited with putting modern nutrition onto a scientific basis. With trade routes dependent upon the health of ships' crews, the problem of scurvy was a major threat: it could devastate entire ships' crews. In 1753 Lind published the results of the first controlled study and established conclusively that scurvy could be prevented and cured by introducing citrus fruit into the diet. This was an early indication of how the science of nutrition could contribute to economic and even military well-being. Napoleon Bonaparte is famously stated to have said that an army marches on its stomach and to have initiated in the late 18th century the search that delivered canning, the means to perfect, portable and long-lasting food. (He also began the French sugar beet industry!).

Two and a half centuries on, nutrition covers a vast field ranging from social nutrition (for example, studying 'at risk' social groups), nutritional epidemiology (plotting the contribution of diet to diseases), biochemistry (the study of the biochemical interaction of nutrients and the body), sports and animal nutritions (optimizing physiological performance) and psychophysiology (including the study of food choice).

Partly fuelled by huge pharmaceutical and food-industry research funds, it is biochemistry that dominates nutrition today, with its researchers seeking profitable health benefits from within the diet. This pursuit began with Sir Frederick Gowland Hopkins' discovery in 1901 that the human body could not make the amino-acid *triptophan*, an essential part of protein, and that it could only be derived from the diet, demonstrating the principle that, without a proper diet, bodily function could be impaired or deficient.[21] Hopkins proved the existence of what he called food hormones or 'vitamines' (sic), most of which had been discovered by the end of the 1930s.

Nutrition, like any study concerning humans, is inevitably framed by social assumptions. Some see the pursuit of better nutrition as a social duty, while others view nutritional science as a tool of greater social efficiency or as an end in itself. Throughout the 20th century, nutrition was a battleground with some forces using it as an opportunity for social control and others arguing that it could liberate human potential. This tension between social control and democracy – 'top-down' science versus people-oriented science – still characterizes the world of food.

W O Atwater, an influential 19th-century American nutritionist, was an early critic of the national diet, but he also pursued a mechanistic approach to understanding food as fuel in physical labour: he calculated how much or little nutrient intake was required by different grades of manual workers, according to whether they were engaged in moderate or heavy work;[22, 23] he produced estimates of the protein, fat and carbohydrate required of workers performing light, heavy or moderate work.[24] His work was taken up east of the Atlantic by B Seebohm Rowntree, scion of a giant UK chocolate dynasty (now owned by Nestlé) and a founding father of UK food and welfare policy. Throughout the first half of the 20th century, Rowntree conducted both domestic and industrial surveys in his home town of York based on Atwater's calculations of nutritional need.[25, 26, 27] Rowntree used Atwater's minimalist approach to nutrition in order to accurately assess the needs of the poor: his minimum criteria made his findings about UK poverty all the more shocking.

In their desire to impose order on food systems, the pioneers of social nutrition developed and promulgated an approach that may seem simplistic today but which was enormously influential. In 1915 TB Wood and Sir Frederick Gowland Hopkins summarized its value:[28]

> *The human body, though doubtless in many of its aspects something more than a mere machine, resembles the steam-engine in two respects. It calls for a constant supply of fuel, and as a result of doing work, it suffers wear and tear. The body must burn fuel in order that the heat it is always giving off may be continuously replaced; and it must burn still more fuel whenever it does work. From this necessity there is no escape . . . It is, of course, the food eaten which provides these fundamental needs of the body; and if we are to understand properly the nutrition of mankind, we must bear in mind the two distinct functions of food – its function as fuel and its function as repair material.*

Arguments about whether diets should be calculated at a minimal or adequate level are inevitably politically highly charged. Not for the first or last time, diets were tinged with morality.

To someone on low income or experiencing food insecurity, it matters considerably whether an expert or an employer or the state is promoting better understanding of their diet in order to shave wages to a minimum or to make improvements in social conditions and individual well-being. Governments and employers might wish for greater influence, but should nutritional scientists collude or differ? During the 1930s' recession, as British wages and social welfare collapsed, a British Medical Association committee courageously argued that state welfare should not be based on an Atwater-type subsistence or bare minimum diet, but on one conducive to maintaining both 'health and working capacity':[29] optimum rather than minimum nutrition. Such historic debates highlighted nutrition's social assumptions and also highlight the fact that they hinged on a view of the food–body interface as an input–output system, with bodies as machines for which food is fuel and from which labour is the output. This somewhat mechanistic model began to be dismantled in the late 20th century with the emergence of biochemistry and the triumph of the doctrine of choice.

A More Sophisticated Approach to Food and Nutrition

An early exponent of a more complex approach to food and nutrition was paradoxically someone whose work might have led him to adopt the ultimate in food-control philosophies. Sir Robert McCarrison was a Director of the Army Medical Service in India, and he questioned the view of food as mere fuel, having been alerted to the impact of poor nutrition by the lamentable health of British army recruits.[30] This had first surfaced as a major issue following the Boer War when the British government set up an interdepartmental Committee on Physical Deterioration to address the eugenic argument that, unless more attention was paid to ensuring that the 'fitter' members of society produced more children and fed them well, it would be radically weakened. In its 1904 report, the Committee adopted a line of national self-interest and promoted optimal feeding, especially in the light of the parlous state of children's nutrition,[31] while McCarrison argued that, although a degree of self-interest was in order, nutrition does not act as an on–off switch, with the consumer having only enough or not enough;[32] gradations must be considered and vitamins introduced to improve the functioning of the body as a whole and not just to prevent specific diseases. McCarrison's view was that optimum nutrition was essential for a sound society, being the lubricant between good agriculture and good health. Education was the key, he argued: science should inform citizens and not control them.

Others, such as Sir John (later Lord) Boyd Orr, the first Director of the FAO, put more emphasis on structural factors. Boyd Orr founded the Rowett Research Institute at Aberdeen, now one of Europe's largest nutrition institutes, in order to explore and promote better scientific links between farming and health. Inclined to favour market solutions, he was gradually convinced of the need for State action. He had conducted a highly influential study of poverty in the 1930s,[33] and concluded that the key solution was adequacy of income: without income above a certain threshold, people could not purchase a nutritionally appropriate diet. He calculated that 50 per cent of the UK population was unable to afford a diet deemed adequate. This argument was taken up with vigour by campaigners for women and families, who argued that, because it was ultimately mothers who controlled food within homes, it was they who should be provided with food-oriented aid and education. The post-War changes in the domestic division of labour fuelled this view, with women entering the labour force and acquiring disposable earnings of their own, yet still retaining control over food in the domestic sphere.

By the end of World War II, the view that income was the key to health had triumphed in Western public policy. Yet the FAO, established in 1945, operated primarily as a production-oriented world body while the WHO, set up in the following year, remained locked in a medical model of health. Boyd Orr had been pioneering the case for better integration of health and agriculture for many years, and by 1945 the idea of joining health and agriculture had already been considered by scientists from all over the world. Such a food policy was reinforced by an earlier Conference on Food and Agriculture in the US in 1943. During World War II, the need to have the agriculture and health sectors collaborate became pressing in order to confront the world's growing nutrition and agriculture problems (such as the legacies of the 1930s 'Dustbowl' crisis and recession in the US). Boyd Orr's vision[34] was global: feed the under-consuming parts of the globe by unleashing the capacity of Western, and particularly US, farmers. His was a classically 'top-down' perspective which now would sit a little uneasily with our new era of community participation and people-led visions: his goal, though egalitarian, was essentially managerialist and Northern-led.

Such food policy vision lost ground after World War II, particularly as the Productionist paradigm took hold, because, first, the nature of production altered both on and off the land; second, there was a radical change in lifestyles, with increasingly affluent and less active proportions of all societies; third, with affluence, people could eat what the food supply chain offered them: 'feast-day' foods in abundance and every day.

Post-World War II Advances in Social Nutrition

The modern epidemiological (that is, population-based) position on the relationship between diet and health had a number of progenitors. Professor Ancel Keys' pioneering research in the 1950s showed that diet was a crucial factor in degenerative disease patterns:[35, 36] in his famous 'Seven Countries' study, he noted that the inhabitants of the island of Crete suffered least from the circulatory diseases. His data concluded that the Mediterranean diet was significantly healthier than the Northern, say Finnish, diet with its higher saturated-fat intake. In fact, it is not the Mediterranean diet alone that is so healthy, but a balance of nutrients and social conditions.[37]

Dr Hugh Sinclair, today scantly remembered in the world of public health nutrition for his work on essential fatty acids, until his death in 1990 promoted a view that is central to this book: namely, that the relationship between food and health requires total food supply chain thinking.[38] In 1961, he argued:

> *[W]e can now see clearly that the nutritional problems confronting the world are more urgent and serious than any others. They can be divided into two broad classes: the provision of adequate food for a rapidly increasing world population, and the disasters caused by the processing and sophistication of food in more privileged countries.*[39]

Yet, decades since scientists like Sinclair and others first voiced their concerns, strategic and policy thinking has continued to go in a different direction. Although nutritional and scientific understanding today is immensely more sophisticated than it was a century ago, the views of people like Keys, Sinclair, Trowell, Burkitt and McCarrison helped map a practicable view of food and health policy in both supply chain and population terms.[40,41] To meet their goals would require a major restructuring of the food supply and threaten many very powerful interests, because their arguments were rooted in – what is sometimes called 'social medicine' – the pursuit of medicine for social good, proposing that life-enhancing nutrition requires good distribution of food within and between populations, good food production and good skills and education.

Today, social nutrition is not regarded as the cutting edge, nor is it seen as a good vocation despite its long pedigree.[42] The academic discipline of nutrition in some respects now lacks public links and pursues a mechanistic (or biochemical) view of health more suited to industrial–pharmaceutical interests; it is in the process of being captured by the Life Sciences paradigm. The gap in the social role has tended to be filled by NGOs and food campaigners in the Food Wars rather than by scientists or dieticians. Nutritionists individually and personally subscribe to the social vision but too often they lack the sympathetic policy networks and skills. The health crisis and the evidence of the nutrition transition discussed in Chapter 2 should surely reinvigorate a social vision for nutrition.

Public Health Strategies: Targeting Populations or 'At Risk' Groups?

Health, in the WHO's 1946 founding charter's definition, is 'a state of complete physical, mental and social well-being, and not merely the absence of disease or infirmity'.[43] While everyone knows when they are not feeling well, many people do not actually think about their health until they are not well or they are reminded of the fragility of life, such as when a relative gets ill or dies. Well-being tends to be a coping issue: we say we are well when we can continue to do what we normally do. From a public policy perspective, the challenge is how to deliver a state of population health which is optimum and permanent and which meets ecological and economic criteria. Health strategies in recent years have tended to place an emphasis upon 'at risk' social groups: that is, people who already do not, or are likely not to, exist in a state of well-being. Studies into CHD in countries such as Finland, Thailand and Costa Rica have shown that getting a dietary improvement over the whole population improves every individual's health.[44, 45, 46] Such whole-population approaches have the capacity to make everyone healthier while retaining the normal diversity and range of behaviour,[47] and pushes the average health of a nation in a positive direction (illustrated in Figure 3.2).

Targeting whole populations provides governments with better chances of public health success, whereas targeting 'at risk' individuals could be socially divisive. This does not mean, as is sometimes assumed, everyone eating the same or a bland diet, but moving overall dietary behaviour *en masse* in a healthier direction. The population approach applies the medical dictum that prevention is better than a cure – for all citizens and not only the ill. This is why the insights of epidemiologists on health and disease patterns are so important. They remind us about public population health, not just individual health.[48, 49, 50]

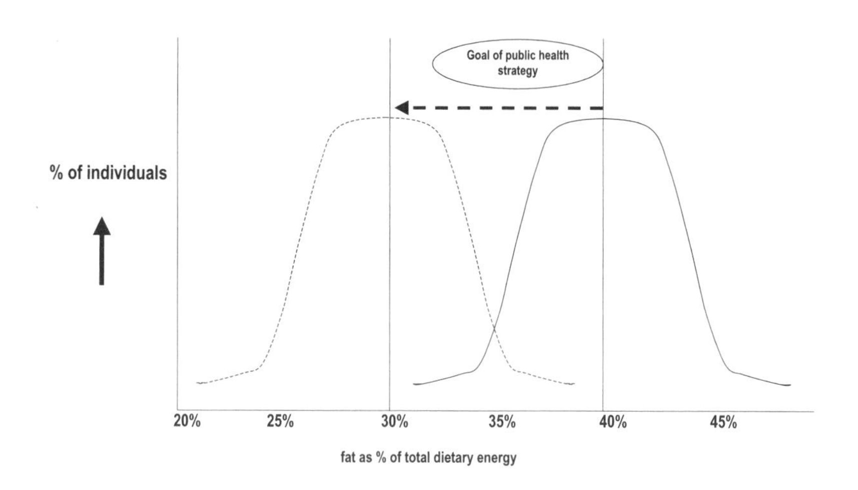

Note: The bell curve represents a total population

Figure 3.2 *Shifting a population in a healthier direction: a hypothetical example of fat intake*

Professor Geoffrey Rose claimed that, if high blood pressure patients are to get well, it is best not to see them as a different category of people but as the extension of normality; they are at one end of the normal statistical distribution curve; there needs to be less emphasis on their sickness than on the wider social determinants of both the healthy and the sick, such as the prevailing causes of high blood pressure. Public policy should aim to deliver social structures that allow individuals to remain well. This insight gives a new impetus for public policy; one which sees health as systemic, something that can be nurtured.

Dietary Guidelines and Goals

International public health and nutrition policy responses have to date been limited to the drawing up of dietary guidelines and goals for populations. These are often translated into giving advice to the public on what constitutes a healthy diet and even suggesting individual changes, alongside product labelling. Dietary guidelines have been both controversial and a major policy battle in the Food Wars, having, in theory, formed the basis of much government and nutrition policy since the 1980s. They may have proved extremely useful in setting goals and raising public awareness, especially through media coverage, but on their own they yield nothing. They need to be accompanied by measures to change supply and to ensure that people meet the set goals.[52] This too rarely happens, and unless implemented down the food supply chain, the guidelines remain only on paper.

Although there were some nutrition guidelines prior to World War II, the modern era of guidelines began in Europe with Norway, Sweden and Finland producing the first recorded governmental dietary guidelines in 1968, with the US following in 1970 and New Zealand in 1971. Initially, with the Nordic exception, these were guidelines produced by expert societies with government approval, rather than by government itself; but gradually most developed economy governments took responsibility for their production. Once created, guidelines inevitably have to be kept under review, given the constant shifts of both population dietary behaviour and scientific knowledge. They must also remain energetic, since reports from around the world continue to document strong grounds for action, yet receive only weary or inadequate policy response. In addition, dietary guidelines can be very threatening to certain corporate interests: to call for a population reduction in sugar, for example, does not go down well with the sugar industry or its main users such as soft drinks or confectionery manufacturers. To sum up, guidelines have to be kept under review and updated; they also need to be linked to actual delivery in the food supply chain. That is why guidelines can be the source of furious lobbying if they offend, for instance, the interests of the sugar, dairy and fats trades.[53, 54]

Today, a general consensus exists, with relatively minor country-by-country variations, on what national dietary guidelines should be.[55, 56] Generally, they promote a variety of foods; maintaining weight within an ideal range; eating foods with adequate starch and fibre; avoidance of too much sugar, sodium, fat (especially saturated fat) and cholesterol; and drinking alcohol only in moderation. People are also strongly advised not to smoke tobacco and to take regular physical exercise. Breast-feeding and pre-conceptual care are also usually recommended.

The early editions of dietary goals were received with due attention, while nutritionists remained alarmed by the problems generated by deficiency diseases. The new dietary guidelines, emerging from the developed world, encountered a certain resistance to the idea that too much of a nutrient might be bad for health, such as when, during the 1970s, high-fat diets came under suspicion.[57] Certain food producing sectors expressed fear at the potential loss of economic growth, particularly industries dependent on purveying fat, salt and sugars.

We need to distinguish carefully between recommended dietary allowances (or intakes) and dietary goals or guidelines: the first RDAs are the levels of intake of nutrients (for example vitamins) considered essential, on the basis of available scientific knowledge, to meeting the known nutritional needs of practically all healthy persons; the second dietary goals or guidelines, more recent than RDAs, aim to reduce the public's risks of developing chronic degenerative disease,[58] and recommend increased consumption of whole foods, foods high in dietary fibres, green vegetables, fruit and fresh produce.[59]

Population-based dietary guidelines have often been converted into consumer-oriented campaigns such as '5-a-day', pioneered in California and replicated in many countries: the consumption of at least five portions of fruit and vegetables daily to deliver the right mix of positive nutrients to protect health. (The Danes argue that this ought to be '6-a-day', and the Greeks, parents of the Mediterranean diet, now recommend '9-a-day'.)[60] Notwithstanding such specific guidelines, the focuses for policy and industry should be on quality, methods of production and processing and overall balance. These are such sensitive matters for the food supply chain that it is no wonder that dietary advice to consumers has been subject to vigorous lobbying and counter-attacks by certain parts of the food industry.

The Dietary Guidelines Battle in the US

The first edition of 'Dietary Goals for the United States', published in 1977, serves as an example of the degree of controversy that dietary guidelines have attracted and has been described as a 'revolutionary document'.[61] Its brief was to set quantitative target levels for reducing fat, saturated fat and cholesterol in the American diet, and the publication created a storm of protest and polarized professional opinion (despite the approval of many of the principal US researchers on diet and atherosclerosis).[62] It was variously seen as being premature; inadequately researched; politically motivated; promising too much; unreliable; puritanical; 'big-brother'-ist; and engendering a 'nutritional debacle'. Predictably, then, by the end of 1977, a second and revised edition of the 'Dietary Goals' had been published, including a foreword which included the following disclaimer: 'The value of dietary change remains controversial, and science cannot at this time insure that an altered diet will provide improved protection from certain killer diseases such as heart disease and cancer.'

The 'Dietary Goals' then, while never an official document, remained in circulation and drew public as well as professional attention to the need for national (and international) guidance on diet and health. For the first time, the American consumer was being called upon to make both quantitative decisions (how much food to eat) and qualitative choices (choosing some foods over others). Despite its deficiencies, it served to divide scientific opinion between those who believe in a 'targeted' approach aimed at those identified as 'high risk', and those who believe in a 'population' approach geared towards general health-promoting behaviours including changes in diet.[63] By early 1980, US nutritional advice was given official credibility with the publication of 'Nutrition and Your Health: Dietary Guidelines for Americans', by the US Departments of Agriculture and (then) Health, Education and Welfare, which is now reviewed and republished at five-yearly intervals.

Acceptance of the national dietary guidelines by many US food producer interests has been hesitant. For example, the US Department of Agriculture (USDA), has consistently promoted diets that 'emphasize consumption of foods from animal sources';[64] lobbying also came from the meat and dairy producers. Its 1992 revised guidelines, the 'US Food Guide Pyramid', suggested eating less meat. Commercial interests exerted considerable energies in undermining dietary advice from the Surgeon-General's department, while bolstering the Productionist mentality of the Department of Agriculture.[65]

In the UK, near-identical battles have been documented.[66] Indeed, the last two decades of the 20th century saw food-industry attempts to stifle health education through dietary guidelines.[67, 68] Even in Scandinavian countries, with their relatively advanced and integrated nutrition policies, industry resistance to dietary advice has been fierce, and progress in implementing the Norwegian nutrition policy of 1976, for example, was slow. Yet Norway seems to have a more integrated food and nutrition policy which has tried to balance the interests of agriculture, fisheries, the consumers and trade as well as of education and research. This is despite initial resistance from the dairy and meat industry, who tried to subvert the policy by producing 'expert evidence' that milk, butter and other dairy products posed no risk factors for CHD onset.[69]

Not all sectors of the food industry are threatened by guidelines; some indeed have responded positively by creating a multi-billion dollar market for 'healthy eating' products, albeit targeted at and premised on an individualized notion of health. However, dietary guidelines have been a rallying point for a new generation of health activists who saw little point in encouraging health education if an avalanche of consumer product choices made it hard to follow.

THE CASE AGAINST THE WESTERN DIET

By the end of the 1980s, eminent nutritional commentators such as Professor Nevin Scrimshaw were able to conclude: 'After years of controversy, a remarkable degree of consensus has developed regarding the kind of nutritional goals most likely to promote good health.'[70] The general 'consensus' on nutrition and public health can be summarized in the following statements:

- During the last half-century, Western diets have become unbalanced. They now contain too much fat in general, too much hard, saturated fat in particular, too much sugar and salt, and not enough fibre.
- Translated from nutrition to food, a healthy diet is rich in vegetables and fruit; bread, cereals (preferably wholegrain) and other starchy foods; and may include fish and moderate amounts of lean meat, and low-fat dairy products.
- The best diet to reduce the population risk of heart attacks is the best diet to protect against obesity, diabetes, common cancers and other Western diseases, and is also the best diet to promote general good health.[71]

However, even though there may be a consensus at public policy level, the scientific debates persist. For example, while dietary guidelines target high intake of fat as an unhelpful component of the Western diet, some scientists argue that the relationship between dietary fat and adiposity, or any other health outcome, is uncertain; breads and cereals advocated as part of a prudent low-fat diet may actually have significant negative effects in an environment of energy abundance, due to their high glycemic index.[72] Further, Professor Marion Nestle has claimed that US fat production by farmers is still excessive and that, once produced, will somehow make its way down consumers' throats.[73] So even if consumers resolve to reduce their fat intake, it may still feature in their diets via hidden routes – in processed foods, when eating out or 'on the hoof'; food processors may invent 'low-fat' products merely to jostle the shelves alongside a plethora of products bursting with hidden fats.

The European Union contains Member States which featured in Ancel Keys' original work as representing both the best (Crete)

Table 3.1 *The Eurodiet Project population guidelines, 2000*[74]

Component	Population goals	Levels of evidence
Physical activity levels (PAL)	PAL >1.75	++
Adult body weight as BMI (body mass index)	BMI 21–22	++
Dietary fat as % of total energy	<30	++
Fatty acids, % of total energy		
Saturated	<10	++++
Trans	<2	++
Polyunsaturated (PUFA)		
n-6	4–8	+++
n-3	2g linolenic + 200mg very long chain	++
Carbohydrates, total % of energy	>55	+++
Sugary food consumption, occasions per day	<4	++
Folate from food, micrograms per day	>400	+++
Dietary fibre, grams per day	>25	++
Sodium, expressed as sodium chloride, grams per day	<6	+++
Iodine, micrograms per day	150 (infants: 50; pregnancy – 200)	+++
Exclusive breastfeeding	About 6 months	+++

and worst (Finland) national case studies of diet and health. It is a matter of some urgency, therefore, that the EU at least considers homogenizing its diet policy to prevent drifts from the best- to the worst-case dietary scenarios. To this end, the European Commission decided in the late 1990s to initiate a project known as 'Eurodiet' to bring together experts in food and nutrition from all Member States to draw up guidelines for nutrition across the EU. Eurodiet worked for two years with a final meeting in Crete in 2000, and its recommendations (Table 3.1) suggest that the attainment of nutrient goals for fat, saturated fat, dietary fibre, complex carbohydrate and many vitamins and minerals would require an overall reduction in the consumption of meat and dairy produce and an increase in consumption of plant-based foods, particularly vegetables and fruit.

A New Approach to the Relationship Between Food, Diet and Health

Building on the experience of creating guidelines and the frustrations of implementing them, some policy-makers are increasingly looking to link guidelines to the food supply chain in a more

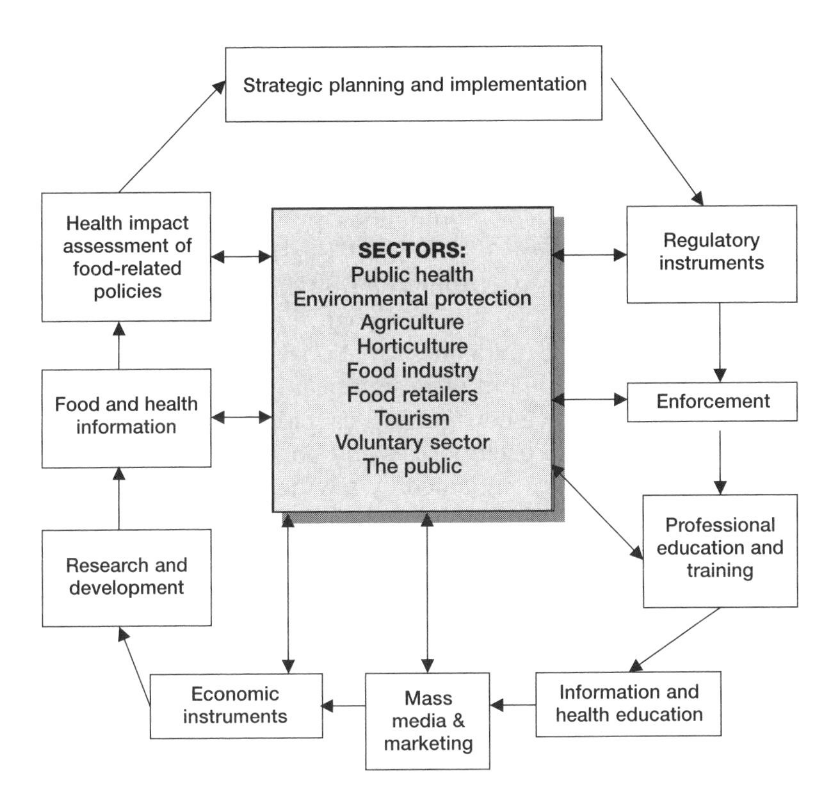

Source: WHO, *The First Action Plan for Food and Nutrition Policy, European Region of WHO, 2000–2005*, Copenhagen, WHO Regional Office for Europe.

Figure 3.3 *An integrated approach to food, nutrition and health*

meaningful way. What is their point if they are not enacted? There is a need to ensure greater transparency on committees, the curbing of commercial funding and the tackling of vested interests. This is a political battle. The approval of politicians needs to be won, if the formal levers of state influence are to be pulled in the public health's favour.

Internationally, there has been much effort to create a positive food policy consensus. The European Region of the WHO representing 51 countries, not just the European Union, has been pioneering an integrated approach to food, nutrition and health embracing the whole food-supply chain, as illustrated in Figure 3.3. The nutrition division of the WHO Europe argues that food policy should not only entail nutrition and food safety, but also a sustainable food supply. Delivering health, it believes, requires partnership.[75, 76] This is a unique and ground-breaking policy approach which requires the WHO and the FAO (sometimes idealogically divergent organizations) to work together. It is an international approach to food and health that is long overdue.[77]

At the global level, in 2000 the WHO also embarked on a major revision of its world guidelines, previously produced in 1990. Its 'Diet, Nutrition and Chronic Disease' report was in 1990 a major policy advance, issuing the first modern set of dietary guidelines, incorporating epidemiological data, compiled from numerous studies and guidelines at national level.[78] A 2003 vision retained the same title but was, significantly, jointly produced with the FAO and gave new guidelines for each major disease – from heart disease to osteoporosis and bone fractures.[79] Just as importantly, it was accompanied by a wider-reaching consultation exercise (2002–2004) involving consumers, the food industry and health professional organizations, and leading, hopefully, to global implementation and political backing with effect from 2005.[80, 81]

This innovation is promising. After a stormy history, public health nutrition and policy could be on the brink of a new era. A new, more integrated policy approach could be emerging in intergovernmental and national circles. But the ground-breaking work of the WHO, either globally or regionally, is only as strong as its national take-up. The WHO and the FAO are Member State organizations; they may have moral authority but little economic or political clout. The success or otherwise of the WHO and others in effectively developing and implementing measures towards a more integrated global food policy will be one of the major battles of the Food Wars over coming years.

At the heart of the war over health, then, is a policy choice between individual or population focus. (Some key features of these different approaches are summarized in Table 3.2.) All three paradigms have features which lend themselves to both individual and population perspectives, but the social policies

Table 3.2 *Individualist and population approaches to food and health*

Policy focus	Individualist public health approach	Population public health approach
Relationship to general economy	Trickle-down theory; primacy of market solutions; inequality inevitable	Health as economic determinant; public–private partnerships; inequalities requiring societal action
Economic direction for health policy	Individual risk; personal insurance; reliance on charity	Social insurance including primary care, welfare and public health services
Morality	Individual responsibility; self-protection; consumerism	Societal responsibility based on a citizenship model
Health accountancy/ costs	Costs of ill health not included in price of goods	Costs internalized where possible
Role of the state	Minimal involvement; resources best left to market forces	Sets common framework; provider of resources; corrective lever on the imbalance between individual and social forces
Consultation with the end user	As consumer; dependent on willingness to pay	Citizenship rights; authentic stakeholder
Approach to food and health	The right to be unhealthy; individual choice; demand will affect supply; niche markets	The right to be well; entire food supply geared to delivering health

which infuse nutrition, as we have seen above, pull the paradigms in one direction or the other. This, too, is a political struggle. Although the Life Sciences Integrated paradigm is heavily subscribed to by corporate interests offering an individualist approach to health, we believe that this paradigm requires a more population-oriented focus is required. The Ecologically Integrated paradigm, by contrast, is more rooted in a social or population approach but it too has individualistic tendencies. In the next chapter we outline the role of food business, the emergence of powerful food companies within and across borders, and we explore tensions between diverse sectors that will settle the shape of these paradigms with regard to health. This will help shape how relatively individualist or population-focused the paradigms become.

Obesity: A Case Study of Battles Over Policy Responses to a Problem

In this chapter we have sketched out the long struggle to inject nutrition into food policy and to bring food into the wider public health arena. There has been a bitter war over how food and health are not just matters of deficiencies and an absence of food, but also problems of inappropriate eating and a warped food supply.

The current obesity epidemic symbolizes the ongoing battles within nutrition policy – battles over targeting population versus 'at-risk' individuals, the delicate balances between government, commercial and 'consumer' responsibilities, and 'medicalized' versus 'social' nutritional approaches to policy.

Probably above all other diet-related diseases that we outlined in Chapter 2, obesity has become the main issue in public discourse now spearheading the case for a return to policy solutions grounded in social nutrition. That said, medicalized solutions to obesity are also being promoted from within the scientific community and industry – for example, solutions such as 'designer foods' for 'weight management', supplements and drugs, as well as full-scale surgical interventions, such as stomach stapling.

The proponents of social nutrition argue that public policy ought to consider, among many actions, taxes (to alter food price signals), transport policy (to build physical activity into daily life), schooling (to give positive messages at an early age) and controls on marketing, advertising and sponsorship (to protect people from unbalanced education). The middle ground is shared by the need to re-think the rights and responsibilities of industry, the state, consumers and key influences such as parents and schools, and particularly by the need to address the issue of overweight and obese children.

But almost all debates within public policy end up by having to face the wider social issues that frame obesity and help to explain why it has emerged so fast in evolutionary terms: what has created the obesgenic environment?

Public Policy Responses to Obesity

Across the obesity debate, policy responses are widely seen to lack integration.[82] In countries throughout the world, there is now public vocalization of the extent of the problem. Evidence has mounted; TV programmes and public health reports have been written; 'wonder diets' and products for losing weight have been promoted; government ministers and health professionals have all agreed on the extent and significance of the problem. For example, in the UK, the normally restrained chief medical officer called obesity 'a health time bomb with the potential to explode over the next three decades . . . unless this time bomb is defused the consequences for the population's health, the costs to the NHS [National Health Service] and losses to the economy will be disastrous' and the UK's Food Standards Agency was urged to 'alert consumers to the risks of regular consumption of foods high in calories, fat and sugars'.[83] The UK Parliamentary Health Committee also launched an inquiry into obesity.[84] In countries across the globe, similar processes were underway, with inquiries, calls for evidence and new action plans being created. In Norway, for instance, the long-established Nutrition Council was given a new emphasis to look at diet (input) and physical activity (output),[85, 86] and in Sweden a national working party was set up, charged to produce a new national policy and action plan.[87, 88] In the US, the Surgeon-General produced a hard-hitting assessment of the obesity crisis.[89]

The conventional policy reflex to offer single solutions for a single problem clearly does not fit the etiology of obesity that lends itself to multifaceted solutions. Key policy tensions in the obesity debate have centred around:

- environmental versus genetic causes;
- targeting 'at-risk' people versus a population approach;
- the relationship and roles between corporate versus public spheres;
- the relative importance of diet versus physical activity;
- fostering 'real' change versus public relations approaches to health education;
- government's role: the 'hands-on' versus the 'hands-off' state;
- people's responsibility: are they simply consumers or 'food citizens'?;
- agriculture and the food industry producing and marketing inappropriate foods and beverages.

As a result of tensions over these issues, and the inevitable sectoral jostling for policy position and influence, there is often no clear-cut or uncontested set of policy options and interventions. Nevertheless, two distinct packages of policy options have emerged. These illustrate the key themes in this chapter, not least individual versus population policies and social versus biochemical approaches to nutrition, in keeping with the paradigms. Political pragmatism – governments trying to be even-handed and not to offend too many powerful interest groups – means that elements of both are likely to be cherry-picked. The packages' differences, however, are important.

The individualized policy package on obesity includes:

- individual responsibility to chose 'healthier' or 'unhealthy' options;
- product branding and market solutions;
- healthy food options (but costs?);
- technical fix: 'designer' foods, supplements, drugs;
- genetic screening;
- gyms for those who can afford them.

The population policy package for obesity includes:

- addressing educational needs;
- regulatory measures, such as food labelling and advertising;
- food assistance programmes;
- health care and training;
- transportation and urban developments;
- taxes;
- governmental policy development;
- changes in food supply chain.

Industry Response

After years of publicly ignoring or denying its role in obesity generation, the food and beverages industry has been brought to account by the rapidity of worldwide public outcry over obesity.[90] It has known the evidence about the problem of obesity for decades,[91] but has tended to expect that it could be dealt with by the diet industry and within the dominant policy model of individualized choice and market forces.

The obesity issue is now, however, changing the food industry's nutritional landscape. The food industry is facing increasing scrutiny over the food and beverages that it produces and there has been a raised risk of increased regulation of advertising, labelling and marketing, notably in Europe. The threat of litigation over obesity in the US, akin to tobacco-style legal actions, has raised the policy temperature.[92]

But this legal threat to US industry interests seemed to have been overcome when the US House of Representatives passed a bill on 10 March 2004 stating that overeating is a problem for individuals, not the courts. The new legislation, formally known as the Personal Responsibility in Food Consumption Act, bars new cases and dismisses pending federal and state suits in which damages are sought as compensation for conditions connected to weight gain or obesity attributed to restaurant food. But the bill's authors said that it would not prevent suits brought because of a restaurant's negligence, false advertising, mislabelling or tainted food. In endorsing the bill, the White House said in a statement: 'Food manufacturers and sellers should not be held liable for injury because of a person's consumption of legal, unadulterated food and a person's weight gain or obesity.'[93]

But the industry is still torn between being defensive about its existing business operations and product offerings, and seeing obesity as a massive new business to produce weight management products targeted at the obese and those suffering diseases associated with obesity, such as diabetes. To give one example, Nestlé, the world's largest food company, in a presentation given to financial analysts in London on obesity in September 2003, outlined its strategy of how it was dealing with 'health and wellness' in the context of 'weight management'. Its presentation indicated a subtle shift in Nestlé's thinking about its product range to address weight-management market issues.[94]

Examples of Nestlé's thinking on weight management in terms of adding value to products ranges from 'reassurance' (such as portion control: claims such as 'contains one serving of vegetables') to 'healthier' (such as 'lower in saturated fat'), through to what Nestlé describes as 'active' (for example, weight and cholesterol control). Product development built around health and well-being, from Nestlé's perspective, involves combining food attributes (such as structure, ingredients, safety, taste, texture, flavour, aroma, satiety and appetite) and matching these to consumer health needs built around the health effects of particular foods or food ingredients, as well as targeting inter-individual differences related to diet and health.

Nestlé summarized its position on obesity thus: proper nutrition and adequate physical activity are integral to maintaining good health; every food has a role to play in achieving a balanced diet. The company also stated that it is committed to responsible communication about all its products, especially those consumed by children, and to clear and user-friendly nutrition labelling. The company plans to encourage nutrition education programmes for the public and to collaborate with public health bodies in both national and international efforts to reduce the incidence of global obesity. Key phrases for Nestlé are apparently now to be 'better taste = better nutrition' and 'easy choices = healthy choices'.

Nestlé is not alone in taking such a position. Other major global companies such as Kraft, Burger King and McDonald's have also reviewed their product portfolios in terms of portion size, composition, labelling, and marketing and advertising practices. Many companies are rapidly addressing the 'obesity risk' to their product portfolios built around fat and sugars, but also focus on business development and innovation efforts on higher-margin nutritional foods.

In all of these cases, part of the food industry's response is to engage and be proactive with public policy to tackle obesity. But it is not a united front. Other sectors of the food industry have taken a different approach and embarked on an aggressive attack against public policy strategies aimed at addressing diet and chronic disease. One of the most high-profile examples of this new open warfare against public nutrition policy has been by the US sugar industry and its response to the World Health Organization (WHO) strategy to help combat the worldwide problem of obesity.[95, 96]

This situation is rapidly changing and the battles over obesity have the potential to transform the food and health policy landscape, possibly in a radical way. They could open up a new awareness and unleash public concerns even further about what is good nutrition. This might herald a rebirth of policy shaped by social nutrition; but, as the brief history of nutrition policy outlined in this chapter has shown, this is likely to be a continuing rather than a quick battle.

CHAPTER 4

THE FOOD WARS BUSINESS

'Take care to drive your cow gently, if you want to milk her comfortably.'

Catherine the Great, Empress of All Russia (1729–1796)

CORE ARGUMENTS

Historically, central focuses of government policy have been agriculture and the facilitation of agribusiness, these industries forming the cornerstone of the Productionist paradigm. But today, in the newly evolved food economy, farming is no longer the driver it was; agribusiness, adding value to raw foods, has become more powerful. The consumption end of the food supply chain, namely retailing, food service and branded food manufacturers, increasingly dictates the terms and conditions of the consumer food market war; they are the brokers for the future but are not in consensus on any one vision for food. Collectively, corporate powers have consolidated both internationally and throughout the food supply chain, and it is corporate policy, as much as public policy, which is now shaping food policy agendas.

THE BATTLE FOR COMMERCIAL SUPREMACY IN THE FOOD SYSTEM

This chapter explores five key issues raised by the contemporary evolution of the food business:

1. Corporate power is now so great within and between national borders that it is redefining what is meant by a 'market'.
2. Food companies now have an increasing interest in health which they apply to marketing and product development, despite the decades spent resisting nutrition policy analysis.
3. There are tensions within the corporate sector over how to approach and deliver health, in part driven by intense business competition.
4. Today's food system is radically different from that of the past: it is fast evolving and full of risks.
5. Corporate policy is becoming more fully engaged in public policy to further its own interests and thus raising questions about accountability.

At the heart of the food economy is the way food and beverages are manufactured and processed. A feature of food manufacturing is its relative conservatism in the sense that many of today's leading manufacturers, such as Nestlé, Coca-Cola, Cadbury and many more, have their origins in the 19th century or early 20th century. With products 50, 60 or even more than 100 years old, some companies could be said to be anachronistic, but the food industry, as a combined sector, as formidable in its range as it is in its influence. It lacks, however, common goals and objectives. In this chapter, we explore the war for supremacy in commerce over food supply by detailing the key links in the food chain and how they are changing, arguing that, for it to provide a health-centred food supply, the food industry itself must undergo a fundamental rethink. We foresee increasing tensions within and between the food sector and food companies over how to address the issues captured by our three-paradigm model: currently the Life Sciences Integrated paradigm is being heavily promoted and major policy battles are being fought between it and the Productionist interests; nibbling away at both are the increasingly articulate proponents of the Ecologically Integrated paradigm.

The Productionist paradigm is reaching the peak of its power, with global corporate 'clusters' now dominating food supply while at the same time having to address a continuing stream of food crises and to implement far-reaching strategies of risk management. These giant clusters are locked in competition between themselves out of which they are adopting two key strategies: the first is a reliance on technology to resolve most problems; the second is alignment of their interests closer to the consumer. Often the two strategies are not mutually compatible: the pursuit of technology (such as genetic modification) is often in conflict with consumers. A newer model of the food economy is centred on technical manipulation, branding (emotional manipulation) and the rise to supremacy of the food retailing and food service industries. (Figure 4.1 summarizes some of the main battlegrounds in the war for commercial supremacy, and how these fit into the paradigmatic model outlined first in Chapter 1.) These key battlegrounds include:

- whether to intensify or extensify production methods;
- how to consolidate market shares and beat the competition;
- how to promote food business growth and development – through internal revenue generation (organic growth), or acquisitions and mergers, or cost-cutting and productivity efficiencies;
- how to expand internationally while still delivering business success locally and nationally;
- which new products and markets to develop;
- the best business strategy with regard to mergers and acquisitions, core competencies, investment processes (returns-on-investment) and geographical control;
- how to move from a primary commodity (or production-led) mentality to a consumer market-led approach and business culture.

The Origins of the Industrial Food Supply

Finding and producing food has been central to the whole saga of human survival and cultural evolution. *Homo sapiens* has been foraging for food for more than 100,000 years; has farmed it for 10,000 years; and has probably manipulated the environment (not farming per se) for considerably longer. But people have been farming and processing food industrially on a mass scale for mass markets across a high proportion of total diet for only 200 years. Now human society is gearing up to change the ancient

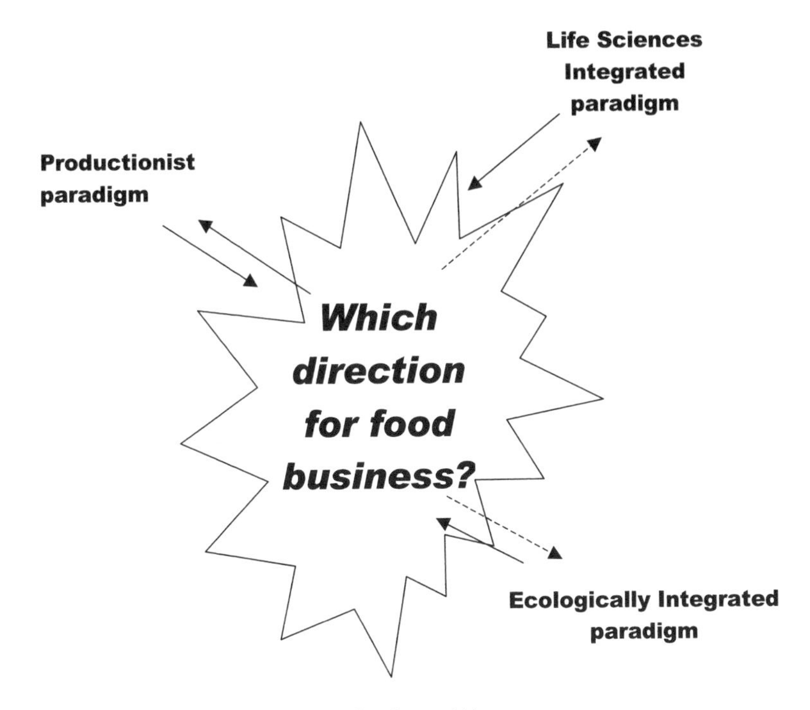

Key Battlegrounds in the Food Business Wars:

- Intensification vs extensification
- Concentration and competition
- Market share
- New products
- Marketing methods
- Business strategy
- Mass vs niche products
- Primary commodity vs consumer-led
- Technology adoption
- Globalization & global sourcing
- Retailer power

Figure 4.1 *Food industry within the paradigms*

and historic genetic make-up of both farm animals and crops in the belief this will secure future food supplies.

The modern food economy has been fed by unprecedented changes in human population growth and demographics. In particular, the decades since the end of World War II have seen

4.2a *United States (1995–1997)*

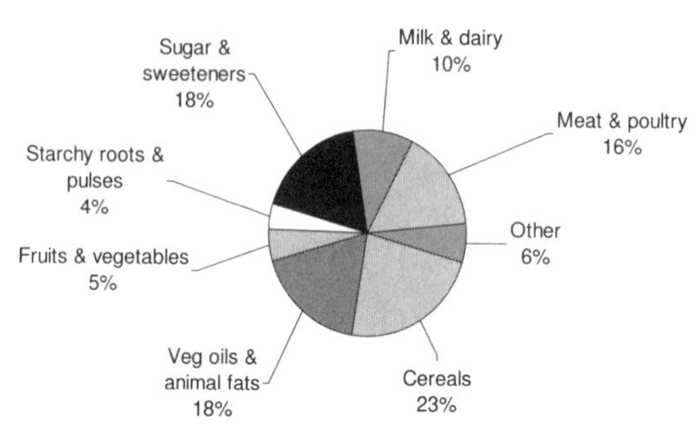

Note: US caloric consumption per person is up 20% since the early 1970s to the world's highest – a third above the global average. Although the US menu reflects the variety typical of industrialized countries, it's lower than the industrialized country average in cereals and higher in meats, sugar and sweeteners (such as found in soft drinks) and dairy products.

4.2b *European Union (1995–1997)*

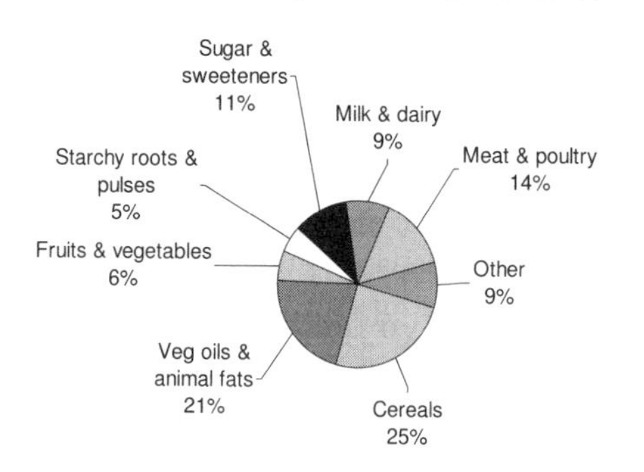

Note: Despite a taste for sausages and cheeses, Europeans consume fewer calories than Americans, and diets more closely mirror industrialized country averages. Compared with the United States, cereal consumption is higher, though down since the 1970s compared to the US trend. Diets have less sugar, but nearly 7% of calories come from animal fats, twice the US level.

4.2c *Latin America & Carribean (1995–1997)*

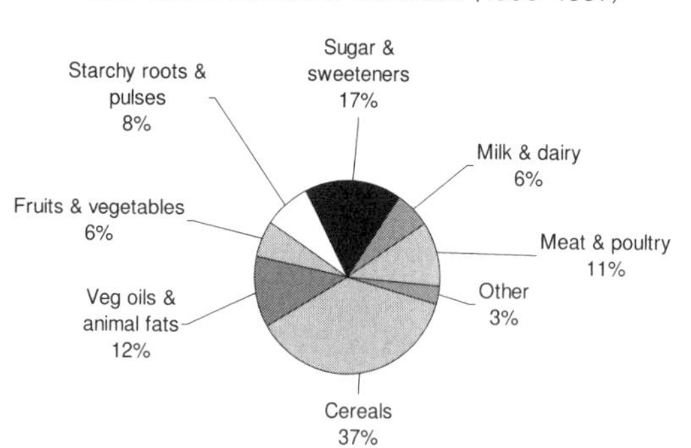

Note: Among major developing regions, Latin America shows the least overall dependence on cereals and the greatest apparent dietary diversity. Since the early 1970s, consumption of cereals and starchy roots and pulses has decreased, and diets now include more meat and poultry, sugar and sweeteners, dairy products and vegetable oils.

Souce: http://fas.usda.gov/info/asexporter/2000/Apr/diets/htm

Figure 4.2a–f *Diets around the world – proportion of energy derived from different foodstuffs determined by different regions' relative stages of development*

4.2d *Developing Asia (1995–1997)*

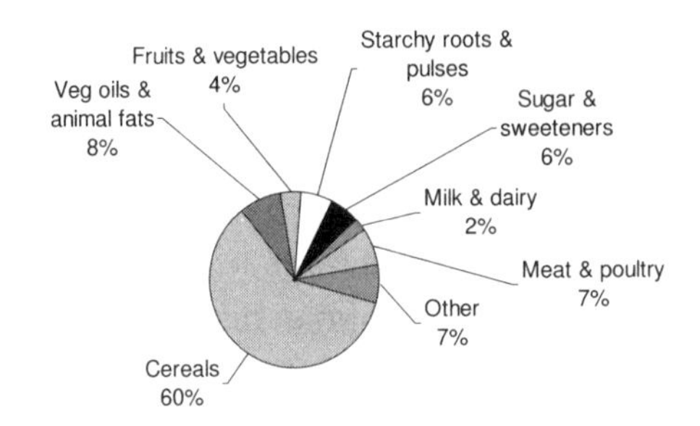

Note: Rising incomes and farm production have altered diets and steadily lifted this region's caloric consumption. Although rice and other cereals still account for 60% of calories, that's down from 67% in the early 1970s. The role of starchy roots and pulses has been halved, while the share of calories from meat and poultry more than doubled.

4.2e *North Africa (1995–1997)*

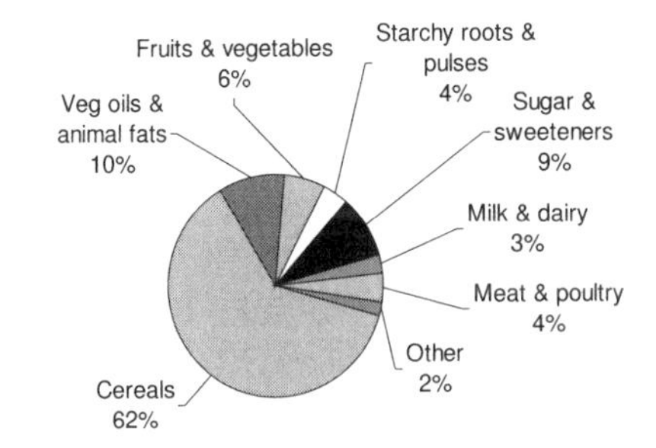

Note: Oil revenues have helped North African countries such as Egypt. Libya and Algeria boost average calorie consumption by nearly 40% since the 1970s. Cereals remain the primary source of calories, but cereal consumption has dropped slightly, while the share of calories from most other food groups has increased.

4.2f *Sub-Saharan Africa (1995–1997)*

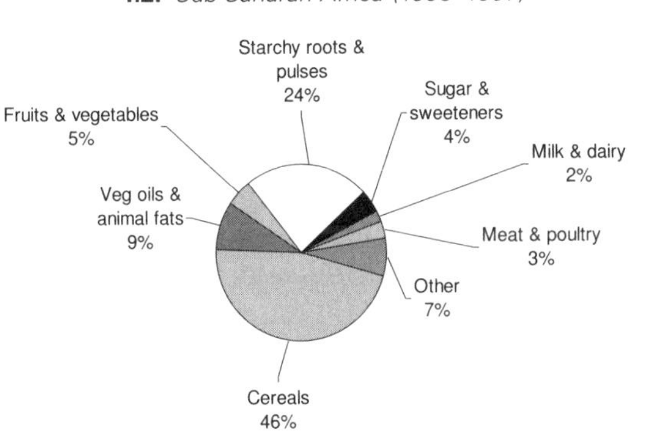

Note: In this generally poor region where tens of millions are chronically undernourished, daily per capita consumption was 2176 calories in 1995–1997, 20% below the global average. Cereals and roots still provide 70% of calories. More costly foods that would add variety and nutrition are consumed at some of the world's lowest rates.

exponential growth in the human population and in our methods to feed ourselves. At the same time world average life expectancy has doubled. Demographic transitions, from rapidly ageing populations in developed countries to the international mix of affluent food consumers, are also having an impact on the food economy. Since the 1950s, proportions of people living in cities increased fivefold from around 10 per cent to 50 per cent; and for the first time in human history the urban population will exceed the rural. Within 50 years, water consumption and global food yields have increased sixfold.

Over the same time-span, consumption of starchy staples has decreased from around 25 per cent or less of total food energy, and fat consumption has increased to 40 per cent.[1] Figure 4.2 illustrates how the proportions of energy derived from different foodstuffs vary between continents suggesting their relative stage of economic development. At one end of the dietary spectrum is the US where cereals account for 23 per cent of dietary energy, sugars 18 per cent and meats 16 per cent; at the other end is North Africa where cereals account for 62 per cent of dietary energy, sugars 9 per cent and meats just 4 per cent. During the past 200 years of industrialization, average individual levels of fat and refined sugar in Western society have increased fivefold and 15-fold respectively due to all-year-round supply of foods.[3]

With growing populations and more mouths to feed, especially affluent ones in the developed world, nationally dominant food and beverage companies have experienced near uninterrupted economic growth and dividend payments for their shareholders. This steady stream of profits and income came under increasing pressure during the 1980s and 1990s, with many food businesses struggling to produce the same levels of growth as in the past, as a new era of mass mergers and acquisitions, industry restructuring, and supermarket power emerged.

Today, the traditional food terrain is fracturing with some interests remaining firmly in the old Productionist camp, while others are marching towards the Life Sciences Integrated paradigm and yet others are turning to the Ecologically Integrated paradigm in their commitment to producing the real health benefits for the planet, societies and for individuals. For example, the German Federal Minister for Agriculture and Consumer Affairs, Renate Kunast, has called for an 'extensification' of food production in place of the 'intensification' model,[2] in 2002 the British government created the Curry Commission into the Future of Farming and Food also proposed the extensification of UK agriculture.[3] Such moves have split opinion about the future of farming, with some seeing a drive for ever more hi-tech farming, and others seeing extensification as a way of getting farmland taken out of production. Either way, the British Chancellor of the Exchequer gave an additional £0.5 billion to implement the Curry Commission's recommendations. At the same time, the European Commissioner for Agriculture, Franz Fischler, announced radical plans to switch funding and support away from commodity regimes towards more support for environmental and rural development.[4] All this change in the direction of agricultural policy followed the unprecedented series of scandals over food quality and confidence in the 1990s, culminating in the 1996 crisis when BSE was shown to have 'jumped' to humans.

This major period of change for the food system has been driven by wider economic forces which include:

- changes on the land which are transforming what agriculture produces and how;
- a rapid industry concentration of control over the food chain;
- labour-shedding and restructuring;
- changes both in the scale and technology of food factories;
- a new emphasis on product development, branding and marketing;
- new levels of control by food retailing and service over the rest of the food economy.

The implications and lessons of these structural changes for public policy are considerable. Public policy often lags behind the restructuring taking place in the food system and is largely reactive; companies are often more ambitious than politicians about the health of individuals; captains of industry are often clearer about the challenges to public policy. In some respects this is to be welcomed, but some of the changes being imposed on agriculture are now largely tailored to the powerful food manufacturing and retail sectors which are content to source foods from anywhere in the world at the expense of their national producers.

Why 'Health' is Important to the Food Industry

There has in the last two decades been an emergence of what the food industry calls 'nutraceuticals' or functional foods which offer health benefits beyond basic nutrition.[5] In the developed world the onus is very much on the individual consumer to choose these functional food options to help prevent or restrict diseases. To this end the food industry has developed its own health and nutrition agenda upon: heart health (in particular cholesterol-lowering products), gut health (to keep the microflora of the human gut in balance) and bone health (a raft of products fortified with calcium); 'energy' products; and foods and beverages fortified with vitamins and minerals. Of course, the reality is that no disease condition or human ailment is more immune by food product development or dietary supplement. Many of these nutraceuticals ('functional foods') require daily consumption over prolonged periods of time in high quantities to deliver similar results to those achieved in clinical trials on the bioactive ingredients, and they constitute a health strategy which is counter-intuitive to modern consumer culture already saturated with food choices and messages The health efficiency of such products to long-term health is not established, and many functional foods usually represent merely a short cut to new markets and big profits. In fact, some consumer advocacy groups have already highlighted the potential for fraud for these types of product, warning, for example:

> *Fortifying conventional foods with physiologically active substances might sometimes be appropriate, but the unbridled marketing in the United States of about $12 billion worth of dietary supplements annually shows the potential for defrauding and sickening consumers. The spread of such mischief to the far larger food industry could prove disastrous. The composition of, and advertising claims for functional foods should be governed by judicious government regulations, not by corporate marketing strategies.*[6]

Despite many examples of successful 'functional' foods and beverages driving this market for products with enhanced health benefits (such as the popularity in the US and UK of soy milk, 'nutrition' bars and dairy-based products containing 'live' bacteria) and the 'promise' that new GM varieties will provide vaccines ('golden' crops) with added vitamins or other health-enhancing bioactive substances,[7] the ethical and practical implications behind this new 'health revolution' are very rarely addressed. A key issue is whether food businesses alone can or should deliver long-term health benefits to individuals or populations, especially in such ad hoc, unevaluated and unproven ways.

The Changing Context for the Global Food Economy

Analysis of the food economy is subject to differing interpretations when viewed from particular economic perspectives. Traditional agricultural economics relies for its sustenance on demand and supply curves, price mechanisms and consumers' willingness to pay for such benefits as food safety or nutritional ingredients. However, this analysis and approach has been confronted by the global nature of food safety crises, consumer unease, and estimates of externalized costs for health and environmental burdens. These are now an incentive for business to change.

The UK BSE epidemic of the 1980s and 1990s, for example – though famously described by the then UK Agriculture Minister as a 'peculiarly British affair' after publication of the BSE enquiry report in 1999 – quickly became a concern for countries around the world: Japan, reported its first case of BSE in 2002, and in Germany, France and Italy, enough cases were discovered for beef sales to be halved within days.[8, 9, 10] More general public concerns about genetic modification and chemical use in food production have also served to raise awareness of major problems inherent in methods of food production.

Ecological imperatives

According to some predictions, the 'health' of our food supply is now in immediate danger. Lester Brown, president of Earth Policy Institute and founder of the Worldwatch Institute, for instance, has spelt out some of these dangers in his book *Eco-Economy*,[11] where he argues that food production has risen on the back of unsustainable use of inputs. On the positive side, since the 1950s: world grain production tripled; world production of beef and mutton increased from 24 million tons in 1950 to 65 million in 2000; and growth in the oceanic fish catch climbed from 19 million tons in 1950 to 86 million tons in 1998. On the down side, inputs also grew to match the production of food: for example, world fertilizer use rose from 14 million tons in 1950 to 141 million tons in 2000. Such food gains will no longer be sustainable if based on past production methods which fail to take global population expansion into account.

Brown sees the very foundations of food production as being in danger. He cites:

- falling water tables, especially in key areas of agricultural production;
- the deterioration of rangelands which supply most of the world's animal protein;
- soil erosion, loss of topsoil and the continuing destruction of croplands; and
- the collapse of global fisheries.[12]

Alongside this pessimistic interpretation of global food and ecological health trends, there also exists a more optimistic interpretation. Professor Tim Dyson of the London School of Economics, while foreseeing huge challenges ahead, argues that grain production, for instance, can keep up with demand.[13] We are not yet sure. Per capita grain production has peaked and is declining (see Figure 4.3). Optimists argue that the 40 per cent rise in soya during 1960 to 2003 compensates, but much of this goes to animals, certainly from US production. Some others argue that one of the key values of biotechnology is its capacity to deliver conservation as a means of addressing ecological degradation and to increase output to meet population growth.[14,15]

An extreme example of the need for the radical reshaping of food systems by putting ecological health limits in place was the collapse of the Newfoundland cod fisheries due to industrial-scale fishing, which resulted in processing factories being closed and tens of thousands of workers being laid off. More than ten

Source: FAO statistics; Colin Butler, personal communication, based on FAO data, March 2004

Figure 4.3 *Grain per person, world, 1961–2003*

years later, neither the cod stocks nor the industry had recovered.[16] Increasing numbers of environmental and agricultural experts are warning about the pressures of similarly intensive food production on natural resources.

Remarkable Changes in Agriculture and Food Production

The challenge for the future of food supply is how to develop sustainable production methods that take into account the health of the oceans, rangelands and croplands of the planet. Professor Marshall Martin, head of the Department of Agricultural Economics at Purdue University in the US, identifies 'environmental quality' as one of the key drivers of change,[17] a second is the industrialization of agriculture as farming continues to shift from a rural lifestyle to an agribusiness sector with a supply-chain mentality; and a third, according to Martin, will be firms that can 'link' the food chain and added value in order to benefit economically. An important aspect of these global dynamics is the way in which global agricultural trade has advanced in recent years. For example, today nearly one-third of US farm income is derived from export sales when, only 35 years ago, the US was a net importer of agricultural products. Land under soy production in Brazil today is 26 million hectares, an increase from just 200,000 hectares 30 years ago.

It is generally agreed by analysts that one of the major changes in world food demand will be in the demand for meat – something the Washington DC-based International Food Policy Research Institute (the IFPRI) has dubbed the 'livestock revolution'.[18] This will be driven by growing demand from developing countries; however, an individual person in the developing world in 2020 will still consume less than half the amount of cereals and just over a third of the meat products consumed by a person in the developed world. The IFPRI points out that people in the developing world collectively consume almost half of the global meat supply and by 2020 will consume two-thirds, as a result of an almost 50 per cent increase in per capita consumption. This presents a major production challenge. The IFPRI cautions: 'The Livestock Revolution will stretch the capacity of existing production and distribution systems and exacerbate environmental and public health problems'.

This projected 'livestock revolution' for animal protein is being fuelled by urbanization and population and income growth in developing countries. By 2020 developing countries will consume 100 million metric tons more meat and 223 million metric tons more milk than they did in 1993, dwarfing developed-country increases of 18 million metric tons for both milk and meat; developing countries will be producing 60 per cent of the world's meat and 52 per cent of the world's milk, led by China for meat production and India for milk production.[19] The revolution will also drive demand for increased cereal production for use as animal feed especially in developing countries. In total, global demand for cereals will increase by 39 per cent between 1995 and 2020 to 2466 million tons, and global demand for meat will increase by 58 per cent to 313 million tons.

Understanding the Modern Food System

Clearly these aggregate levels for cereals, meat (and for other foodstuffs) make it possible to build a model for projected demand and to start to weigh up the consequences of what growing so much food over the past 50 years and into the next 15 years means for farming and agriculture and for the processing industries.

The 20th century witnessed arguably one of the most significant food revolutions since settled agriculture began around 10,000 years ago. Although the roots of this revolution lay in the 18th and 19th centuries, it was in the 20th century that the application of chemical, transport, breeding and energy technologies transformed food supply. Beginning in the UK and US agricultural heartlands, radical changes to how food was grown, processed, distributed, and consumed were experimented with, applied and marketed.[20, 21] These countries entered the 20th century already well endowed with processing technology, such as the giant roller mills used in grain milling, and with industrialized baking machinery enabling, for example, the production of biscuits by the million. But these changes were nothing compared to what was to follow. Giant machinery soon began to replace human labour and 'Fordist' thinking was applied to both plant and animal production.[22] Large-scale experimentation was expended on trying to reduce nature's unpredictability.[23] Agrichemicals replaced the hoe; feedlots replaced grazing; monoculture replaced smallholdings. This growth owed much to the spread of fossil fuel culture, in particular the use of oil to drive machines.[24]

If the first half of the 20th century was marked by the industrialization of both agriculture and processing, the second half will surely be enshrined as the decades of retailing industrialization.[25, 26] New ways of packaging, distributing, selling, trading and cooking food were developed, all to entice the consumer to purchase. These power shifts in the food economy have contributed to the contemporary conflict within the food system between the 'Productionist' and 'consumerist' sectors: agribusiness versus consumer business; primary producers versus traders; food processors versus food retailers; and even production interests versus public health goals. Other changes saw year-round varieties of foods available from all corners of the world and the virtual elimination in the seasonality of fruits and vegetables by global sourcing.[27, 28] However, when reduced to its bare essentials, the global food economy is revealed as having a relatively simple base: most consumers eat foods from a core group of about 100 basic food items, which account for 75 per cent of our total food intake.[29]

From this simple base a food economy of immense size and economic power was created. Professor Marion Nestle notes that of the $890 billion food and beverage sales in 1996 (total US food and beverage sales were estimated at more than $1 trillion in 2002), nearly half was spent on items consumed outside the home but only 20 per cent of food expenditures went to food producers; the remaining 80 per cent constituted added value in the form of labour, packaging, transportation, advertising and profits. In that year, the US marketplace, featuring 240,000 packaged goods from US manufacturers alone, saw 13,600 new food product introductions: 75 per cent of these were candies, condiments, breakfast cereals, beverages, bakery products and dairy products. These and other foods were advertised to the tune of $11 billion spent on electronic and print media, with another $22 billion or so on coupons, games, incentives, trade shows and discounts. In the 21st century, food and beverage product launches internationally are occurring at the rate of more than 20,000 a year.[30] An important consequence of such economic activity is that the American food supply is providing 3800 kcal per day for every man, woman and child in the country, an increase of 500 kcal per day since 1970. This level is nearly twice the amount needed to meet the energy requirements of most women, one-third more than is needed by most men, and far higher than is needed by babies and young children. Professor Nestle concluded: 'These figures alone describe a fiercely competitive but slow-growing food marketplace, one in which food companies compete to sell more of more profitable foods'.[31]

Thus, the 20th century saw a food supply chain revolution characterized by integration, control systems and astonishing leaps in productivity, as measured in labour and capital use. Its more recent restructuring has seen key changes in:

- how food is grown, such as mass use of agrochemicals and hybrid plant breeding;

- how animals are reared: factory farms, intensive livestock rearing, prophylactic use of pharmaceuticals to increase weight gain;
- the emergence of biotechnology applied to plants, animals and processing;
- food sourcing: a shift from local to regional and now global supply points to monoculture;
- the means of processing, such as the use of extrusion technology, fermentation and cosmetic additives;
- the use of technology to shape quality and to deliver consistency and regularity;
- the workforce: labour-shedding on developed world farms; a retention of cheap labour and a strong push to 24-hour work;
- marketing – a new emphasis on product development, branding and selling;
- the retailers' role as the main gateways to consumers;
- distribution logistics, such as the use of airfreight, regional distribution systems, heavy lorry networks and satellite tracking;
- the methods of supply chain management – centralization of ordering and application of computer technology;
- the moulding of consumer tastes and markets – mass marketing of brands, the use of product placement methods, investments in advertising and marketing and the targeting of particular consumer types;
- the level of control over markets – rapid regionalization and moves towards globalization, and the emergence of cross-border concentrations;
- the growing importance of tough, legally backed intellectual property rights.

Table 4.1 plots some of these changes.

The Emergence of Food Company Clusters

Much discourse in food studies assumes the presence of a linked 'food chain'. Figure 4.4 details a simple food and grocery chain,

Table 4.1 *Directions of change in diet, food supply and culture*

	Issue	Direction of change
Dietary change (nutrition transition)		
	meat & animal products	up
	edible oils	up
	cereals	down but variable
	fruit and vegetables	Varies
	soft drinks	up
Food supply		
(concentration)	farmers	down
	processors	up
	retail (supermarkets)	up
	eating out ('fast food')	up
	long-distance foods	up
	seasonal dependency	down
	import–export trade	up
	capital intensity	up
Socio-cultural change		
	range of foodstuffs	up (urban)
	price relative to income	down
	packaging and appearance	up
	physical activity	down
	food inequality	up

in this case for the UK, indicating how value is accrued along its links and the sort of activities taking place.

At the end of the 20th century there was, amongst so many other things, a rebirth of academic interest in food studies. One rich seam which we can draw upon is known as agrarian political economy – or a 'food-systems' – an approach pioneered by rural sociologists trying to understand the deepening rural and farm crisis and the squeeze on farming.[32, 33] This approach tries to take further the idea of the food supply chain or 'value chains' in order to understand how and why different parts of that supply chain have an impact on one another,[34] in the context of the rural

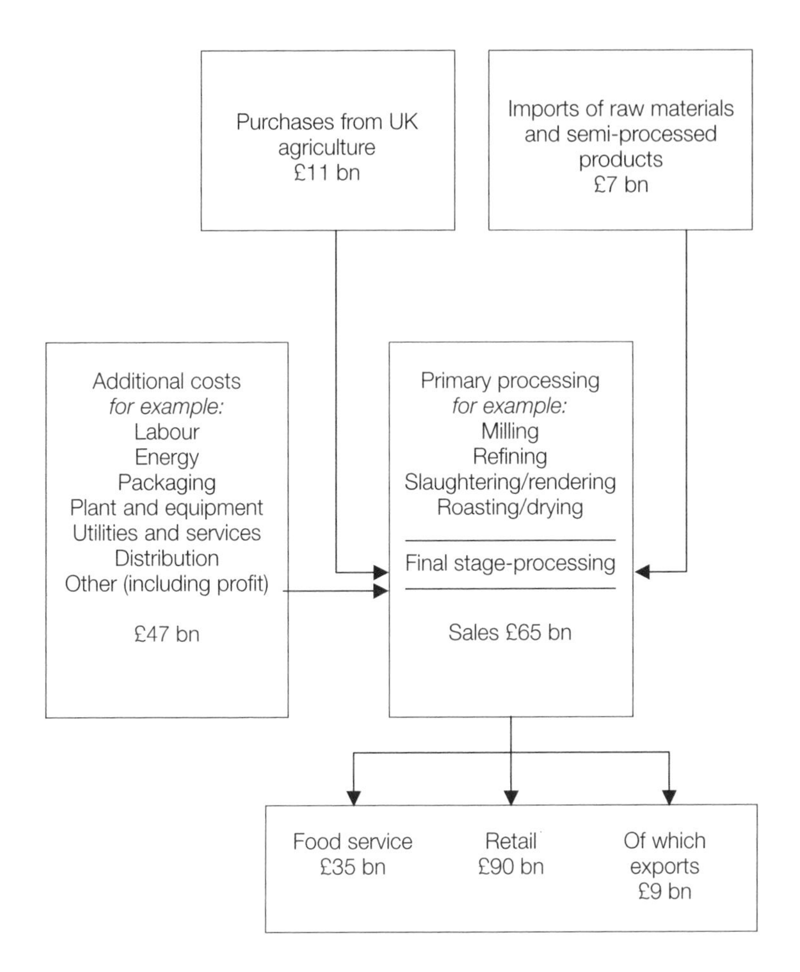

Source: IGD data in UK Food and Drink Federation (2001) *Submission to the Curry Commission*

* *Note*: The figures represent the monetary value of different stages as food passes through the supply chain from farming to point of sale.

Figure 4.4 *The UK food added-value chain, 2001**

economy. During the 1990s, social scientists interested in this approach focused on four major areas:[35, 36]

1 how agrarian structures and state agricultural policies developed over time in both the developed and developing world and in the growth of globalized food regimes;

2 detailed empirical analyses of particular agricultural commodity regimes, with an emphasis on the structures and strategies of multinational firms;
3 the role of regulation: how state practices and rules governing food systems are changing and how they shape agrifood systems;
4 how key players and networks of interest work together to formulate policy and define the workings of the food supply chain.

Although these four areas have been identified as distinct areas of theoretical and empirical study, in practice there are areas of overlap. From our perspective a weakness of this work, as in other areas of food studies, is that it too rarely acknowledges or includes health as either an outcome of the food supply chain or as a separate topic worthy of research and analysis.

Throughout the 1990s, a group of rural social scientists in the US led by Professor Bill Heffernan developed an analysis of how the food supply is being controlled and concentrated in the hands of a few 'food clusters': that is, firms that control the food system all the way from gene to supermarket shelf.[37, 38] They suggested that the changing nature of the food system means that relationships between food corporations are becoming much more complex and bound together. Acquisition is still the common method of combining two or more firms, but mergers, joint ventures, partnerships, contracts and less formalized relationships such as agreements and side agreements are also used. Professor Heffernan and his colleagues use the concept of 'clusters of firms' to represent these new economic arrangements – predicting the development of five or six dominant food clusters among US food companies, citing such links as between Cargill and Monsanto, or ConAgra and subsidiary companies along most parts of the food chain. This analysis reveals that control in the food system is exercised by a handful of firms, and US examples of Heffernan's findings are given in Table 4.2.[39]

Just 20 feedlots feed half of the cattle in the US and these are directly connected to the four processing firms that control 81 per cent of the beef processing, either by direct ownership or through formal contracts. In this sort of food system the farmer becomes nearly a contractor, providing the labour and often some capital, but never owning the product and never making the major

Table 4.2 *Concentration in the US food processing sectors*[40]

Sector	Concentration ratio	Companies involved
Beef packers	**81%**	Tyson [IBP], ConAgra Beef Cos Cargill [Excel] Farmland National Beef Pkg. Co
Pork packers	**59%**	Smithfield, Tyson [IBP] ConAgra [Swift] Cargill [Excel]
Pork production	**46%**	Smithfield Foods Premium Standard Farms [ContiGroup] Seaboard Corp. Triumph Pork Group [Farmland Managed]
Broilers	**50%**	Tyson Foods Gold Kist Pilgrim's Pride ConAgra
Turkeys	**45%**	Hormel [Jennie-O Turkeys] Butterball [ConAgra] Cargill's Turkeys Pilgrim's Pride
Animal feed plants	**25%**	Land O'Lakes Farmland Feed LLC/Purina Mills Cargill Animal Nutrition [Nutrena] ADM [Moorman's] JD Heiskell & Co
Terminal grain handling facilities	**60%**	Cargill Cenex Harvest States ADM General Mills
Corn exports	**81%**	Cargill-Continental Grain ADM Zen Noh

Table 4.2 *(continued)*

Sector	Concentration ratio	Companies involved
Soybean exports	**65%**	Cargill-Continental Grain ADM Zen Noh
Flour milling	**61%**	ADM Milling ConAgra Cargill General Mills
Soybean crushing	**80%**	ADM Cargill Bunge AGP
Ethanol production	**49%**	ADM Minnesota Corn Producers [ADM has 50% equity stake] Williams Energy Services Cargill
Dairy processors	**n/a**	Dean Foods [Suiza Foods Corp.] Kraft Foods [Philip Morris] Dairy Farmers of America Land O'Lakes
Food retailing	**38%**	Kroger Co. Albertson's, Safeway Wal-Mart Ahold

Note: The percentage figure is the concentration ratio, relative to 100 per cent of the top four firms in specific food industries

management decisions. The pricing mechanisms and hence market so dear to traditional agricultural economists are all but hidden; through contractual relationships the value is distributed up the food chain and not at the farm level, as the production stage is integrated into the larger food system. If, as another example, 95 per cent of US broilers are produced under production contracts with fewer than 40 firms, there is essentially no price discovery for chicken feed, day-old chicks or live broilers; the food producer does not sell at these stages and the price is not revealed until sold to the final consumer. Basically, in the economic sense, there is no 'free market' for live broilers, just one seamless supply chain. Professor Heffernan concludes:

> *Today, most rural economic development specialists discount agriculture as a contributor to rural development. The major reason why agriculture contributes so little to the community is because of the emerging structure of the food system.*[41]

Food clusters give unprecedented power to a relatively small number of companies, and it is difficult to establish dialogue with vast corporations such as the US conglomerate ConAgra. In 1996 it accounted for 25 per cent of US sales in feed and fertilizer, 53 per cent of sales of refrigerated foods and 22 per cent of all grocery products. Much food production in the developed world, however, is still in the hands of smaller companies in a local or regional context: food is still produced and consumed by rural populations. It is estimated that 1.9 to 2.2 billion people are today still excluded, directly or indirectly, from modern agriculture technologies. In fact, only about one-fifth of the world's 6 billion people are able to participate in the cash or consumer credit economy that modern food capitalism thrives on.[42]

Farming Becoming 'Irrelevant'

One of the tragic consequences of the restructuring described above is that the loss of farmers and family farms in the US and Western Europe, for example, has little impact on transnational corporations: corporations are often mainly concerned about sourcing a product with the least cost and then move the product where it can be sold for the highest price. In many poor countries, workers in rural areas receive less than $5 a day, with health and environmental regulations unlikely to be enforced, again helping to drive down costs. It is remarkable how cheap labour characterizes the supposed efficiencies of the food supply chain. Behind low-cost food can be even lower cost labour.[43] Transnational corporations are experts at reaping the economic benefits of globalization while pushing the economic, social and environmental costs onto the public.[44] In turn this makes developed world farmers 'high cost' producers. Some neo-classical economists now openly argue that consumers in the developed world no longer need their own farmers because countries can import food from poorer countries more cheaply. The solutions offered by right-thinking analysts emphasize global market efficiencies; this is why an estimated $35 billion is spent on agricultural research annually.[45] It is not clear that this money is well spent. As in other areas of research, we would argue that the approach of such farm economics has ideological framing assumptions embedded within the science. Half a century ago, economists were justifying and promoting a decentralized agriculture production and processing system, while today, much economic theory defends a highly centralized monopolistic or near-monopolistic system.[46]

Industrial farming or agribusiness – seeing fewer farmers, on bigger tracts of land, producing a greater share of the total food supply – is shaping food production into monocultures (sometimes referred to as 'green concrete'): either the same crop is grown year after year in the same field or very simple rotations are used (such as corn–soybean–corn–soybean). The shift towards monoculture and homogeneity in food production can only happen with a technology 'bundle' consisting of mechanization, the manipulation of crop varieties, the development of agrochemicals to fertilize crops and control weeds, insects and other crop pests, as well as antibiotics and growth stimulants for agriculture animals.[47] Further up the production chain it means the increasing fractionation of foodstuffs into smaller and smaller biological components and ingredients and then the recombining of these fractions into 'value-added' retail food products. Such activity has spawned a massive food technology industry whose practitioners have been increasingly involved over the past ten years in 'adding' health 'benefits' to foods and beverages.

Food monoculture gives rise to the agricultural 'treadmill': there is downward pressure on prices for farmers and upward pressure on the inputs needed for production; farmers are forced to adopt new technologies and increase their scale of production or go out of business; over time, the geographical concentration of production is into narrower and narrower locations or centres of production.[48] By 1994, 50 per cent of US farm products came from 2 per cent of farms, while 9 per cent of produce came from 73 per cent of farms. For many food technologists, monoculture means battling to make retail food products cheaper to manufacture while maintaining their integral qualities.

This shift of the food dollar from farmers to other parts of the food chain has gone hand in hand with intense concentration at every other link in the food chain – from farm inputs (such as fertilizers and pesticides) to food processing and retailing. In the US, the share of the consumer's food dollar that gets back to the farmer has dropped from around 40 cents in 1910 to 7 cents in 1997; in the UK, the same has occurred and the gap between farmers' prices and the retail prices has become a very contentious issue. In July 2002, for example, supermarkets were accused by the UK National Farmers' Union of overcharging consumers for milk but underpaying dairy farmers: the price paid to farmers for milk had fallen over the previous year from 20 to about 14 pence a litre, while consumer prices had remained constant at around 45p.[49] The fall in the monies received by US farmers was graphically illustrated in March 2000 at a farm rally in Washington DC when farmers served legislators a 'farmers lunch'. Typically an $8 lunch – of barbecued beef on a bun, baked beans, potato salad, coleslaw, milk and a cookie – they charged only 39 cents, reflecting what farmers and ranchers actually received to grow the food for such a meal.[50]

The economic pressures of the treadmill have led to a sharp decline in the numbers of so-called 'inefficient' farms, with smaller family farms being particularly badly hit. For example, in the US there were close to seven million farms in the 1930s, but less than 1.8 by the mid-1990s; in France 3 million farms in the 1960s, yet fewer than 700,000 in the 1990s; 450,000 farms in the UK in the 1950s, half that number in the 1990s. Over the past 50 years the number of actual farmers has declined by 86 per cent in Germany, 85 per cent in France, 85 per cent in Japan, 64 per cent in the US, 59 per cent in Korea and 59 per cent in the UK.[51]

For economists these long-term trends represent greater efficiencies from economies of scale. From this perspective, however, the remaining number of large farms still remains high, especially in comparison with concentration in other industrial sectors, and many economists predict a continuing reduction in farm numbers, as though this is natural and desirable.[52] While the number of farms remains stubbornly high, in many global primary commodity markets there already is considerable concentration and control of markets: five corporations, for example, market 60–90 per cent of all wheat, maize and rice; three corporations control 80 per cent of banana trade; three corporations control 83 per cent of cocoa trade; three corporations control 85 per cent of tea trade.[53] It is a similar story with agrochemicals: in the late 1980s, the top 20 firms worldwide accounted for around 90 per cent of sales; by the late 1990s, ten firms controlled this much of the market; in 2003, it was just seven.[54]

To use Canada in more detail, while in 2000 there were 276,548 farms supporting the Canadian food chain (down from 430,522 in 1966), today:

- three companies retail and distribute gasoline and diesel fuel;
- three produce most nitrogen fertilizer;
- nine make pesticides;
- four control the seed market;
- three produce most of the major farm machinery;
- nine grain companies collect all Canadian grain;
- two railways haul it;
- four companies dominate beef packing;
- four companies mill 80 per cent of Canadian flour; and
- five control food retailing in Canada.[55]

In a 2003 study the Canadian National Farmers Union (CNFU) argued that, over the 14-year period since the signing of the 1989 Canada–US free trade agreement (CUFTA) (and the implementation of the North American Free Trade Agreement in 1994 and the World Trade Organization Agreement on Agriculture in 1995) 'free trade' is simply not working: while Canadian farmers have more than doubled exports with agrifood exports growing from CAN$10.9 billion in 1988 to CAN$28.3 billion in 2002, Canada is facing its worse farm income crisis since the 1930s: 'Free trade agreements may increase trade but, much more importantly, they dramatically alter the relative size and market power of the players in the agri-food production chain . . . much more significant . . . may be the effect these agreements have on the balance of market power between the farmers and agribusiness corporations.'[56]

But even the economics of the remaining farms is distorted towards the largest. In the US, the 122,000 largest farms, representing only 6 per cent of the total number, receive close to 60 per cent of total farm receipts. These large farms have also been able to reap a disproportionate amount of government support payments, receiving over 30 per cent of the payments for commodity programmes. In 2002 President Bush announced a $180 billion support package for farming, while in the EU the financial strain of the $40 billion Common Agriculture Policy is prompting calls for radical reform. Although calculations of farm subsidy figures need to be treated with caution, it has been estimated that in 1998 agricultural subsidies in OECD countries totalled $362 billion – 2.5 times the combined gross domestic product of all the world's least developed countries.[57]

A New 'Health' Colonialism?

A more recent feature of agricultural trade has been the development of new food commodity chains, especially for fresh produce, built upon production in developing countries for consumption in rich countries. This trend is giving rise to a different form of competition not only between 'logistic' chains such as bananas and exotic fruits, but also whole food production systems. Table 4.3 illustrates the banana logistics chain for a Caribbean producer and a UK supermarket, showing product flow from field to store. Bananas were in 2003 a significant contributor to UK supermarket profits, suggesting health and high turnover at the point of consumption; the sad reality is that

Table 4.3 *The banana supply chain*[60]

Stage in product flow	Time taken
Field to leaving the packhouse	3 hours
Packhouse to loading in the ship	11 hours (maximum)
Shipment to Zeebrugge	13 days
Transport to storage in Kent, UK	12 hours
Storage in Kent, UK	5–6 days
Delivery and store in supermarket depot	12 hours (maximum)
Delivery to store	1–2 hours
Maximum shelf life in store	2 days

most banana workers in Ecuador earn wages that are below half of what is required to feed an average family. Of every £1 retail price for Ecuadorian bananas, the plantation worker receives just 1.5 pence, whereas the plantation owner receives 10 pence, the trading company 31 pence (of which the EU tariff is 5 pence), the ripener/distributor 17 pence, and the final retailer 40 pence.[58] For a 40lb box of Costa Rican bananas sold in the UK supermarket for the equivalent of £14.69, the grower would have received a maximum of £2.22, only 15 per cent of the resale value.[59]

The banana story is but one part of what has been described as a disaster for small farmers who trade in tropical commodities such as coffee, cocoa, rubber and sugar where the price per tonne in US dollars is lower today than it was in the early 1980s.[61] This hard analysis presents a challenge to the conventional liberal argument that world trade rules need to be redesigned to allow 'fair' prices for such goods. The new analysis suggests that encouraging developing countries to pour out such commodities creates over-production and is not the panacea for the ills of economic development.

Horticultural exports from developing countries have also become a major growth sector in international trade, driven by supermarket policy to offer consumers year-round supplies of fresh produce sourced from different areas of the world. A team led by Dr Hazel Barrett and Professor Brian Ilbery has documented the horticultural trade for Kenya where 85 per cent of the country's horticultural trade is destined for just four EU countries: the UK, The Netherlands, France and Germany. Most of this business is through supermarkets. By the mid-1990s horticultural exports accounted for 10 per cent of total Kenyan export earnings (and are the third most important agricultural export after tea and coffee),[62] supplying green beans, mange-tout, runner beans, okra, chillies, aubergines, avocados, mangoes and cut flowers to Europe, 93 per cent of which is delivered by air. Barrett and her colleagues have documented how, to meet the growing demand for consignments of produce to EU supermarkets, Kenyan export production concentrated in the hands of the larger and highly capitalized producers. There are now signs that China could enter this export trade. Historically associated with intensifying its agriculture to keep pace with a burgeoning population, China's food policy is poised to reposition itself as a potential exporter of labour-intensive foods; it has low production costs, amenable to some intensive but high-value exports. One estimate of onion production, for instance, found that China had costs of US$3605 per hectare, compared with onion growing costs of $5110 in the US and $20,901 in Japan.[63]

The mechanics of the Productionist paradigm have altered, both figuratively and literally, the landscape and practice of farming and agribusiness. As Productionist-style agriculture struggled to find acceptable levels of profitability, it turned increasingly to global trade liberalization as one route out. But as the limits of chemical farming technologies are reached, the new business model is aiming to capture value throughout the food chains, starting with the seeds to grow food, and to impose a new homogeneity on agricultural production. Control is the crux of the Life Sciences Integrated paradigm. The Ecologically Integrated paradigm, however, aims for different kinds of efficiencies through reducing inputs, and re-aligning of production and land to smaller-scale, biodiverse and localized farming practices.

The Global Scope and Activity of Food Processors

In food manufacturing, as well as in agriculture, big business is getting bigger, despite small-scale food companies continuing to survive through 'niche' and specialty markets. The world's top three food processing companies at the turn of the millennium were Nestlé with global food sales of US$44,640 million, Philip Morris (Kraft) topping $26,532 million and ConAgra Foods Inc with sales of $25,535 million, giving combined global food sales for these three alone of a huge $97 billion. Table 4.4 lists the world's 50 largest food groups based on food sales. It is important to recognize that, compared to other industrial sectors such as aerospace or the automotive industries, food manufacturing is not nearly as concentrated; the food industry remains highly fragmented.

Another survey, this time of the world's leading 50 grocery manufacturers, found that only four food companies were big enough to make it into the Fortune Global 100, a US business index of the biggest corporations in the world. The same survey

Table 4.4 *World's top 50 food groups, 2000*[64]

Rank	Company	Total sales $ millions	Food sales $ millions
1	Nestlé	47,489	44,640
2	Philip Morris Co Inc	80,356	26,532
3	ConAgra Foods Inc	25,535	25,535
4	Cargill	47,602	22,500
5	Unilever Bestfoods	41,403	20,712
6	The Coca-Cola Company	20,458	20,458
7	PepsiCo Inc	20,438	20,438
8	Archer Daniels Midland Company	20,051	20,051
9	IBP Inc	16,950	16,950
10	Arla Foods	15,824	15,824
11	Diageo	16,938	15,584
12	Mars	15,300	15,300
13	Anheuser-Busch	12,262	12,262
14	Danone	12,576	12,123
15	Kirin Brewery Co	12,826	11,928
16	HJ Heinz Company	9430	9430
17	Asahi Breweries	11,351	9194
18	Suntory	10,485	8823
19	Snow Brand Milk Products	9256	8516
20	Sara Lee Corporation	17,511	7705
21	Dairy Farmers of America	7600	7600
22	Nippon Meat Packers	7389	7389
23	Tate & Lyle	8058	7252
24	Tyson Foods	7158	7158
25	General Mills Inc	7078	7078
26	Kellogg Company	6955	6955
27	Interbrew	7038	6898
28	Cadbury Schweppes	6515	6515
29	Parmalat	6466	6466
30	Maruha	7232	6292
31	Campbell Soup Company	6267	6267
32	Eridania Beghin-Say	8625	6038
33	Smithfield Foods Inc	5900	5900
34	Heineken	6175	5891
35	Associated British Foods	6289	5660
36	McCain Foods Ltd	5603	5603
37	Yamazaki Banking	5919	5504
38	Suiza Foods Corporation	5756	5410
39	Ajinomoto	7372	5308
40	The Quaker Oats Company	5041	5041
41	Meiji Mild Products	5748	4943
42	Dole Food Company Inc	4763	4763
43	Cenex Harvest States Co-operatives	8571	4714
44	Procter & Gamble	39951	4634
45	Lactalis	4609	4609
46	Novartis	20,879	4593
47	Tchibo	8819	4410
48	Danish Crown	4318	4318
49	Uniq	4249	4249
50	Hershey Foods Corporation	4221	4221

found that the global top 50 grocery companies also accounted for less than a quarter of the consumer expenditure on food, drink and tobacco in the developed world – although just 20 of the 50 accounted for 69 per cent of combined turnover.[65] So even within the 50 largest players, there is considerable concentration and differentiation.

The food industry is the leading industrial sector in the European Union (with 15 Member States), with production estimated at €572 billion accounting for 13 per cent of the total EU manufacturing sector.[66] It is the EU's third-largest industrial employer with over 2.5 million employees, or 11 per cent of employment. Although dominated by a relatively small number of giant food firms, the food sector is characterized by a relatively low concentration. In 2000, 99.3 per cent of all food and drink companies were defined as small- and medium-sized enterprises, or SMEs (that is, less than 250 employees). Of the EU's 257,807 food and beverage companies, SMEs account for one half of total EU production and 62 per cent of employees. Large food companies – just 0.7 per cent of all EU companies – produce half of all EU's production. It is a diverse industry, reaching across a wide range of production, covering both first-stage processing (agricultural produce) as well as second-stage processing (fabricated products). It is also displays relatively stable overall growth of around 2 per cent a year, albeit with significant variations between the various branches that make up the food and drink industry. Four sectors dominate the EU food industry: beverages, 'various products' (which includes bakery, chocolate and confectionary products), meat products and dairy products (by production 16, 27, 20 and 15 per cent, respectively; by added value 19, 37, 17 and 10 per cent, respectively). France, Germany, Italy, the UK and Spain are the leading producers of foods, representing 80 per cent of total EU production (when 15 Member States).

Long-term Structural Change in Food Manufacturing and Processing

In 1981, the OECD identified in a seminal study of the food industry a range of pressures on resources employed in the food economy and outlined the choices facing the food industry:[67]

- improving efficiency (lowering unit costs);
- diversification of product ranges and costs;
- remaining within the food economy, but accepting lower returns on resources;
- transferring resources to activities providing higher returns;
- persuading governments to subsidize the various parts of the food economy.

These OECD 'predictions' have proved remarkably prescient. In our view, the key components of business strategy led by food manufacturers today now include:

- restructuring and concentration through factory closures;
- cost savings through massive reduction in numbers of employees;
- increasing scale of production in the remaining factories;
- mergers and acquisitions and other intra-firm alliances;
- focusing on core brands;
- focusing on key product categories.

Food manufacturing locks companies into a never-ending stream of new product introductions – the vast majority of which fail.[68] For example, annual new food and beverage product launches between 1994 and 2000 in North America averaged 14,358 per annum over the seven-year period: that is 100,506 products over seven years.[69] Food companies strive to introduce products with health benefits in order to differentiate their product from others within this flood of product launch activity; product differentiation strategy is not always supported by science and products too often go under.

For many branded food manufacturers there has recently been renewed focus on brand value. Even the leading consumer-brand companies, often as a result of their histories, are still wedded to a Productionist and 'economy of scale' business model: selling masses of commodity-style products and foodstuffs to keep factories churning at near maximum capacity, which maintains 'production-led' rather than 'consumer-led' marketing. Brands and brand marketing remain one of the central pillars for the future of food companies.

Changing Company Cultures for the 21st Century

In the modern economy food manufacturers are something of an anachronism: 'big business' but not very exciting to stockbrokers; they are not just 'old economy', but positively 'ancient economy' in terms of the history of capitalism. A brief history of Nestlé (Box 4.1), founded in 1866, illustrates the long-lasting and conservative nature of the food industry:

Box 4.1 A brief history of Nestlé[70]

1866 Company founded
1905 Merger between Nestlé and Anglo-Swiss Condensed Milk Company
1929 Merger with Peter-Cailler-Kohler Chocolats Suisses SA
1947 Merger with Alimentana SA (Maggi)
1971 Merger with Ursina–Franck (Switzerland)
1985 Acquisition of Carnation (US)
1988 Acquisition of Buitoni–Perugina (Italy)
1988 Acquisition of Rowntree (GB)
1992 Acquisition of Perrier (France)
1995 Acquisition of Victor Schmidt & Söhne, Austria's oldest producer of confectionery
1997 Through the Perrier Vittel Group, expands its mineral water activities with the outright acquisition of San Pellegrino
1998 Acquisition of Spillers Petfoods of the UK and the Carnation Friskies brand

To achieve its near $50 billion of annual sales, Nestlé products at the start of the 21st century included:

- soluble coffee
- roast and ground coffee
- water
- other beverages
- dairy products

- breakfast cereals
- infant foods
- performance nutrition
- clinical nutrition
- culinary products
- frozen foods
- ice cream
- refrigerated products
- chocolate and confectionery
- food services and professional products
- pet care
- flavours for the food industry
- pharmaceutical products.

With such diversity, it seems incredible that even a giant of the global food stage should have to re-invent itself for the new century as we saw at the end of Chapter 3. The Coca-Cola Company, too, dates back to the 19th century, its product being founded in 1886 as a health food drink (initially a hangover cure);[71] it has grown to become one of the world's leading manufacturers, and has operations in 200 countries, yet is not immune from the new tensions in modern food supply. With revenues of US$20,458 million in 2000, the company embarked on a major restructuring process that included reducing its workforce by a fifth (that is, by 6000 jobs) and recuperating from a multi-million dollar product recall in 1999. Worse, Wall Street in 2000 was marking the company's stock down heavily from the glory days of the 1980s and early 1990s. Its competitors were also being seen as a lot more viable in the carbonated soft drinks market as consumers thirsted for drink alternatives.[72, 73] The company was facing a strategic impasse and, worst of all for a consumer brands company, was seen as out of touch.

From Globalization to Localization

Coca-Cola's strategy for recovery was to go for a new scheme it called 'localization': defined by Douglas Daft, Coca-Cola's then CEO, as a promise to 'think locally, and . . . act locally . . . making sure we are operating as a model citizen . . . so those two principles – brands, and citizenship – are key to our business'. He highlighted people and new opportunities as the key to new success, and claimed that, 'instead of seeing an admirable stock-price as the ultimate driver of our actions . . . we must see admirable actions as the ultimate driver of our stock price'.[74]

Many other major food processors are rethinking their business strategy along similar lines. A change in company approach and culture is being addressed by Unilever, with worldwide sales of £47.5 billion in 2000 and a diverse portfolio of products ranging from soap powders to foods. Unilever's new 'Pathway to Growth' strategy resolved to concentrate product innovation and brand development on just 400 brands, down from around 1200 in its original portfolio; to revise knowledge and information systems; and to reorganize or sell off under-performing businesses.[75] Unilever estimated this restructuring strategy would reduce its workforce of 295,000 in 2000 by £3.3 billion in total and lead to a reduction of 25,000 jobs worldwide over a five-year period. The objective is to achieve £1 billion annualized savings by 2004, to increase annual growth to 5 per cent and operating margins to 15 per cent.

Coca-Cola and Unilever are particularly visible and articulate expressions of thinking about change in response to health and social demands, and about restoring their intangible assets of brands and trust: that is, their intellectual capital, not just production efficiencies alone. The future for food, according to these giant companies, lies in a composite strategy that includes: implementing knowledge management; exploiting imagery and brands more successfully; clarifying the emotional appeal of products; how a company is perceived or trusted by consumers; and as Douglas Daft of Coca-Cola put it, through reconnecting with the 'local'. Biotechnology might open up new opportunities but, above all, health is central to this composite approach.

Thus, the world's major food manufacturers are redefining their vision, mission, values and strategic focus for the future. They will need to wrestle with the implications of the competing frameworks we define as the Productionist, Life Sciences Integrated and Ecologically Integrated paradigms. The redefining of food manufacturers is in part a response to the changes they see ahead in food retailing and food service over the next five to ten years.

Rapid Consolidation and Concentration in Food Retailing

Part of the rationale for this major rethink by branded food manufacturers is the need for a competitive response to the power of international food retailers. It is predicted that just a handful of – perhaps six – food retailers will dominate global food sales in the next few years.

The power of the food retailer is illustrated by the US corporation Wal-Mart – America's most 'admired' company of 2003, according to *Fortune* magazine.[76] With total annual retail sales of US$250 billion, it operates 4300 stores in nine countries. Within ten years of entering the food trade it drove down market prices by 13 per cent. The impact on food producers has been immense: in the US, 10 per cent of Kraft sales now go through Wal-Mart (17 per cent of Procter & Gamble's annual revenue), and some analysts suggest that the success of Wal-Mart has influenced food industry mergers such as Kellogg's purchase of Keebler in 2001, and the merger of Kraft and Nabisco in 2000.[77]

By 2010, the top seven US retailers are forecast to control close to 70 per cent of the food retailing environment, with Wal-Mart driving its grocery share to a massive 22 per cent. Similar concentration pressures are occurring in European markets, as Figure 4.5 shows. The growth in market share being anticipated for the top ten European grocery retailers to 2010 is expected to increase from 37 per cent to 60 per cent in just one decade. Their combined European grocery turnover will grow from €337 billion in 2000 to €462 billion by 2005 and €670 billion by 2010. Table 4.5 identifies the world's leading food retailers in 2002, and Table 4.6 identifies Europe's top food retailing companies in 2001.

Dr Bill Vorley in a 2003 report for Development NGOs – anxious about the squeeze on primary producers of commodities – has suggested that the European market also exhibits 'bottlenecks' where a tiny number of powerful firms can control the dynamics of the supply and value chain.[80] The European retailing market illustrates this perfectly. A report in the same year by CAP Gemini estimated that the EU 2002 retail market prior to EU enlargement comprised:[81]

- consumers and customers: 249 million
- outlets: 170,000

Table 4.5 *Leading global food retailers, 2002*[78]

Rank	Company	Country	Turnover (million $)	No. of countries	% Foreign sales	Ownership
1	Wal-Mart	US	180,787	10	17	Public
2	Carrefour	Fr	59,690	26	48	Public
3	Kroger	US	49,000	1	0	Public
4	Metro	Ger	42,733	22	42	Public/ family
5	Ahold	Nl	41,251	23	83	Public
6	Albertson's	US	36,762	1	0	Public
7	Rewe	Ger	34,685	10	19	Co-operative
8	Ito Yokado (incl Seven Eleven)	Jap	32,713	19	33	Public
9	Safeway Inc	US	31,977	3	11	Public
10	Tesco	UK	31,812	9	13	Public
11	Costco	US	31,621	7	19	Public
12	ITM (incl Spar)	Fr	30,685	9	36	Co-operative
13	Aldi	Ger	28,796	11	37	Private
14	Edeka (incl AVA)	Ger	28,775	7	2	Co-operative
15	Sainsbury	UK	25,683	2	16	Public/ family
16	Tengelmann (incl A&P)	Ger	25,148	12	49	Private/ family
17	Auchan	Fr	21,642	14	39	Private/ family
18	Leclerc	Fr	21,468	5	3	Co-operative
19	Daiei	Jap	18,373	1	0	Public
20	Casino	Fr	17,238	11	24	Public
21	Delhaize	Bel	16,784	11	84	Public
22	Lidl & Schwartz	Ger	16,092	13	25	Private
23	AEON (formerly Jusco)	Jap	15,060	8	11	Public
24	Publix	US	14,575	1	0	Private
25	Coles Myer	Aus	14,061	2	1	Public
26	Winn Dixie	US	13,698	1	0	Public
27	Loblaws	Can	13,548	1	0	Public
28	Safeway plc	UK	12,357	2	3	Public
29	Lawson	Jap	11,831	2	1	Public
30	Marks & Spencer	UK	11,692	22	18	Public
	Total		**930,537**			

Table 4.6 *Europe's leading retailers, 2001*[79]

Retailer	Rank (ERI)*	Rank (Turnover)	European total grocery market share	European status
Carrefour	1	1	7.2%	Leading pan-European
Metro	2	2	1.9%	Leading pan-European
Auchan	3	5	2.9%	Major
Aldi	4	7	2.9%	Major
Lidl & Schwarz	5	14	1.7%	Major
Ahold	6	10	2.4%	Major
Tesco	7	6	3.3%	Major
Rewe	8	4	2.3%	Major
ITM	9	3	2.9%	Major
Casino	10	15	1.7%	Major
Tengelmann	11	12	1.3%	Major
Wal-Mart	12	13	1.9%	Major
		Total	**32.4%**	

Note: ERI is IGD's composite index measuring both 'hard' factors such as turnover, and 'soft' factors such as culture, strategy and internal learning

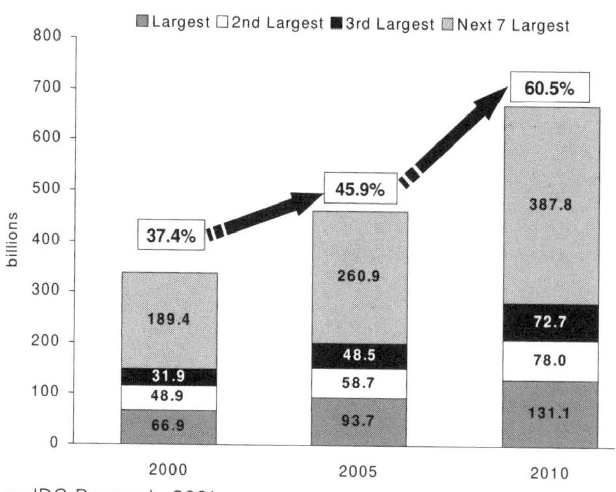

Source: IDG Research, 2001

Figure 4.5 *Anticipated growth of concentration in European food retailing (by sales), 2000–2010*

- supermarket formats: 600
- buying desks: 110
- manufacturers: 8,600
- semi-manufacturers: 80,000
- suppliers: 160,000
- farmers/producers: 3.2 million.

The power of the 110 buying desks will shrink with concentration and cross-border mergers. The eventual impact on consumers will depend significantly on the nature of the competition between the large retail chains and on whether buying groups (of large retailers) allow small retailers to compete.[82] In the US, 'horizontal integration' – retailers buying each other out – has been an increasingly common strategy.

Supermarket power and the trend to 'hypermarketization' are not just limited to rich, developed countries. Recent studies by Professor Tom Reardon and colleagues have shown the remarkable rise of supermarket power in developing regions such as Africa, Latin America and Asia.[83] In Latin America, where four in ten people live in poverty, the supermarket share of food retail markets rose from 10 to 20 per cent in 1990 to 50 to 60 per cent by 2000.[84] In China, following a change of state policy in 1992 to favour supermarket development and privatization, there has been very fast emergence; by 1994, 150 chains owning 2500 stores; by 2000, this had grown to 2100 chains with 32,000 stores.[85] The supermarket share of urban Chinese food markets was 48 per cent in 2001, up from 30 per cent two years earlier in 1999.[86]

Although much 'supermarketization' is supported by 'national/home' finance, foreign joint venture activity accelerated remarkably from the 1990s. Companies such as Carrefour of France (the world's second largest food retailer after Wal-Mart) achieved 37 per cent foreign sales out of its 1999 total sales. In the same year, Wal-Mart had just 13 per cent foreign sales, The Netherlands-based Ahold had 76 per cent and Tesco in the UK was rapidly building on its 10 per cent foreign sales. With home markets saturated, joint ventures with foreign chains (perhaps leading to ultimate takeovers) is perceived as making good financial and marketing sense. Foreign direct investment has overwhelmingly come from the three giants of global finance capital (the US, the EU and Japan)[87] but other regions are generating their own powerful retail forces. In South Africa, for example, 1700 stores provide 55 per cent of food sales for 35 million people, and its leading supermarket chains such as Shop Rite have crossed borders and become regional powers. Although concentration and supermarket penetration is advanced in some countries, the picture is not universal. Nigerian supermarkets have only 5 per cent market share; Bolivia has a low supermarket rate while Brazil's is high; India has few supermarkets compared to Korea. Such differences may occur because supermarkets are traditionally associated with packaged rather than fresh foods which in developing countries at least still tend to be sold in markets or street stalls or specialist stores and to be central to food culture.

But food retailers do not make their money solely from selling foods and beverages to the consumer. To win market share and to control the flow of foodstuffs to the consumer, retailers charge manufacturers fees for 'slotting' (paying for a product 'slot' on the shelf), display and presentation, and pay-to-stay and failure fees. It has been estimated that the large US retailers generate some $9 billion through such methods.[88] These retailer fees present barriers to smaller processors and farmers, and also to consumers and communities. Because inner urban and rural areas are no longer profitable for global food clusters, they get left behind in retail developments; in addition, consumer choice, in the sense of where and what to buy, becomes more restricted. French-based retailer Carrefour, after its merger with Promodés in 1999, became overnight the biggest retailer in France, Spain, Belgium, Portugal, Greece, Brazil, Argentina, Taiwan and Indonesia, with stores in 26 countries.[89]

The world business future for food retailing is that there will be local companies or global companies and not much in between. Three forces are pushing the top retailers to further globalization: first the growing sophistication of consumers,[90] second, capital intensification to extract ever-tighter financial returns; and third, the need to get the best price from suppliers in order to stay competitive, globally sourcing while appearing local.

Food Retailers and Their Suppliers

Prior to the EU growing from 15 to 25 countries in 2004, the combined populations of Britain and Germany accounted for 40 per cent of the EU. With more than 208,000 stores, the two countries owned 30 per cent of supermarkets, a market share of 44 per cent of total EU-15 grocery sales. That is why Wal-Mart decided to establish its first European base camps in both Britain and Germany.

The UK is dominated by four large food retailing companies – Tesco, Sainsbury, ASDA and Safeway – which account for 47.3 per cent of spending in the sector. Since the entry of Wal-Mart into the UK with its acquisition of ASDA, price competition has become a key characteristic of UK food retailing. But a rare insight, before Wal-Mart's influence in the UK, into the inner workings of retail domination of the food economy was revealed by the UK government's 2000 Competition Commission inquiry[91] which identified 24 supermarkets which fell within its terms of reference. Five of these had 8 per cent or more of the market each: Tesco (24.6 per cent), Sainsbury (20.7 per cent), ASDA/Wal-Mart (13.4 per cent), Safeway (12.5 per cent) and Somerfield (8.5 per cent).[92] The Commission questioned whether price trends in the UK compared with abroad and recent falls in wholesale prices, especially in the livestock sector, were being fully reflected in prices charged to consumers. The Commission identified 27 practices by supermarkets that were against the public interest and recommended that a code of practice between retailers and suppliers be implemented. These business practices, the Commission said, gave the UK's five major supermarkets substantial advantages over other, smaller retailers whose competitiveness was likely to suffer as a result. The Commission found the major supermarkets operated 'against the public interest' but they could find no market remedy for this other than to suggest that a voluntary 'code of practice' be developed by the retailers to guide their dealings with their suppliers. The Commission found that the UK's major supermarkets engage in a number of business practices against the public interest that distort competition, but still said the industry is 'securing a good deal' for the British consumer.

The Commission also uncovered regular selling, by all the five main retailers, of some frequently purchased products below cost: the so-called Known Value Items (KVIs). 'Price flexing' (that is the way supermarkets sell the same product, but at different prices depending on the location in the UK) was found to be practised by Safeway, Sainsbury and Tesco, thus distorting competition in the supply of groceries.

The Competition Commission report detailed the extraordinary relationship between supermarkets and their suppliers: a one-sided business relationship in favour of the supermarkets that has created a climate of fear to such an extent that the Commission found it hard to communicate with suppliers. They feared being de-listed and the weakening of the terms or values of orders. Indeed, the Commission found very low margins among suppliers to supermarkets, noting the 1999 average operating margin as 4.3 per cent, with many as low as 2 per cent, and some. The Commission's examination of trading found even negative margins.

Another interesting finding was on product promotions. Promotional activity in supermarkets is considerable. Sainsbury, for example, offered promotions over the course of a year on 10,000 items out of its 20,000 main product lines; but one of the effects of this culture of constant promotions is that it makes it very difficult for consumers to work out comparative or competitive prices, and it constitutes 'a lack of full transparency'.

Other negative business practices pursued by the supermarkets include:

- over-ordering goods at a promotional price from a supplier which are then sold into retail at a higher price without compensating the supplier;
- requesting compensation from a supplier when the multiple's profit on a product is less than it expects;
- seeking discounts from suppliers retrospectively which reduce the price of the product agreed at the time of sale;
- requiring suppliers to purchase goods or services from designated companies, such as hauliers and labelling companies,

Table 4.7 *UK food retail market share (%), 1900–2000*[93]

	Independents	Multiples	Co-ops
1900	80	5	15
1960	60	20	20
1970	44	37	19
1980	25	57	18
1990	10	76	14
2000	6	88	6

from whom the supermarkets often also received a commission.

Through global expansion and consolidation, food retailers are looking to wrest even greater control over pricing and promotions away from food manufacturers. For the foreseeable future, food retailers that grew strong within the framework of the Productionist paradigm and helped shape its workings, will be the key to its future. They will also be key whether the Life Sciences Integrated or Ecologically Integrated paradigms have any future. They will determine the conditions under which health is delivered or swamped by unhealthy arguments. Their power is immense over the health of food supply as well as over their suppliers. When the major UK retailers decided, for example, not to stock GM products, reacting to consumer concern, this effectively blocked the European market and, similarly, their late embrace of organic produce has stimulated rapid market growth, albeit from very low levels. And in 2003, the Competition Commission, asked to adjudicate on whether a proposed takeover of Safeway, the fourth largest chain, by Morrison's a smaller but highly capitalized and profitable chain was against the consumer interest, gave the takeover the go-ahead, effectively triggering yet another round of consolidation.[94]

The Scale of the Food Service Industries

Even more poorly acknowledged in food policy until recently than the role of food manufacturing are the food service industries, partly due to their fragmented nature, ranging as they do from 'greasy' roadside cafés to up-market restaurants, and from schools to the workplace. But the totality of food manufacturing and processing and food retailing needs to be seen within the context of the remarkable growth in food service and catering over the past 30 years. Up to half of consumer expenditure on food and drink in developed countries is now spent outside the home, and the global market for consumer food service is forecast to be worth around $1.6 trillion by 2004 (see Tables 4.8 and 4.9).[95] In the US, the 89 publicly traded US restaurant companies had

Table 4.8 *World food service – outlets by region, 1995–2000*[96]

Region	1995	1996	1997	1998	1999	2000	% growth 1995/2000
Asia–Pacific	3,547,430	3,755,754	3,970,980	4,196,713	4,399,359	4,594,772	30
Latin America	1,182,852	1,235,240	1,253,937	1,272,202	1,289,165	1,307,439	11
Western Europe	1,088,546	1,109,939	1,140,642	1,163,172	1,182,989	1,201,801	10
Africa and the Middle East	360,906	408,085	454,878	495,472	583,337	592,905	64
North America	384,758	395,729	406,065	416,055	426,314	436,100	13
Eastern Europe	249,852	254,842	266,610	279,498	284,518	291,757	17
World	6,814,343	7,159,590	7,493,112	7,823,112	8,120,681	8,424,774	24

Table 4.9 *World food service outlets, by value and type, 2000–2004, US$*[97]

Outlets	2000	2001	2002	2003	2004	% growth 2000–2004
Full-service restaurants	887,665	915,729	952,402	990,289	1,027,033	16
Quick-service restaurants	351,689	365,595	381,440	397,630	415,315	18
Cafés/bars	164,317	169,591	176,189	183,134	190,318	16
TOTAL	1,403,671	1,450,916	1,510,031	1,571,053	1,632,667	16

sales of just under $100 billion in 2002. This accounted for 36 per cent of total restaurant industry sales. These 89 companies operated 121,000 outlets.

In terms of type of outlet and value, full-service restaurants dominate the world with more sales than quick-service restaurants, cafés and bars, stand-alone sites accounting in 2000 for almost 80 per cent of total food service sales. The primary driving force of unit expansion has been the overseas growth of major US chains such as McDonald's, Yum! Brands (formerly Tricon) and Burger King. One market study suggests that such chains performed well 'due to their fashionable image, reputation for good hygiene, strong marketing power and ability to tap into the increased demand for swift, convenient meal solutions at affordable prices'.[98] Not surprisingly, there is growing competition between food service and food retailing, with supermarkets striving to develop what they call 'home meal replacement' (HMR) product offerings.

Growth in fast food has been phenomenal. While in 1970 Americans spent $6 billion on fast food, by 2000 they spent more than $110 billion; they now spend more money on fast food than on higher education, computers or new cars; more than on movies, books, magazines, newspapers, videos and recorded music combined.[99]

Table 4.10 illustrates the growth in the high-profile fast food company Burger King, a US company which is now truly international and relies on consumers and countries outside the US for its business success. The renamed Yum! Brands (the new holding company for Kentucky Fried Chicken, Taco Bell, and Pizza Hut) also has a new global reach (see Figures 4.6a and 4.6b).

Table 4.10 *Burger King global presence, 1 January 2002*[100]

Region	Number of countries present in	Number of outlets
North America	2	8248
Europe	27	1657
Latin America	21	561
Asia Pacific	10	598

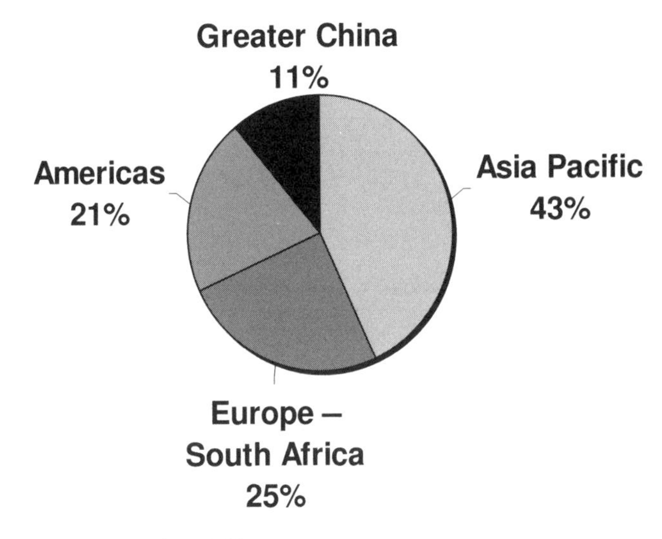

Source: Company website, 2002

Figure 4.6a *Yum! Brands, by region, 2002*

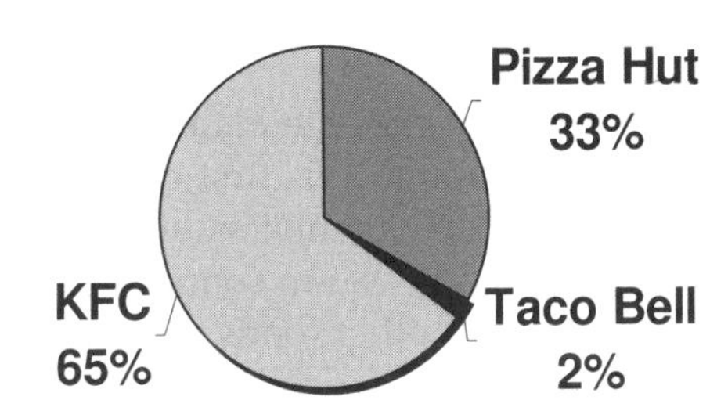

Source: Company website, 2002

Figure 4.6b *Yum! Brands, global sales by fascia, 2002*

But the global icon of the fast food industry remains McDonald's, which more than most companies carries the public profile that personifies the Productionist paradigm. It is the world's largest-ever fast food chain with more than 30,000 outlets serving 46 million customers a day in more than 100 countries, and in 2002 it created system-wide revenues of $41.53 billion. (Table 4.11 lists the top five countries for McDonald's by number of outlets.)

But as the Productionist paradigm begins to destabilize, so McDonald's has looked less formidable. From a share price of just

Table 4.11 *Global top five countries by number of McDonald's outlets, (2003)*

1	United States	13,609
2	Japan	3773
3	Canada	1339
4	Germany	1244
5	Britain	1235
	Total worldwide	31,129

Source: McDonald's *Annual Review 2003*, http://www.investor.macdonalds.com

over $48 in 1999, the company's stock fell 72 per cent to around $13.50 by February 2003, when the company announced its first-ever quarterly loss in its 47-year history. In addition, McDonald's was in 2002 at the centre of high-profile litigation on behalf of a class of obese and overweight children. It was alleged in a New York suit, that the fast-food chain 'negligently, recklessly, carelessly and/or intentionally' marketed to children food products that are 'high in fat, salt, sugar, and cholesterol' while failing to warn of 'obesity, diabetes, coronary heart disease, high blood pressure, strokes, elevated cholesterol intake, related cancer,' and other conditions.[101] Although this first case was thrown out of court, in February 2003, such regulatory activity around fast food and childhood obesity looks set to continue. The healthiness of products like burgers and other foods associated with fast-food culture such as snacks and soft drinks is likely to be increasingly in the health firing line, and will present major challenges to the food sector. It is why healthier options and some portion size reductions are appearing in chains – not before time.

Meanwhile, food service, in its totality, continues to be one of the most dynamic sectors of the food supply chain, despite the evidence that food consumed away from the home is generally higher in fat, and saturated fat, and lower in fibre and calcium and overall nutrition than food eaten in the home. A US study showed that, with eating out playing an increasingly large role in the American diet, more nutritional activities are needed to focus on improving food choices and quality.[102] This 'cultural' dimension of new trends is central to any understanding of the business dynamics of food service and its competition with retailing. A tension is already apparent about whether people eat at home or on the streets, whether they eat in public or private worlds, and whether they live to eat at home or eat and then go home. The issue, as with elsewhere in the food supply, is who does the cooking. We discuss this cultural dimension in more detail in Chapter 5.

The Politics of GM Biotechnology and the Growth in Organics

To illustrate further the far-reaching effects of the Food Wars on shaping the fundamental scientific framework which is redefining the Productionist paradigm, and the dynamics of the Life Sciences Integrated and Ecologically Integrated paradigms, this section briefly traces the rise of both genetically modified and organic agriculture. What is surprising about both is how recently they have come into prominence. Although the modern-day practice of organic agriculture has its origins in the 1930s, it is only since the early to mid-1990s that there has been an exponential growth in the development of land under organic management in areas such as the European Union. Likewise, it was only from the mid-1990s that the world was introduced to GM technologies and that wide-scale planting began, starting in the United States.

We must make it clear here that in our three-paradigm model we are in no way suggesting that the Ecologically Integrated paradigm only means agriculture has to turn 'organic'. There are other ecologically integrated practices such as agro-ecology, low-till methods and so on. Equally, the Life Sciences Integrated paradigm does not mean that all production will be through GM crops. Both paradigms are wider in scope than those narrow conceptions. The paradigm concepts are used to suggest the key drivers of a whole food system from inputs through to consumption and to help identify human and environmental health implications. Both are founded on the science of biology, and biological sciences can have as much of a role to play in the Ecologically Integrated paradigm as the Life Sciences Integrated paradigm. 'Soft' biotechnologies could perhaps have a role to play in the Ecologically Integrarted paradigm; but it is hard to see how GM foods as currently being introduced from an environmental, social and political perspective would fit in.

It is the Ecologically Integrated paradigm which best embraces the broader goals of the organic movement in its aspirations to protect environmental health and biodiversity, and to develop traditional skills and techniques to improve agriculture. It also espouses the application of low-till or 'zero-till' (ZT) agriculture currently being pioneered in South Asia, which increases harvest yields while reducing the use of water and herbicides.[103] This technique has been approved by the International Maize and Wheat Improvement Centre, the International Rice Research Institute and also Dutch and Chinese agencies, where herbicide usage has been cut by over 50 per cent, water consumption by 30–50 per cent and yields improved. The defining characteristic of this paradigm is that, from local to global level, it attempts to find genuinely sustainable solutions within ecosystems.

At the same time as organic products have set record-breaking sales, organic production itself has come into head-to-head conflict with the proponents of GM foods, as consumers increasingly come to see 'organic' as a label for non-GM foodstuffs, over contrasting methods of agricultural production. The battle has become bitter since, under organic certification schemes, produce cannot be labelled as 'organic' if it is found to be contaminated by GM material. The global spread of GM crops has the potential to make organic standards null and void through such contamination. But the battle is deeper than this, with organics squaring up to GM foods over the future direction and shape of food supply.

Although there is not yet a consumer market for GM-produced foods, a growing consumer market already exists for organic fresh produce and processed foods: GM foods entail the more worrying technology, particularly for environmental reasons and their impact on human health has so far mainly been analysed on safety terms: that is, for immediate toxicity or for allergic reactions. In this section we briefly consider both organics and GM from a commercial perspective.

Organic production

One of the more dynamic and fast-growing food markets since the mid-1990s has been the demand for 'natural' products and an explosion in the rate of demand for organic produce, and market analysts are predicting that it will constitute one of the most sustainable health-driven markets worldwide for decades to come.[104] In the US, for example, more than one-third of supermarket shoppers list organic foods, and more than half list additive- and preservative-free foods, among the products they wish to purchase to maintain health, with sales of just under $13 billion in 2001: organic foods at $6.95 billion (up 19.9 per cent over 2000) and natural foods at $5.95 billion; and expected to reach $25.8 billion in total by 2010.[105] However, in 2002 no more than 1.5 million hectares in the developed world were managed organically, representing only a 0.25 per cent share of the total agricultural area. As Table 4.12 shows, total area under organic production in the same year was just 2.99 per cent in the EU and there were just under 143,000 organic farms (Austria and Italy are the European countries with the largest land areas devoted to organic production at 8.68 per cent and 7.14 per cent respectively, followed by the Scandinavian countries).

In terms of world markets, as Table 4.13 illustrates, sales of organic foods in the US, Europe and Japan were less than 2 per cent of total food sales in 2001, but significant growth is forecast in the medium term to 20 per cent of total food sales: around $21 billion in global market value. Figures suggest an organically managed area of around 22 million hectares in total, with retail sales expected to grow to US$29–31 billion by 2005.[107] To put the organic market in context, in 2003 it was valued at less than half of the total sales of Nestlé.

A strong selling point for organics is that they have been confirmed to contain fewer chemical residues than other foods. A review comparing foods grown under three regimes – organic, Integrated Pest Management (IPM) and conventional farming – found that: organically grown foods contained one-third fewer residues than conventionally grown foods; conventionally grown and IPM grown foods were more likely to contain multiple residues than organic foods; residues in organic foods were consistently lower than the conventionally or IPM-grown foods; and conventional foods contained the most residues, IPM foods the next and organic foods the least. Such potential health benefits of organic foods, however, do not prevent them from being contested. An interesting tussle went on in the UK, for example, between the organic movement and the national Food Standards Agency (FSA) shortly after the FSA was created in 2000 and in the

Table 4.12 *Organic farming in Europe, 2000–2001*[106]

European Union	Date	Organic area (ha)	Agr utilized area (000)	%	Organic farms	All farms (000)	%
Austria	31/12/2000	271,950	3497	8.68	19,031	226	8.42
Belgium	31/12/2001	22,410	1391	1.61	694	76	0.91
Denmark	31/12/2001	174,600	2646	6.20	3539	63	6.40
Finland	31/12/2000	147,423	2192	6.73	5225	91	6.60
France	31/12/2001	420,000	30,150	1.50	10,400	680	1.53
Germany	31/12/2000	546,023	17,389	3.14	12,732	434	2.93
Greece	31/12/2000	24,800	3465	0.72	5270	821	0.64
Ireland	31/12/2000	32,355	4415	0.73	1014	148	0.69
Italy	31/12/2000	1,100,000	15,401	7.14	60,000	2315	2.59
Luxembourg	31/12/2000	1030	127	0.81	51	3	1.70
Netherlands	31/12/2001	38,000	1961	1.94	1510	94	1.61
Portugal	31/12/2001	70,000	3812	1.84	917	417	0.22
Spain	31/12/2001	485,079	29,272	1.66	15,607	1208	1.29
Sweden	31/12/2000	171,682	3107	5.20	3329	90	3.70
UK	31/12/2000	527,323	15,858	3.33	3563	233	1.53
Total EU		**4,032,675**	**134,683**	**2.99**	**142,882**	**6988**	**2.04**

Note: Provisional results of an FiBL (Forschungsinstitut für biologischen Landbau) survey, June 2002; continual updating

Table 4.13 *World markets for organic food and beverages*[108]

Markets	Retail sales (million US$) 2000	% of total food sales	Expected growth – medium term %	Retail sales (million US$) 2001
Germany	2100–2200	1.6–1.8	10–15	–
UK	1100–1200	1.0–2.5	15–20	–
Italy	1000–1050	0.9–1.1	10–20	–
France	800–850	0.8–1.0	10–15	–
Switzerland	450–475	2.0–2.5	10–15	–
Denmark	350–375	2.5–3.0	10–15	–
Austria	200–225	1.8–2.0	10–15	–
Netherlands	275–325	0.9–1.2	10–20	–
Sweden	175–225	1.0–1.2	15–20	–
Belgium	100–125	0.9–1.1	10–15	–
Other Europe*	400–600	–	–	–
Total Europe	7000–7500	–	–	8500–9000
US	7500–8000	1.5–2.0	20	9000–9500
Japan	2000–2500	–	–	2500–3000
Total	17,500	–	–	21,000

Note: Official trade statistics are not available. Compilations are based on rough estimates. Retail sales in US$ are based on average exchange rates. The figure for Japan is particularly uncertain. (This figure also includes non-certified products, for example some 'Green Products')

* Finland, Greece, Ireland, Portugal, Spain, Norway

wake of the BSE and other food scandals. The FSA launched a public attack on organics, arguing that they were poor value for money; there was furious riposte from the organic movement and surprise from independent observers who felt that absence of evidence is not valid bargaining power. The FSA has modified its attack but persists that 'there is not enough information available at present to be able to say that organic foods are significantly different in terms of their safety and nutritional content to those produced by conventional farming.'[109]

By contrast, supporters of organics argue that organic agriculture provides more than just raw food and fibre; it is good for the landscape and for rural development. This position has been built into the multifunctional approach to the reform of the Common Agriculture Policy (CAP): under reforms announced by EU Commissioner Franz Fischler on 10 July 2002, there will be a gradual switch under CAP from direct payments to farmers for food production, to payments to farmers for their part in conservation and rural development. But, as one review has noted, there are limits to what can be achieved by organics if it works within the dominant food supply chain.[110] The impact of commercial contracts and specifications could outweigh the gains of organic systems, and, given the distinct possibility of an intensification of organics, key issues for the organics industry will be:

- how to grow the market from its niche position in the developed world market, currently dominated by a very small number of dedicated consumers who purchase the majority of organic produce;
- how to make organic produce more affordable;
- how to extend the range of products from fresh produce to more prepared foods;
- local versus export-driven production;
- extending the organic movement's 'value' proposition, for example to fair trade as well as to organic management;
- demonstrating and evaluating the human health benefits of the consumption of organic produce.

GM foods: key themes and issues

Biotechnology – in all its forms – promises to be one of the defining technologies of the 21st century, with GM foods proving to be one of the most controversial issues in the future of food supply. GM biotechnology allows science to manipulate the natural world at its most fundamental level: in comparison to the organics industry, characterized by mainly small and medium-sized companies, the early pioneers of the biotechnology industry are global corporations: Pharmacia (Monsanto), DuPont, Syngenta, Bayer and Dow dominated agricultural biotechnology in 2002.[111] As a result, the rise of GM has been rapid. The first commercial GM crop – tomatoes – introduced as recently as 1994, was soon followed by insect-resistant potatoes and cotton in 1995, then insect-resistant corn, herbicide-tolerant soybean, canola, insect-resistant maize and oil-seed rape in 1996.

From an agricultural point of view the commercialization of GM crops so far has been spectacular, rising from 1.7 million hectares in 1996 to more than 52 million hectares planted worldwide by 2001 (see Figure 4.7). In 2002, there were around 58.6 million hectares of GM crops cultivated in 16 countries;[112] 68 per cent of all plantings taking place in the United States, 22 per cent in Argentina, 6 per cent in Canada and 3 per cent in China. One company's GM seed technology – Monsanto's – accounted for 91 per cent of the total world area devoted to commercial GM crops in 2001, and two genetically engineered traits (herbicide tolerance and *Bacillus thuringiensis* (*Bt*) insect resistance) accounted

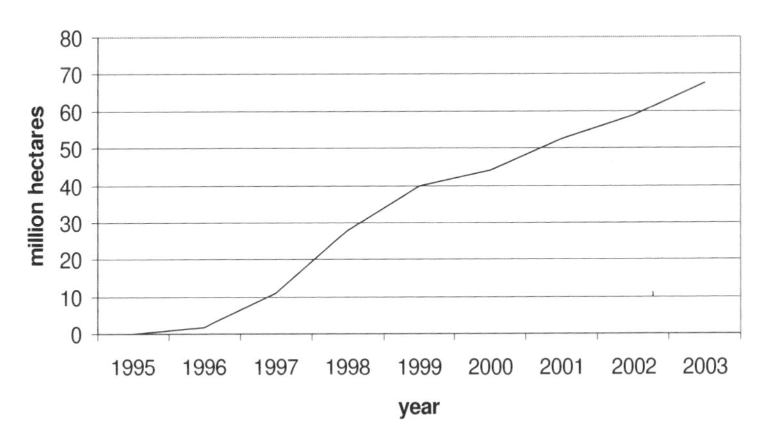

Source: Clive James, ISAAA Brief No 30, *Global Status of Commercial Transgenic Crops*, 2003

Figure 4.7 *Globalplanting of transgenic crops, 1995–2003, million hectares*

for 52.6 million hectares planted to GM crops in 2001.[113] For the GM seed companies the technology is proving something of a bonanza; they notched up sales of US$3.67 billion in 2001 with forecast growth by more than 50 per cent to $5.57 billion in 2005.

Advocates of GM biotechnology point out a whole range of benefits, including their potential to produce crops and foods with enhanced health benefits. A report from the US Institute of Food Technologists on biotechnology and food summarized many of these; they included:[114]

- a more abundant and economical food supply for the world;
- continued improvements in nutritional quality, including foods of unique composition for populations whose diets lack essential nutrients;
- fresh fruits and vegetables with improved shelf-life;
- foods with reduced allergenicity;
- the development of functional foods, vaccines and similar products that may provide health and medical benefits;
- further improvements in production agriculture through more efficient production practices and yields;
- the conversion of non-productive toxic soils in developing countries to productive arable land;
- more environmentally friendly agricultural practices (for example, through improved pesticide usage).

It should be noted that many of these claims lack substantiation and that some remain theoretical, whilst others are already reality or approaching commercialization.[115] Despite the widespread introduction of GM crops, many research and scientific questions remain to be satisfactorily answered. The 2003 report from the International Council for Science (the ICSU) showed that, hitherto, there had been no attempt to draw together the global research base to establish where consensus lay and it pointed to areas of disagreement and to gaps in knowledge requiring additional research.[116] The main area of disagreement was the long-term environmental risks, which, in the UK, NGOs have challenged, the most vocal group of which have been environmentalists who argue that GM may irreversibly upset the balance of nature, and that more ought to be known about its potential impacts on flora, fauna and human beings. Additional issues shaping the debate on GM include:

- the impact of biotechnology on agriculture, the environment and eco-systems (including intellectual property rights over genetic materials);
- the role of experts in debates about science and technology;
- how to regulate new and emerging technologies;
- what is meant by consumer choice, including consumer information, whether through labelling or other means; and
- how food production models are to be exploited under a development banner.

Arguments for ecological caution are being expressed not only by environmental campaigners but also from within mainstream biological science. For example, in a measured review of GM plants for food use published by the Royal Society of London in October 1998, it was recommended that a 'super regulator' be appointed to monitor the wider issues associated with GM development,[117] and the UK government's official advisers from English Nature wrote to Prime Minister Tony Blair in 1999 as follows:[118]

> *We are . . . very concerned about the effects that introducing herbicide tolerant crops would have on biodiversity. These new varieties would give farmers the ability to eliminate wildlife in crops where they currently cannot easily do so. This type of genetic modification will make farming even more intensive and is undesirable in the British countryside where farming and wildlife must co-exist. We must learn the lessons from the previous intensification of agriculture and its disastrous effects on biodiversity.*

In response, the UK government's GM Science Review Panel has now claimed that there have been no verifiable untoward toxic or nutritionally deleterious effects of GM foods, despite remaining gaps in their knowledge.[119] Notwithstanding persistent consumer, environmentalist and (some) expert concerns, the biotechnology industry is quietly making sure that GM crops are almost a *fait accompli* in world food supply: their rapid spread is making it difficult for consumers to avoid them and the trend is possibly irreversible. As biotech crops proliferate, the GM industry is determined to win consumer acceptance. Monsanto, in May 2004, withdrew its GM wheat in the face of consumer resistance and in order to concentrate on less high-pofile consumer GM markets than bread, such as cotton or oilseed rape. The industry was boosted in 2001 when the UN Development Programme's *Human*

Development Report came out in support of GM technologies as helping to solve the developing world's food problems, and reduce world poverty;[120] it stressed the unique potential of GM techniques for creating virus-resistant, drought-tolerant and nutrient-enhanced crops.[121] The Report also pointed to an urgent need for developing modern varieties of millet, sorghum and cassava, which are staple foods for poor people in many developing countries, asserting that current debates over new biotechnologies mostly ignore the concerns and needs of the developing world. The Report stated that Western consumers 'naturally focus on potential allergic reactions and other food safety issues. People in developing countries, however, may be more interested in better crop yields, nutrition, or the reduced need to spray pesticides . . .' Some critics of GM in the developing world were irritated by such a patronizing tone and reassert GM as irrelevant to their needs which are focused on access to land, securing local markets and poverty alleviation.[122, 123, 124, 125, 126]

Many in the biotechnology industry, while acknowledging that part of the public concern about GM crops is that they deliver benefits to farmers only (agronomic traits), express great faith that, once 'second-generation' GM crops come onto the market, negative public opinion will fade. 'Second-generation' GM foods, they argue, will deliver consumer benefits through the genetic modification of fruits, root and leaf vegetables and grains, with enhanced nutritional and health-promoting properties for both humans and eventually even animals.

Summary and Conclusion

In this chapter we have tackled the historical forces, goals and challenges of production, distribution and processing, highlighting how corporate interests provide a main focus of the Food Wars. The emergence of unprecedented power within the supply chain shapes how business diverges in its approach to the emerging paradigms, and how key trends shaping the food economy during the early 21st century point to health being a key concern for some business. For many companies, health has become purely a marketing tool but for others it is a commitment. For the future of food, we foresee the following dynamics:

- There will be a continued consolidation of food power into the hands of supermarket chains. They currently dictate the competitive environment in which the food chain operates and they are extending their influence globally, and this can be contained only by political intervention.

- Food service is growing in importance and absorbing more consumer expenditure. This growth will blur the business boundaries between manufacture, retailing and catering.
- 'Sustainable' issues are of mounting importance and will climb up the food economy agenda as evidence of environmental challenges deepens.
- There will be continued consolidation of agriculture and agribusiness, a continuing decline in the numbers of small farmers and widespread but contested use of GM technology.
- The 'localization' of global food corporations will grow. As food transnationals become more dominant in key food categories, they will be increasingly challenged to meet 'local' demands and consumer needs.
- There will be escalating pressure on the food sector to innovate and add value in order to deliver returns for financial markets as well as to meet consumer demands.
- There will be continuing fragmentation of markets with a mass market for 'niche' products.
- Health and nutrition will become more important across the food supply chain, although the meaning of health will be defined differently to suit different interests.
- There will be a 'green shift', away from intensification of food supply to extensification schemes such as the promotion of organic and natural products.
- New technologies, especially in respect of GM, will offer to alleviate the shortcomings of the Productionist paradigm.
- There will be consumer resistance to what is seen as the 'corporate takeover' of food and growing unease about low-waged labour and other ethical issues.

While not wishing to downplay the importance of science and technology to the future of food supply, we suspect that cultural and social values will play an increasing role in how the food supply chain develops, and the role of science in society will be more critically assessed and evaluated. The last two decades of the 20th century witnessed consumers flexing their opinionative muscles and reasserting their rights, but whether consumers really exert effective collective power as forces within the food supply chain remains to be seen.

Chapter 5
The Consumer Culture War

'There is hardly anything in the world which some man cannot make a little worse and sell a little cheaper, and the people who buy on price only are this man's lawful prey.'

John Ruskin (1819–1900)

Core Arguments

In an ideological world where the consumer is ostensibly sovereign, the notion of a 'food culture' is a more robust way of understanding people's beliefs and behaviours regarding food: the concept is central to food and health and to which ideas about diet will dominate the consumer's mind. Business spends huge sums of money trying to mould and respond to consumer aspirations: by contrast, governments deliver huge amounts of rhetoric but very little money on urging consumers to change their diet. The 'consciousness' industries are now competing with public health education to shape food culture and food policies. That a homogeneous 'food culture' simply no longer exists is becoming a new reality for food policy.

The Battle for Mouths and Minds

Food companies today all want to describe themselves and their activities as 'consumer-led', and the Food Wars are played out in the arena of public consumption and the imagery and media that help shape it. The battles here are not just for the mouths of consumers, but increasingly for their minds. As we saw in the last chapter in the case of Coca-Cola and Unilever, there is considerable emphasis by food companies on getting the consumer to form an emotional bond with products and companies.

A struggle is underway between food processors selling their branded dreams, caterers luring consumers to their food offerings often through strategies that owe more to the entertainment industries than to food (a trend called 'eatertainment' in the US), and supermarkets trying to outdo both. There is now a belated rearguard action by some farmers who are attempting also to inject their imagery and produce into this consumer mix.

This chapter maps out the new consumer landscape as it applies to our core theme of food and health and the policy responses. This is difficult terrain. Whereas conventional marketing and management theory focuses on narrow market definitions of a 'food consumer' and the pursuit of consumer sovereignty, we think it more valuable to investigate and understand a wider notion of food culture and to situate health within it. It is only by understanding how food culture is being shaped by powerful food supply chain forces, and their responses to consumer understanding of food and health, that policy-makers can hope to create realistic strategies. The current policy emphasis on health education as the means for tackling diet-related ill health is out of touch with both business realities and food culture.

The notion of 'food culture' is also useful for expressing how food beliefs and behaviours are socially framed: it refers to a constellation of socially produced values, attitudes, relationships, tastes, cuisines and practices exhibited through food. To explore food culture requires us to acknowledge the meanings that individual people and population groups attach to food.[1] When people sit down to a meal together, every action they take – sitting on chairs, using plates and cutlery, the order of food, the type of food they choose, where and what they eat and how much – is an indicator of food culture.[2] Food culture comprises both social 'cement' – binding groups together with shared assumptions – and opportunities for difference and distinction:[3,4] people express their identities, and classes through food and derive cultural meanings from it.[5]

From the point of view of our present concerns, we reiterate that food culture is not homogeneous. Consumers often hold opposing food beliefs: they may welcome some innovations but resist others; they say that they are concerned about health, but do not always want to change diet-related behaviour for health reasons alone; they judge food by its price but also by its quality. Consumer food culture is immensely complex, being multi-dimensional and constantly changing: food patterns, tastes and beliefs are now in a permanent state of transition; existing and older food cultures are semi-permanently under assault,[6, 7] partly driven by willing change and partly by changes stemming from the restructuring of the food economy and its major players (see Chapter 4).

More than any other factor including science, we see the current food culture as driving the demise of the Productionist paradigm and shaping the future success or otherwise of the Life Sciences Integrated and the Ecologically Integrated paradigms. Food culture is in tension with business philosophies and state policy, as well as with the ways people have traditionally purchased, prepared and consumed. Our thesis here is that culture is a key to understanding the dynamics between the supply chain and health; but culture is not a consistent entity.[8, 9]

Food and Health: a Done Deal for the Consumer?

According to some market research, consumers already see the connections between food and health, and in many cases make choices based on their understanding of diet and health. In the 1990s, a large study for the European Union of 14,331 people in 15 countries found that, while overall, respondents saw quality, price and taste as the main factors affecting their food choice, health was close behind.[10, 11, 12, 13] The same survey found that the majority of EU citizens are happy with their diets and see no need to change them, despite a third of respondents reporting that they had already made changes for health reasons; that young people see convenience as the greatest influence on their choice of food, while older people are most interested in food's health attributes; that women are more interested in health than men; and that Northern Europeans tend to focus on fat, fruit and vegetables and the need to maintain variety and balance, while Southern Europeans express a preference for fresh and unprocessed foods.

Such findings are also supported by more commercial market research. The US company HealthFocus, for example, surveyed consumers in both the US and selected European countries.[14] Their results suggested that 74 per cent of European shoppers were 'health active', to use their terminology, (those who see a connection between their dietary choices and their health and believe that they can influence their health condition by the choices they make, both for themselves and for their families), and that 94 per cent of American shoppers were 'health active'. The survey found that four out of five Western European shoppers are eating healthy food because they want to and feel that their eating habits are already healthy and do not require major improvement or changes to their diets. HealthFocus found that Europeans are unwilling to make diet or food compromises for health benefits. They want to eat healthfully, they feel they do eat well, but they won't go out of their way to make any major changes in their diets or activity levels in order to improve their long- or short-term health. HealthFocus advised marketers: 'Don't expect consumers to change their behaviour. Fit your solutions to their current eating habits and lifestyle needs.'

Linda Gilbert, president of HealthFocus, said:

> *One lesson is to speak to the internal benefits that shoppers realize from making healthy choices, more than external rewards. In the U.S. we have found the most compelling benefits are often anchored in emotions and lifestyles. Clearly consumers recognize that nutritional needs differ by individual based on gender, age, health condition and more. Product developers need to customize products and marketers need to personalize communications so that shoppers can purchase products that are 'right for me'.*[15]

The same HealthFocus research showed that European shoppers are concerned about pesticides in food, food irradiation and genetically modified organisms; also about food preservatives, highly processed foods and food biotechnology. Such concerns over quality and adulteration suggest consumers awareness that is often not acknowledged within the policy (and business) context. Policy-makers too readily subscribe to the view that all that consumers are concerned with are macro-issues such as price, convenience and time, when, in fact, consumer research has suggested that they have a complex set of demands and aspirations for food in which health plays a significant role.

A report commissioned by Food Science Australia in 1999 presented a 'snapshot' of opportunities for the development of food products which had health- and vitality-enhancing properties. It predicted, up to 2010, the following market opportunities:

- greater consumer interest in Asian, herbal and alternative health and vitality products as societies become more multicultural;
- a shift in consumer perception from food as fuel to food as preventative medicine;
- growing consumer interest in personal health advisers, both human and digital;
- greater emphasis on 'organic' production as a source of competitive advantage;
- increased regulation of health claims on food products;
- the increased privatization and politicization of health care;
- new technologies, such as nutraceuticals, emerging as the focus of a food industry 'bandwagon'.[16]

With the results of such studies around the world pointing to health as one of the key consumer concerns driving food markets – from what they eat to how it is produced – policy must address the adequacy of the food industry's response of developing one-off products containing 'health attributes'. Much of the food industry's answer to the health challenge is short term and simplistic, merely adding more vitamins and minerals, fibres or various proteins, or so-called nutraceuticals to products, thereby treating health as a subordinate goal and a discrete entity, rather than as intrinsic to food culture and the entire food supply chain. Health still tends to be seen as an individual corporate characteristic rather than a goal for the entire food culture. Truly to meet the needs of consumer health and build an effective public health strategy, the food industry will have to develop a new business culture of health. Just producing new one-off products with health attributes, however successful in the marketplace, cannot constitute a public health strategy.

Consuming Wants and Needs

The relationship between the consumer and food can be fraught. Affluent societies are now understandably concerned about developing obsessions with body images:[17] media messages extol the beauty of waif-like fashion models on the one hand, and the dangers of anorexia on the other. The new food and health *zeitgeist* is an eternal triangle linking 'health', 'food' and 'beauty'. Food, beverage and cosmetic companies are now busy dreaming up new products for consumers that will deliver the perceived benefits and imagery represented by this new consciousness. Debates have, of course, raged for decades about the morality of consumerism,[18] its impacts on global divisions[19] and the environment,[20, 21] its myths and so on.

Global food marketing is putting before people an awesome offer: endless food choice for very little effort, and now at last, the new global food machines – 'intellectual property' as well as actual techniques – are meeting resistance: after decades of fast food, there is a fast-growing 'slow food' movement;[22] after decades of the legal adulteration of food, there is now a burgeoning market for natural foods; after decades of enticing consumers to eat world cuisines, there is now a counter-move to return to localism, regional foods and real cooking. It is not fanciful to see the revolt over GM foods, for instance, as a popular uprising against imposed new technologies.[23] There is growing consumer resistance to the perceived corporate and scientific arrogance of companies patenting a rice, for example.[24] With the food companies' global brands so ubiquitous, it is not surprising that they are becoming sometimes subjects of a citizens' backlash.[25, 26] In addition, today's consumption is tomorrow's waste, both in the form of excrement and packaging. As we will show in the next chapter, energy used in food production is astonishingly wasteful in itself. Consumption affects climate change because of use and abuse of fossil fuels. No civilization ever before has had or expressed this capacity to threaten its own biosphere so extensively.

'Burgerized' Politics

Food is intensely personal: it is, and it conveys, our identity, which is why these deep meanings are so carefully monitored and tracked by food companies, building on the thinking first developed by Sigmund Freud (an explanation of why so many psychologists are employed in the consciousness and marketing industries). Everyone likes to think that they control their own diet, that it is theirs by choice and tradition; we like to think that the tomato is Italian and that rhubarb is English, for example. Yet even a cursory review of food history reminds us that there is nothing new about food globalization: food culture follows trade, which, in turn, follows people and culture. Take the potato, which came from the Andes:[27] it met a combination of resistance and apathy when introduced to Europe by Sir Walter Raleigh in the 17th century, but now the chip or 'fries' drive global fast-food production so closely associated with America. In fact, the US love affair with fries came back from World War II with returning troops enamoured with 'French fries' – hence the name.[28] The potato like many foods was projectiled by trade into the mass food culture worldwide: the British introduced the potato to India where it is now a staple alongside rice; they also introduced it to Ireland, where its crop failures due to monocropping and bad weather engendered the dreadful Irish Famine of 1845–1846; it was the culinary 'technical fix' for British oppression, grown by peasants on smallholdings while absentee British landlords, who milked the land and put little back, relied on them doing so: potatoes gave high yield on low acreages. When grown in monocultures, however, potatoes are highly vulnerable to decimation by the blight.[29]

Few major food commodities are heavily cultivated and eaten solely where they originate, their so-called 'Vavilov' centres (named after the Russian scientist Vavilov who first mapped the botanic origins of species to centres of the world). Culture and production have a symbiotic relationship. Cane sugar, a native of Polynesia, spread to India, China and elsewhere but only became a global commodity and staple ingredient and food in the industrial diet through the slave trade in the notorious 'golden triangle' between the Caribbean, Africa and Europe.[30,31] Wheat originated in Africa, but it is US and European surpluses via the GATT (the Global Agreement on Tariffs and Trade) which are now entering Asia to win over the rice-eaters of Asia to its charms. Food in this sense is both cultural ambassador and 'fifth columnist'.

What is new today about the role of food and its 'cultural dissemination' is the pace and scale of corporate influence in the process of change. Colas and burgers are now synonymous with America, and 'Mcdonaldization' – what we call 'burgerization' –

used as shorthand for the modern bureaucracies of society.[32] The burger has become a symbol of modernity and the triumph of an American mode of eating, a metaphor for how rational, bureaucratic society can command the production of any good and the fulfilment of any need. The burger symbolizes a society on the move, constantly seeking new markets, travelling large continents, forced to eat 'on the hoof' and at speed.[33] In cultural terms, sweet, bland foods are usurping gustatory diversity; palates are kept at the pre-adolescent stage of development. 'Comfort' foods sell. Sugary drinks triumph.

Food's capacity to be changed – its 'plasticity' – is why it has been such a perennial threat to those in control, whether governments, companies or land-owners. The relationship between consumers and food is uneasy. Companies know only too well that consumers can be fickle but customer loyalty is the dream of any firm; governments know that it is worth paying more than mere lip-service to consumer opinion. At the same time, if consumers led opinion, there would be no room for politics. The legacy of the French Revolution in which all citizens vote, not just the free as in Athenian democracy under which slaves had no rights, is replaced by a marketplace 'ballot box' which occurs every time the consumer buys and pays for a good or service. The notion of citizenship is thus being redefined as a purchaser whose '(purchasing) ballots . . . help create and maintain the trading areas, shopping centres, products, stores, and the like'.[34] Buying becomes tantamount to voting, market surveys the nearest thing that exists to expressions of the collective will.[35] It follows that the more wealth or purchasing power the consumer has, the more 'votes' she or he gets.

Indeed the parallel between workers, who without organization (trades unions) are easily outmanoeuvred in the wage market, and consumers has been much commented on. The French sociologist Jean Baudrillard, for instance, stated acidly in 1970:

> *Overall . . . consumers as such are lacking in consciousness and unorganized, as was often the case with workers in the early nineteenth century. It is as such that they are everywhere celebrated, praised, hymned by 'right-thinking' writers as 'Public Opinion', that mystical, providential, sovereign reality. Just as 'the People' is glorified by Democracy provided that it remains the people (and does not intervene on the political and social stage), so consumers are recognized as enjoying sovereignty [. . .] so long as they do not attempt to exercise it*

> *on the social stage. The People are the workers, provided they are unorganized. The Public and Public Opinion are the consumers, provided they content themselves with consuming.*[36]

Is that the role of consumers? To consume and not to 'intervene'? We think their role has developed significantly and that certainly in the food world, consumers have begun to 'exercise' their muscles and to threaten.

THE NEW CONSUMER WEB AND COMPETING MODELS

Most countries have had a long struggle over who controls food: between 'democratizing' food culture and the imposition of a top-down control system.[37] Much depends upon whether one is rich or poor, living in a developed or developing country. The reality of modern food consumption is that the dominant food culture is pulling one way, while reason and caution urge other policy goals: the mantra of 'choice' pulls one way, the necessity of environmental sustainability another. The language is of consumer sovereignty; the reality is of increasing market concentration. These contrary patterns are shown in Table 5.1 which contrasts the 'dominant' model with 'alternatives'. Even though the term 'alternative' is used, these are just as real, as the 'dominant' models and have policy ramifications; it remains to be seen which pattern emerges as dominant overall. The challenge for policy-making is to reconcile these contradictory demands and to make a definite choice. We see this list as key indicators for which paradigm is adopted within food and health policy and thinking.

At its most basic level, consumerism offers a dream of total choice. The neo-liberal economic model has ushered in an era in which consumers, rich ones in the main, can 'browse the world' in their local supermarket without even having to travel; the world's food is literally coming to their shopping basket and table. The reach of retail giants, as we showed in the last chapter, is a new food colonialism: instead of owning the land, as the great European imperialists did, the 21st century food empires own the trade routes and marketing channels. At the same time as this

Table 5.1 *Competing models for patterns of food consumption*[38]

The dominant model	The alternative model
globalization	localization
urban/rural divisions	urban–rural partnership
long trade routes (food miles)	short trade routes
import/export model of food security	food from own resources
intensification	extensification
fast pace and scale of change	slow pace and scale of change
non-renewable energy	re-usable energy
few market players (concentration)	multiple players per food sector
costs externalized	costs internalized
declining rural workforce	vibrant rural population
monoculture	biodiversity
one-track agriculture	multifunctional agriculture
science replacing labour	science supporting nature
scientific farming	organic/sustainable farming
biotechnology	indigenous knowledge
processed (stored) food	fresh (perishable) food
food from factories	food from the land
hypermarkets	markets
de-skilling	skilling
standardization	'difference' and diversity
niche markets on shelves	real variety on field and plate
people to food	food to people
fragmented (diverse) culture	common food culture
created wants (advertising)	real wants (learning through culture)
brands	local distinctiveness
'burgerization'	local food specialities
microwave re-heated food	cooked food
fast food	slow food
individualized food	commensality/shared food
private intellectual property	common goods
'new' food economy	old ways of food economy
footloose production	fixed location for production
food from anywhere	bio-regionalism
rapidly changing artificial nature	slowly changing nature
global decisions	local decisions
top-down controls	bottom-up controls
dependency culture	self-reliance
health inequalities widening	health inequalities narrowing
social polarization and exclusion	social inclusion
consumers	citizens
de- or self-regulation	state/public regulation
food control	food democracy
policy segmentation	policy integration

food trade has grown, so has the travel of people away from their homes: an estimated 600 million people travel abroad each year (and run an estimated 20–50 per cent risk of contracting a food-borne illness when they do so).[39]

Modern food culture thus offers huge choice for some, but also little for many:[40] it promises information on labels but provides little education with which to interpret them.[41] In the far north of Russia, one of us has had discussions with public health specialists wanting to know what these Western European labels mean. As Western food products flood eastwards over the former Iron Curtain, the vocabulary needed to interpret the packaging devalues existing cultures and imposes new constraints: for example, an 'E' label designates EU approval of an additive; ingredients are listed in order of decreasing quantity by weight; and so on. These questions indicate the cultural drift being pushed from West to East by powerful food companies looking for new markets, and often aspired to as modernity by consumers in the East. Wants are being subtly shaped while needs remain unmet. Local foods become culturally suspect while foreign and particularly Western food becomes chic.[42] Food is becoming a commodity which allows whims and fashion to determine its consumption; consumer culture is implanting what environmentalists call its 'footprint':[43] elite food consumer culture is leaving a deeper and deeper imprint in the sands of history. Bottled water is an example of a commodity which travels the globe and which is hugely energy-wasteful, while modern lifestyles threaten local water sources and water shortages are turning the goal of clean water for all into a commodity scramble.[44, 45] The paradoxes and idiocies of the emerging global food culture are legion.

'Schizophrenic' Consumers?

In theory, modern consumers can now choose to remain victims or to be active citizens, finding meaning and enchantment from food in an otherwise disenchanted world.[46] In reality, however, only affluent consumers have significant food choice; middle-income consumers have rather less, and the poor next to none. Table 5.2 gives a conceptual classification of these new global

Table 5.2 *World consuming classes*[47]

Category of consumption	High	Middle	Poor
Population	1.5 billion	3 billion	1.5 billion
Diet	meat, packaged food, soft drinks	grain, clean water	insufficient grain, unsafe water
Transport	private cars, air	bicycles, bus	walking
Source	long-distance foods; hypermarket and delicatessen/ specialist shops	some long-distance food; local shops and markets	local food; local shops and markets
Materials	throwaways	durables	local biomass
Choice	big choice; global horizons	sufficient, regional horizon	limited or absent, local horizon
Environmental impact	high	considerable	low

consuming classes. (Ironically low-income consumers can have a considerable impact on the environment, chiefly by burning wood to cook.) The purpose of the table is to highlight how the more affluent, by dint of purchasing and consuming more, exhibit the greater environmental impact.

Immigration is another well-established mechanism of food cultural change. Immigrants bring their food tastes and cultures with them. It is why setting up restaurants selling their food culture and cuisine is such a common first or second generation commercial enterprise. It is why one could eat good Ethiopian food in Washington, DC, not long after the famine.[48] France has long had a range of excellent African food. The UK is a second home to marvellous Indian sub-continental foods. In fact, the ethnic food market in the UK had a retail value of more than £900 million in 2002,[49] and today English food is defined as much by 'chicken tikka masala' as by the formerly ubiquitous but now declining British 'fish and chips'.

In the public sphere, the food industry may appear to bow before the consumer, yet this facade is thin: the marketers, opinion pollsters and focus group consultants who daily probe consumers' tastes, fetishes and aspirations will privately admit that there is no one such type as 'the consumer'; market research highlights how, over and above demographic givens such as gender, age and ethnicity, consumers are divided by income, social class, ideology and desires; they are fragmented in their tastes in ways that make mass markets hard to corral. The public wants good safety and environmental practices in theory but is often unprepared in practice to pay the extra costs that tougher standards entail.

A product specialist for a top UK food retailer conducted a detailed tracker survey on food integrity which showed consistent rises in concerns about issues such as pesticides, GM foods, additives and health issues generally: echoing the US psychologist Abraham Maslow's hierarchy of human needs (that humans cannot move on to more mature emotional states until 'lower' and more basic needs have been fulfilled),[50, 51] the survey produced a pyramid model to explain rising consumer aspirations with regard to food (see Figure 5.1). At the base is the demand for food to be safe – the *sine qua non* of contemporary food supply management. But, as people get more affluent, they demand features further up the pyramid, features which also

Figure 5.1 *A model of consumer food aspirations used by a leading UK food retailer*

have a strong generational dimension. This hierarchy, also recognizing young consumers' demands, is being planned into that company's long-term strategy.

All medium to large food companies ceaselessly research the consumer and incorporate findings into their product planning and strategy. There is a veritable army of market researchers, consumer monitoring services and consultants whose entire existence centres on tracking shifts in consumer consciousness, aspirations and needs. Most large companies conduct and/or buy such research on a continuing basis. Market researchers and advertisers are very good at (re)packaging such work, producing ever-more clever categories of consumers in market-led groups to tailor products to suit their differences. They assume the standard socio-economic differences and overlay aspirational group differences to yield categories. The homogeneous 'consumer' is a myth that the food industry realizes no longer exists.

Moulding Food Culture

One of the most remarkable changes in consumer food culture is the sources people use for their information about food. Home experience may still be the initial 'fount', but there are now many competing channels of information: advertising, sponsorship, formal education, the media and the internet, and peer groups. The media has often been the butt of blame from the food industry if their particular industries are not portrayed in a favourable light. Notwithstanding, the global food industry spends hundreds of millions of dollars annually on advertising, public relations and other promotional activities. The battle over GM foods is an obvious example of where things have unravelled in the management of consumer consciousness. At the 2000 International Food and Agribusiness Management Association conference, for example, speakers claimed that more money needed to be spent on persuading consumers of GM foods' attractions. 'Perhaps the greatest challenge we face lies not in the area of technology but in marketing,' David Rowe of Dow AgroSciences, a unit of Dow Chemical Co, told a session on food technology;[52] he called for more investment in consumer needs and education. Seven companies and the Biotechnology Industry

Organization formed the Council for Biotechnology Information, which has developed a $50 million, three- to five-year agenda for building public support for high-tech foods. Another delegate told the conference of the need to 'lose the term GMOs' and to use 'food biotechnology' or 'agricultural biotechnology' instead.[53] Similarly, in the area of functional foods and nutraceuticals it is now a common assumption that it is not more science that is needed for market success but better communications and being able to 'talk' the language of consumers.[54] The reality of this approach, of course, in many cases is that it is putting marketing hype over health substance.

Food Advertising and Education

There has over recent decades been persistent concern about the targeting of young people by advertisers;[55] the advertising industry has responded by sponsoring health research in order to dispel its negative image.[56, 57] In the UK, however, around half a billion pounds a year continues to be spent on food advertising, most of it for sugary, fatty, 'fun' foods. Table 5.3 shows how the healthier foods – fruit and vegetables – receive negligible advertising support, and Table 5.4 details the UK advertising

Table 5.3 *Adspend by food category, UK*[58]

Food category	£000s
Bread and bakeries	11,124
Biscuits, cakes, pies and pastries	14,420
Cereals: ready-to-eat	69,219
Dairy products	55,489
Fish: canned, fresh and frozen	5,040
Frozen ready-to-eat meals	18,600
Fruit: fresh, canned, dried and frozen	3,506
Margarine	23,148
Meat and meat products	24,041
Potato crisps and snacks	34,221
Sauces	40,187
Vegetables: fresh, frozen and canned	13,255
Total advertising for all foods	**471,497**

expenditure on leading food and drink brands. (It is interesting to note that, although the brands are diverse, it is a narrow range of agencies that shape the advertising and communication messages.) Table 5.5, by way of contrast, gives examples of advertising expenditure by leading US food corporations.

The success of businesses – and particularly of brands – depends upon consumers being predictable. A single industry is staffed by an army of professionals, from psychologists to home economists, dedicated to studying, analysing and squeezing consumers into 'pigeon holes' – or 'niches' in business parlance – better to predict and manage them. As Sir Michael Perry, former head of Unilever, said in 1994 in a keynote presentation to the UK Advertising Association:

> *Our whole skill as branded goods' producers is* in anticipation of consumer trends. *In earlier appreciation of emerging needs or wants. And in developing a quality of advertising which can interpret aspirations, focus them on products and lead consumers forward* [our emphasis].[61]

The food industry was a strong supporter of, and beneficiary from, the European Single Market, legislated in the Single European Act of 1987. The cost savings it promised from coordinating from diverse national food standards was calculated to save the chocolate industry 30 per cent, the beer industry 22 per cent, the ice cream industry 12 per cent and the pasta industry 9 per cent.[62] (It is not certain whether these cost savings were ever passed on to consumers because these industries immediately entered a period of mergers and acquisitions to take benefit of massive new markets.) Meanwhile, implementation of the other side of the Single Market 'bargain' – giving consumers more information and better labelling – was remarkably slow; it was not for another 12 years, for example, that quantitative ingredients labelling (QUID) was introduced.

The food industry is highly skilled in using diverse media to present itself in a favourable light, from producing educational materials for schools to linking up with charities.[63] This type of promotion is now global. In 2002, McDonald's and UNICEF launched a fund-raising initiative called 'World Children's Day', held on 20 November, the anniversary of the UN adoption of the Convention on the Rights of the Child in 1989. On that day in 121

Table 5.4 *Adspend by food companies in top 50 UK brands*[59]

Brand	Manufacturer	Sales year to April 2000 (£m)	Creative agency	Adspend (£)
Coca-Cola	Coca-Cola Company	635–640	Publicis	**26,575,202**
Nescafé	Nestlé	335–340	McCann-Erickson	**22,662,100**
Weetabix	Weetabix	90–95	Lowe Lintas	**16,631,619**
Budweiser	Anheuser-Busch	100–105	BMP DDB	**13,135,168**
Birds Eye Wall's Ice Cream	Birds Eye Wall's	130–135	McCann-Erickson	**11,712,507**
PG Tips	Van den Bergh	125–130	BMP DDB	**10,600,844**
McCain Chips	McCain Foods	115–120	D'Arcy	**9,092,710**
Kit Kat	Nestlé Rowntree	175–180	J Walter Thompson	**9,077,997**
Fosters	Scottish Courage	90–95	M&C Saatchi	**8,904,286**
Müller	Müller	230–235	Publicis	**8,608,144**
Tango	Britvic Soft Drinks	90–95	HHCL & Partners	**8,289,474**
Tetley	Tetley GB	110–115	D'Arcy	**8,103,598**
Sunny Delight	Procter & Gamble	140–145	Saatchi & Saatchi	**8,006,084**
Mars Bar	Mars	110–115	D'Arcy	**7,968,903**
Walkers	PepsiCo	455–460	Abbott Mead Vickers BBDO	**7,672,949**
Pepsi	PepsiCo	190–195	Abbott Mead Vickers BBDO	**7,508,165**
Dairylea	Kraft Foods	90–95	J Walter Thompson	**7,109,456**
Lucozade	SmithKline Beecham	120–125	Ogilvy & Mather	**6,762,215**
Flora	Van den Bergh	110–115	Lowe Lintas	**6,527,470**

Brand	Manufacturer	Sales year to April 2000 (£m)	Creative agency	Adspend (£)
Ribena	SmithKline Beecham	170–175	Grey	**5,601,488**
Stella Artois	Whitbread	255–260	Lowe Lintas	**5,539,885**
Carling	Bass Brewers	145–150	WCRS	**5,178,819**
Pringles	Procter & Gamble	135–140	Grey	**4,055,268**
Robinsons	Britvic Soft Drinks	170–175	HHCL & Partners	**3,769,187**
Galaxy	Mars	80–85	Grey	**3,725,219**
Anchor	New Zealand Milk (UK)	90–95	Saatchi & Saatchi	**2,782,055**
Cadbury Dairy Milk	Cadbury	160–165	TBWA GGT Simons Palmer	**2,780,558**
The Famous Grouse	Highland Distillers	85–90	Abbott Mead Vickers BBDO	**2,772,168**
Smirnoff Red Label	Diageo	90–95	J Walter Thompson	**2,441,745**
Lurpak	Arla Foods	100–105	BMP DDB	**2,269,430**
Heinz Baked Beans	Heinz	120–125	Leo Burnett	**2,198,289**
Heinz Canned Soup	Heinz	130–135	Leo Burnett	**1,901,585**
Ernest & Julio Gallo	E&J Gallo Winery	85–90	Mountain View	**1,447,893**
Birds Eye Frozen Vegetables	Birds Eye	105–110	McCann-Erickson	**899,824**
Bells 8 Year	Diageo	110–115	Court Burkitt & Co	**3,600**
Birds Eye Frozen Poultry	Birds Eye	110–115	N/A	N/A

Table 5.5 *Adspend by leading food companies, US*[60]

Rank	MegaBrand	Total Spend $000s
1	McDonald's	655,743
2	Wal-Mart Stores	375,331
3	Coca-Cola Company	345,090
4	Kraft foods (Altria)	315,715
5	Burger King (Diageo)	306,147
6	Pepsi-Cola (PepsiCo)	273,135
7	Big G Cereals (General Mills)	265,749
8	K-mart	250,284
9	Quaker Oats Co (PepsiCo)	249,484

countries, McDonald's mounted a variety of activities and promotions in aid of local children's organizations. This serves as an example of the 'global going local' strategy outlined in the last chapter.[64]

One of the serious points of tension between commerce and government has been food labelling. Labelling is a mantra for consumerism: its role is to provide consumers with information upon which to be able to make informed judgements and choices, thereby encouraging market efficiencies yet there is little internal consistency in its application to food. In practice, what goes on a label becomes a subject of intense politicking. One British senior civil servant went on record as saying that in effect the UK government in the 1980s and early 1990s regarded food policy as no more than what information could or could not appear on a label.[65] Within Europe, for instance, the presence of food additives has to be declared, but not pesticides.[66] If a food has been irradiated, a declaration on the label is required only if the ingredient is more than 1 per cent by weight of the final product. This allows irradiated pepper and spices (commonly irradiated ingredients) not to be declared. Labelling of GM foods has been a continuous battle zone, with consistency only emerging when civil disobedience forces retailers and the supply chain to implement traceability systems. Labelling, in short, is not a resolution to the Food Wars but another battlefield within them. Yet politicians regularly cite labelling both as a principle of and a means for consumer sovereignty.[67]

One of the major battles within the warzone of labelling has been what can and cannot be said about the nutrient content of foodstuffs. While the US has now introduced mandatory nutrition labelling for most food products, in Europe such labelling is still voluntary unless a particular health claim is made. Food labelling and its effectiveness in consumer transactions are likely to continue to be contentious for as long as some critics argue that complex information on a label in tiny writing across dozens of purchased goods does not enable the consumer to deliver for him or herself a health-enhancing diet, and others argue that without such information, the consumer remains in ignorance.

Whilst vast resources are now put into this area of policy, with national and international meetings proliferating on all aspects of labelling, we remain to be convinced that labelling is an effective health strategy. Unfortunately, infringements of such regulations as exist which involve intervention by food law enforcement officers mean that labelling issues are likely to remain a battlefield, set to become an even more hotly contested area through the development of 'health claims' regulation which will allow for disease-related labels on certain foodstuffs and ingredients.

Obesity: Redefining Food Marketing

The obesity epidemic outlined in Chapter 2 will become a factor in reshaping the future of food and beverages marketing throughout industrialized countries. While the problem of obesity has been often been classed as one of individual responsibility, not least by the food industry, the rapid rise in obesity in low-, middle- and rich-income countries is repositioning the problem as a population one – with severe societal impacts. In the UK, for example, the Chief Medical Officer described obesity in 2002 as a 'health time bomb' bearing with it risks to the population's health, costs to health care services, and 'disastrous' losses to the economy.[68]

But it has taken factors other than public health concerns alone to focus the attention of the food industry on the true extent of the obesity problem and force them to reassess their food and beverage marketing practices. Instead, it has been the twin threats of legal action against US fast-food restaurants in relation to their products contributing to obesity in some individuals, and dire warnings from financial analysts about slow economic growth and weakened profit margins of some leading companies, that has put the food industry on the defensive. In a truly revealing piece of global equity research published by UBS Warburg in 2002, the authors warned that the share prices of many of the world's biggest food companies are at risk because of 'consumers' exposure to 'unhealthy' foodstuffs.[69] They told investors: '. . . we believe there are risks associated with obesity that have not yet been factored into share prices.'

Further, an equity research report published by JP Morgan in April 2003 predicted that food manufacturers faced risk of increased regulation, notably in Europe, in labelling, advertising and distribution of some of their key growth product areas.[70] The authors described this as 'not good news for volume growth or margins', and predicted long-term risks to producers of fast foods, soft drinks, confectionery and snacks. As the Warburg report put it: 'Reducing obesity implies reducing calorie consumption. If the [obesity] epidemic is to be tamed, the major purveyors will have to see sales of many of their traditional product lines fall.'

As part of the JP Morgan analysis, the authors drew up a list of the product portfolios of those companies most exposed to the 'obesity risk' (based on percentage of total revenue), also predicting growth opportunity for those companies focused on healthy products and able to make the transition to 'healthier' options. They singled out French company Danone as one such positive example. Table 5.6 gives the figures.

The weak spot for the food industry in relation to obesity is its marketing practices aimed at children.[72] Indeed, trends in childhood obesity are starting to appear frightening: prevalence of the condition escalated in the US from 5 per cent in 1964 to 14 per cent in 1999; in the UK, in 2001, one in six 15-year-olds were classified as obese; in Australia, one out of every five children is overweight. Obesity-linked 'adult onset' diabetes mellitus is for the first time being reported in children and adolescents in countries like the UK.

In addition to this the public health voice is now speaking up in ways that are unprecedented. For example, at the start of 2003 the World Health Organization went public about its concerns that products like sugary drinks are contributing to obesity, especially in children. The WHO urged governments to consider

Table 5.6 *Leading food companies exposed to 'obesity risk, 2003'*[71]

Rank	Company	% of 'not so healthy' food	% of 'better than'
1	Hershey	95%	5%
2	Cadbury	88%	12%
3	Coca-Cola	76%	24%
4	PepsiCo	73%	27%
5	Kraft	51%	49%
6	Kellogg	38%	62%
7	Wrigley	35%	65%
8	General Mills	35%	65%
9	HJ Heinz	32%	66%
10	Campbell	23%	77%
11	Unilever	23%	35% (43% non-food)
12	Nestlé	22%	60% (18% non-food)
13	Danone	20%	80%

clamping down on TV ads for 'sugar-rich items' to children. The International Obesity Task Force in its September 2002 report 'Obesity in Europe' was unequivocal in its concern about childhood obesity, calling for a European Union-wide restriction on advertising targeting the young, including pre-school children, to consume inappropriate foods and drinks.

What is more, the idea of taxes on unhealthy foodstuffs is again being openly discussed. For example, state legislators in many US states have considered legislation to tax soda and snack foods. Schools throughout the US are looking at the types of product they stock in vending machines or sell on school premises with the goal of reducing or banning sugary and fat-laden products.

Campaign groups like the International Association of Consumer Food Organizations (IACFO), an alliance of NGOs that represents consumer interests in the areas of food safety, nutrition and related matters, reports that the health of the children around the world is put at risk by the marketing of junk food. In the IACFO report, 'Broadcasting Bad Health: Why food marketing to children needs to be controlled',[73] it is calculated that:

- the food industry's global advertising budget is $40 billion, a figure greater than the Gross Domestic Product (GDP) of 70 per cent of the world's nations;
- for every dollar spent by the WHO on preventing the diseases caused by Western diets, more than $500 is spent by the food industry promoting these diets;
- in industrialized countries, food advertising accounts for around half of all advertising broadcast during children's TV viewing times. Three-quarters of such food advertisements promote high-calorie, low-nutrient foods;
- for countries with transitional economies (such as in Eastern Europe), typically 60 per cent of foreign direct investment in food production is for sugar, confectionery and soft drinks; for every $100 invested in fruit and vegetable production, over US$1000 is being invested in soft drinks and confectionery;
- over half the world's population lives in less-industrialized countries such as Russia, China and India and they are now suffering a rising tide of diet-related diseases as food companies export their products and their advertising practices.

Research officer Kath Dalmeny, co-author of the report and a member of the London-based Food Commission, part of the IACFO alliance, continued: 'Junk food advertisers know that children are especially susceptible to marketing messages. They target children as young as two years old with free toys, cartoon characters, gimmicky packaging and interactive websites to ensure that children pester their parents for the products.' The report calls for international controls on the marketing of high-calorie, low-nutrient food to children. In particular, the IACFO recommends that the WHO should:

- ensure children's health protection is put before trade concerns;
- support international controls on food marketing, including cross-border television, websites and email marketing;
- coordinate a statement of responsibility outlining the rights and responsibilities of food and beverage manufacturers and advertisers; and
- monitor industry marketing practices and develop global and regional targets.

In addition, a report by the UK Food Standards Agency in 2003 has prompted calls for a ban on advertising food to children and for health warnings on snacks and sweets. The report singled out what it called the 'Big Five' of bad children's food – sugary breakfast cereals, soft drinks, confectionery, savoury snacks and fast foods – as indicating a direct link between the amount of advertising seen by children and the food they eat.[74]

The food industry, in turn, has tried to respond to this onslaught on their food products and marketing practices with the publication of a number of company and public policy statements. One of the most high-profile early responses was from Kraft Foods, announcing a new series of steps to strengthen the alignment of its products and marketing practices with societal needs. The Kraft strategy promises to focus on four key areas: product nutrition, marketing practices, consumer information and public advocacy and dialogue. To this end, Kraft has formed a global council of advisers to structure a response to obesity and to develop policies, standards, measures and timetables for its implementation.[75]

Companies, such as Kraft and McDonald's, aimed to be ahead of many national health ministries in putting forward new policy initiatives relating to the rapid rise in obesity and by taking such initiatives to pre-empt regulation, taxes or other 'burdens on business'.

Cooking and Food Culture

Television chefs are not a new phenomenon; they are as old as television itself. Over the decades, TV cooking and food programmes seemed to mushroom almost in inverse proportion to the amount of cooking that actually occurred in the home, and went mass, mainstream and practically daily. Books on cooking and cuisine are currently in best-seller lists worldwide, and some research has suggested that the greatest strength of the phenomenon is its entertainment value;[76] in rich societies, actual cooking is often now only an occasional or 'hobby' pastime rather than a daily necessity. A 1997 European survey suggested that, although the British public spent less time in the kitchen than their European neighbours, 42 per cent viewed cooking as an enjoyable occupation, 14 per cent saw it as a creative activity and 11 per cent used it as a 'de-stressing activity'.[77] A UK government-funded study, however, found that, while some young people were interested in learning to cook, they lacked actual skills.[78]

Cooking is a universal point at which the natural world meets food culture.[79] In this respect, the matrilineal thread to food culture may be fraying: as family structures alter, mothers and grandmothers are not the important purveyors of food culture – teaching what to cook and how – they once were. In mobile and media-dominated cultures, there are other means for transferring food knowledge. Traditionally, and almost without exception across cultures, cooking has been a predominantly female task, but the skills base, in industrialized societies at least, is changing. Young people, for instance, may not cook but they have different skills from their predecessors.[80] The question is: does it matter if cooking is marginalized? In one corner are those who argue that the arrival of pre-processed foods is a liberation for women generally;[81] in the other are those who argue that by not being taught to cook, people lose a basic means for taking control of their diet. People cannot be expected to control their diet as health educators exhort them to do if they do not know what goes into food; cooking is a life skill rather than a high moral pursuit. If people do not know how to cook, they are entirely dependent upon whatever others or the food industry provides for them. Food culture then becomes a dependency culture – a far cry from the rhetoric of consumer sovereignty.[82]

One in five Americans eats in a fast-food restaurant each day;[83] 48 cents of every US dollar spent on eating in the US is spent out of the home.[84] Fast food is only one expression of the American love of being on the move; the 150 million cars on the roads by 2000 created a pressure for fast eating,[85] which in turn fed the US 'long hours/few holidays' work ethic and broke the link with the cultures that immigrants brought from Europe. Convenience is now also regularly cited in the EU as a key influence on shifting food patterns. Yet, ironically, the concept of 'convenience' also implies leaving the responsibility for food consumption choices to someone else – the food and beverage processing industries. UK consumers, perhaps the most Americanized of Europeans, are now adopting 'grazing' – constant snacking – rather than eating at set times of the day.

Research on the British suggests their cooking patterns are determined by structural factors such as longer working hours, the increase in waged female work and longer shopping travel times.[86] Pre-prepared meals and better cleaning products mean the average person spends 2 hours and 41 minutes less time doing domestic chores per week.[87] This is, however, balanced by an increase in the time spent shopping and on associated travelling to food suppliers, which has risen by two hours and 48 minutes per week, largely because of the growth of out-of-town supermarkets.[88, 89] In evolutionary terms, an entirely new food culture is being induced. Mobility of course is not new; traditionally almost all cultures have developed forms of wrapping food either to preserve it or make it transportable. For example, the ubiquitous sandwich – so named after an English earl, Lord Sandwich,[90] who in 1762 was so addicted to his gaming table that he refused to leave it and stuffed meat between two slices of bread in order to carry on gambling – is now a multi-billion dollar industry across many countries.

Shopping, Spending and Food

Although food is a basic need, it is only one of many spending areas, and the food shopping culture is, like other shopping, increasingly driven by the retail giants. Hypermarkets tend to have a higher turnover per square metre of sales space than do small shops, so that encouraging consumers to travel to a hypermarket in pursuit of small cost savings is a very successful marketing strategy. As a result, food shopping becomes a near-identical experience everywhere in the world: aisles in bricks-and-mortar-malls with car parks and open 24 hours; instead of the consumer having at least eye contact with the retailer and perhaps at best a chat with him, there is an alienated exchange with a harassed, low-paid worker.

The net impact of this emerging shopping culture is externalized social and environmental costs: the bread or beans might be cheaper at the hypermarket but the food travels further and the consumer travels further to meet it. In the UK, it has been estimated that half the food consumed by the country's 60 million mouths is sold from just 1000 stores,[91] with village stores and overall diversity threatened and with some communities becoming 'food deserts', where there are no or few shops.[92, 93] (And this in a land that, more than two hundred years ago, first Adam Smith and then Napoleon famously called 'a nation of shopkeepers'!) The retail concentration trend is widespread internationally, and within this food culture, it is often difficult to see a place for a health culture.

Food Activism and the Role of NGOs

A relatively new player shaping food culture and driving many food trends has been food activism and, allied to this, NGOs campaigning on behalf of food and health. In the 1980s and 1990s, some extraordinarily creative and effective NGOs emerged in the food policy world: in Brazil, for example, groups like IBASE (Instituto Brasil rode Análises Sociais e Econômicas) in Rio and the Global Forum on Sustainable Food and Nutritional Security in Brasilia both lent their weight to the mass campaigns for the right to eat and against hunger; in India, groups like the Delhi-based Research Institute for Science, Technology and Ecology worked with farmers on sustainable agriculture and in defence of home-bred seeds, while consumer groups like the Jaipur-based Consumer Union and Trust Societies (CUTS) worked on food standards and globalization; and in Australia, the Australian Consumers' Association, a member of Consumers International, worked on labelling, while the Eco Consumers Network linked human and environmental health and the Landcare network encouraged positive systems for protecting our fragile soils and ecosphere. In Canada, too, the Toronto Food Policy Council pioneered a new forum for stakeholders in the local food system, while at a national level the Council of Canadians tackled governmental sale of Canada's ample water assets. In Malaysia, the Consumers' Association of Penang campaigned against pesticides and for the quality of chicken feed, while the Third World Network pushed back the boundaries of consumerism by being the earliest to recognize the significance of the Uruguay Round of GATT.[94]

All this activity laid the foundations for the demonstrations in Seattle in December 1999,[95, 96] an unprecedented coming-together of NGOs from all around the world. Unfortunately, although there was strong representation from proponents of environmental health, there was next to none from the public health world.

In the UK, partly due to the vacuum in food governance in the period of neo-liberal Conservatism from 1979–1997, NGOs opened up new fronts and popular discourse on food and health issues such as labelling, the introduction of new food technologies, food safety and adulteration, food poverty, the morality of marketing targeted at children and nutritional standards of public food. NGOs throughout the world have been adept at building not just national, but international coalitions and networks, and have become a powerful force behind food and health concerns, but mostly as 'single issues' rather than as public health.

The core themes of this hybrid movement of food activism – sometimes rather pompously referred to in UN or international circles as 'civil society' – are beliefs that:[97]

- consumers have rights which must be fought for rather than assumed;
- human and environmental health go hand in hand;
- there is no such thing as the average consumer and consumption is socially fragmented;
- what matters is not just *what* is eaten, but *how* it is produced and distributed;
- policies can be changed for the better but this requires imagination, coalitions and focused effort.

A central tenet of these organizations, espoused also by the academics who work with them and identify with them, is that the ecological public health cannot improve or be protected simply by exhortation and that health education is limited as an effective policy strategy.[98] Due to their capacity to generate publicity and to bear witness, NGOs have become powerful voices within government and effective researchers into how food is currently made, processed, distributed and marketed. They are moving beyond the health education model that advises people to take better care of themselves, arguing that individuals are hampered in the marketplace by systematic biases and unequal power. They lay down a challenge to business and governments to reframe food governance and to set the rules of food engagement in a way that biases health.

Formal reviews of consumerism suggest that consumer NGOs tend to have limited industrial 'muscle'.[99] Consumer boycotts – pioneered in the US against the British as early as 1764–1776[100] – have had limited success; only when coupled with mass political movements (as happened with Gandhi's salt tax boycott when leading his revolt against the British) do boycotts have an industrial-style leverage. The international boycott of Nestlé in the late 20th century for breaking of the UN code against the marketing of breast-milk substitutes may have been considered effective but it did not break the company: it only embarrassed it and on occasions dented the share price.[101] Instead, the greatest weapons that NGOs bring to the Food Wars are their speed and flexibility, their ability to challenge carefully crafted commercial images, a high level of public trust, international networks and media credibility, and quality and independence in their research and policy proposals.

An opposing force to the growth of NGO influence has been the number of global public relations companies and lobbyists who have recruited teams to develop 'dialogue' with NGOs on behalf of major corporate clients. An outcome of this activity has been the emergence of counter-groups such as the fast-food industry's Center for Consumer Freedom, or the American Council for Fitness and Nutrition, a coalition set up in 2002 by leading food and beverage groups to counter charges that suggest that the industry is answerable for the obesity problem confronting the US population.

NGOs are now a clear and positive force in shaping food and health culture, but, because they are neither governments nor commerce, they lack ultimate power. The strengths of NGOs lie in their abilities to build political pressure, to forge the links between food supply chains and health, to integrate human and environmental health, and to put these on the policy and political agenda. Ultimately, however, only government can be a forum for an integrated food and health policy.

In this chapter we have set out the notion of a 'food culture' in order to capture the diversity of influences within the food economy that shape food consumption and the food consumer, who in turn feeds new ideas and demands back into food supply. It is within the dynamics of what we argue is a new food culture emerging for the 21st century (as part of our paradigmatic shift) that food policy-makers have to grapple with food governance and reconciling the long-term needs of policy strategy with the short-termism of much business, media and consumer activity. The food culture challenge is heightened further by bringing the health needs of individuals and populations within the sphere of food policy. Not least is the challenge of linking both human and environmental health demands as they apply to food. In the next chapter we set out the scope of these environmental challenges.

CHAPTER 6

THE QUALITY WAR: PUTTING PUBLIC AND ENVIRONMENTAL HEALTH TOGETHER

'In an unequal agricultural society, with primitive techniques, where men were at the mercy of nature and starved if the harvest failed; where plagues and warfare made life uncertain; it was easy to see famines and epidemics as punishments for human wickedness. As long as the level of technique was too low to liberate men from nature, so long were they prepared to accept their helplessness before a God who was as unpredictable as the weather. Sin, like poverty and social inferiority, was inherited.'

Christopher Hill, historian, 1912–2003

CORE ARGUMENTS

Natural resources such as water, soil, land, plants and biodiversity which underpin food consumption, have been placed under growing strain by the modern food economy, both by misuse of the land and by increasingly wasteful systems of distribution. The inputs and outputs of food production from chemicals to energy, are creating major problems in their own right. Because human and environmental health are so inextricably connected, solutions for the future of food supply have to address environmental quality and human health goals simultaneously. Policy initiatives, while limited, collectively have awesome implications for the redesign of the food supply chain and for reframing consumer culture.

Introduction

This chapter turns to another war zone where the conflict between how food is an environmental determinant of health and how the environment determines or affects food is fought out. Food *quality* is put into the front line: more than just cosmetic issues such as appearance and colour, but hidden features such as toxins or residues and the external implications of the food supply chain (such as how much water is used and wasted in growing or processing food and in transporting it).

Too many policy-makers still believe that they can merely 'bolt on' an environmental safety valve or eco-friendly niche market to address the crisis of food and the environment. In any other business but the food market, to despoil the natural capital upon which one depended would be self-evidently folly and managers would be disciplined or dismissed. Yet daily, poor environmental decisions are taken in the name of so-called efficiency to produce and distribute food. Food production depends upon a sound and healthy environment in which quality of input and output must be central. Food managers are caught between the need dramatically to alter how food is produced and distributed – for instance, to slash energy use, ozone depleters and waste – and their inability to do so because the supply chain is so competitive and cost-cutting. In addition, for example, eating fish products in the belief that ingesting fish oils is beneficial for human health is 'unproductive' if landing or farming the fish eventually destroys the very base of its own production.

Market research into new product design or creation seldom asks whether new food products are really needed in the first place; instead, the economic assumption is that, if a product can be made and sold, therefore it must fill a need. It is possible, for example, to grow lettuce hydroponically – on water in greenhouses, fed by a judicious mix of basic nutrients, but it is not viable to build the expensive greenhouses needed if lettuces can be grown in season out of doors, or organically in the soil in the open air, fertilized by composts that build the soil structure. When food quality and environment quality are considered in such a case, there are two health issues at stake: the first is the actual health infrastructure of the food production system – its soil, carrying capacity, ocean, access to water, and so on; the second is the potential human health impact of inputs used during production – such as pesticides, residues from drugs and animal production, pollution and microbiological toxins. In this chapter, we document both of these types of environmental health impact, and we discuss in addition: biodiversity, the use of water, soil degradation, climate change, urbanization, energy use, food miles, fish and the seas, meat production and the use of antibiotics, in order to underscore the intimate connection between human and environmental health.

Consumers often say that they want to act more responsibly towards the environment, yet they consume food that goes against that principle; they say they are prepared to pay full costs of production but they demand cheaper foods; the nature of production in the food supply exercises them. As a result, issues of food quality and their environmental links now demand detailed management and more insightful policy. As food supply chains have extended and become more integrated along their links, the capacity for breakdown to spread rapidly through these highly integrated logistic systems presents unprecedented difficulties for managers. A 'buzz' phrase in contemporary supply chain management is 'efficient consumer response' (ECR), a management belief that inefficiencies remain in long chains unless all the management operate as one corporate whole, and whose goal is to weed out weak links and inefficiencies in the chain and to cut costs.

While ECR is being initiated throughout the Western food supply chain, the consequences of its global reach are being analysed by a group of social scientists meticulously mapping the socio-geography of modern food supply chains, often by studying particular commodities such as tomatoes, bananas and tea.[1, 2, 3, 4, 5] They reveal that, in the name of efficiency, social and environmental degradation is often kept conveniently out of sight of the Western consumer, but that damage happens nonetheless. In the Life Sciences Integrated paradigm, for instance, there is a control ethic governing nature, plants, animals and ultimately consumers; in the Ecologically Integrated paradigm, the principles are more holistic: what matters are linkages rather than mechanisms.

Environmental hazards facing the world extend beyond food alone. Figure 6.1 gives a diagramatic representation from the

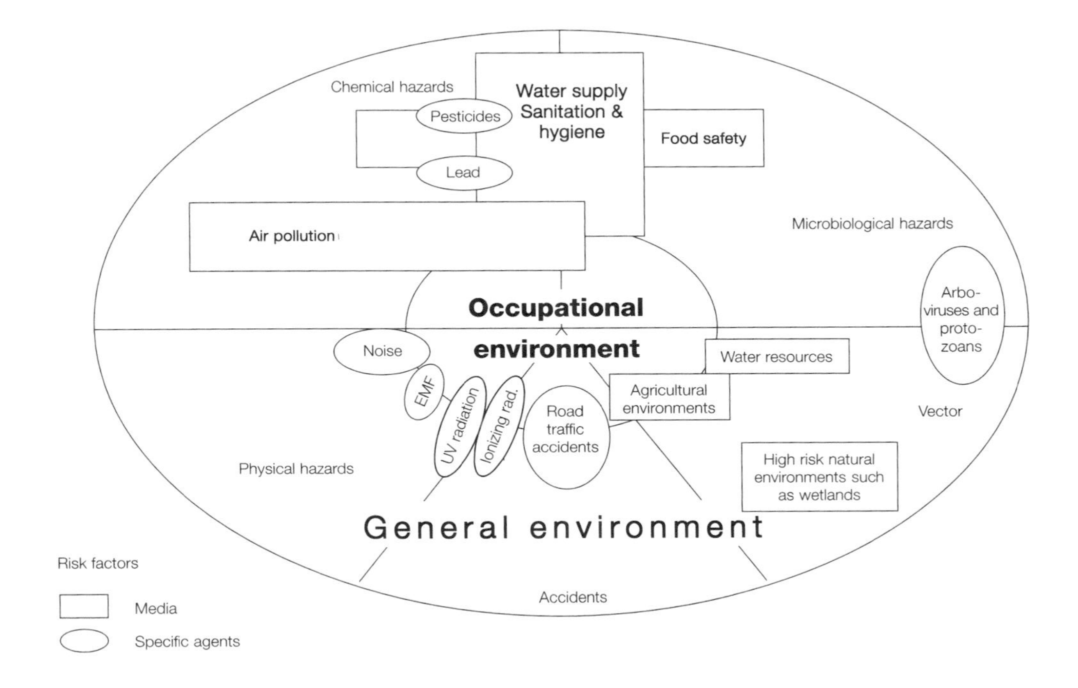

Source: WHO (2002) http://www.who.int/peh/burden/hurdenindex.htm

Figure 6.1 *WHO environmental hazards and risk factors*

WHO of how food risks fit into this wider environmental context. Note in this figure how the hazards that dominate contemporary food policy – food safety and pesticides – while important are by no means the whole picture. (Indeed, food safety's contribution to the alleviation of ill health is, as we noted in Chapter 2, very small in the West but more important in the developing world, largely due to poor water hygiene.) The new agenda of food and public health has to be seen within this wider environmental context.[6, 7, 8] Public health could in the future be more appropriately renamed 'ecological public health'.[9]

Can Consumers Save the Planet?

In the light of 'new' environmental-protection factors such as the limits to growth,[10, 11] the cycle of pollution,[12] and a holistic understanding of the earth's life-support systems,[13] a key question has presented itself to public policy: what sort of lifestyle can the planet sustain? To promote global human development as though all can aspire to the lifestyle of the presently better-off assumes wrongly not just that there are unlimited resources,[14, 15] but also that there are no better ways of redistributing those that exist.[16] The UN Development Programme has calculated that it is the richest 20 per cent of the world's populations which already accounts for:

- 86 per cent of all total private consumption;
- 58 per cent of the world's energy;
- 45 per cent of all meat and fish;
- 84 per cent of all paper;
- 87 per cent of all cars;
- 74 per cent of all telephones.

Conversely, the poorest 20 per cent of the world consume 5 per cent or less of all of the above goods and services.[17]

Proponents of maintaining current policies argue that such statistics and claims fail to take account of human ingenuity or that the poor are being denied the opportunities of the world's affluent. There is some validity in these positions but there is also much evidence-based concern about the earth's carrying capacity and the unequal distribution of its output. We need to construct a new ecological wealth for all nations.

To meet projected population growth and consumer demand, global food production will need to double by 2020,[18] and the FAO is optimistic that food supplies can and will grow faster than the world population. Even with 8 billion people by 2030, '. . . more people [will have] an adequate access to food than in earlier times . . . Growth in agriculture will continue to outstrip world population growth of 1.2 per cent up to 2015 and 0.8 per cent in the period to 2030.'[19] This optimism, whilst welcome, is, however, based on total food supplies and takes little account of distribution or other ecological constraints. Food sits at the apex of many biological givens: if good food requires clean water, soil, air and healthy biodiversity, the purpose of any public policy ought to be to nurture that interdependence of factors in a virtuous rather than a vicious circle.

Intensification

Intensification refers to the re-ordering of the food production process to make it more capital and labour efficient.[20] Under the Productionist paradigm intensification had strong public health proponents for the need to produce more food:[21,22] investment in raising output would surely yield health, since under-production and poor distribution were the key factors in food-related ill health. On the land, there is an increasing reliance upon non-farm inputs such as fertilizers and agrochemicals; labour is squeezed to produce more per person and with fewer numbers, and made more productive by the introduction of machinery and new farm technology which replaces animal motive power. Such concentration occurs at both an enterprise and a regional level: in the dairy sector, for example, there has been a remarkable increase in milk output from fewer farms, farmers, labourers and animals. There is also an agricultural lean towards specialization and standardization, which lead to monocropping and a decline in the number of traditional mixed farms practising crop rotation.

Intensification has been the motor of 20th-century consumerist prosperity in general and the increased output of agriculture in particular, but it comes at an environmental cost: food is a biological system, not the same as say, textiles or motor industrial production, and intensification of foodstuffs can fundamentally alter the nature of the final product. It affects, above all, the nature of production and the dynamics of capital accumulation;[23] it externalizes environmental costs. Pollution, waste and health care costs can all be more fully costed by economists on the 'balance sheet' of food production.

There is also growing concern about the implications of intensification on health in regard to pesticides, for example, and their potential health risks to humans and the ecology,[24,25] as well as potential hazards of chemicals in farming (which, research now suggests, may be implicated in disrupting the endocrine system).[26]

Intensification also means that, if there is a system breakdown, the ramifications spread quickly and extensively. In January 1999 in Belgium, for example, two synthetic compounds (PCBs and furans) were mixed with animal oils and fats by feed processors and sold as animal feed,[27,28,29] which travelled from just ten feed makers to 1700 farmers and thence to millions of poultry, pigs and cows. The resulting dioxin-contaminated feed was the focus of a huge political scandal that toppled the government and further dented European consumer confidence in big farming.[30,31] The impact on Belgian food production was immense, declining by 10 per cent in June 1999. The overall cost to the Belgian economy was in excess of £750 million.[31]

Perhaps the greatest illustration of the health risks of the intensification approach was the BSE crisis in the UK. Although the official government report made it clear that in its view intensification was not a cause,[33] the intensification of the feedstuffs, dairy and meat processing sectors meant that it was almost impossible to contain the outbreak easily. The freefall of public confidence in national and European safety protection sent shock waves throughout the continent and world and contributed to strong pressure to reform the Common Agricultural Policy. The BSE crisis' resulting call to de-intensify agriculture weakened the grip of the Productionist paradigm on policy-makers, and led British and European policy-makers to increase support for the greening of agriculture.[34]

One controversial argument is over whether the nutritional composition of food is affected by the way it is produced and how this effects the nutritional value of the diets of humans. There is good evidence of the extraordinary changes in the composition of meat and other animal products as a result of intensification. A series of studies took as its base healthy relatives of animals used in food production and observed their differences from domesticated animals from the perspectives of intensive or extensive rearing respectively. The study concluded that:

> *In the past it would have taken over six years for a steer to reach about 500kg body weight; with the feeding of high protein and energy feed it takes less than 20 months to attain the same body weight. A 2kg broiler is now produced in six to seven weeks instead of about 14 weeks. Similarly, a dairy cow now produces about 9,000kg of milk a year as compared to about 2,000kg about 40 years ago.*[35]

Such meat, reared on a high-protein diet, has shifted from being a 'protein-rich' to a 'fat-rich' food; the fat, in turn, is high in saturated fats, and the weight gain in both beef and chicken is partly due to the animal depositing excessive amounts of fat. Wild species provide more than three times as much protein as fat, compared to twice as much fat as protein in domesticated animals. In addition the proportion of polyunsaturated to saturated fatty acids is about 1:50 in domesticated animals compared to around 1:23 in wild animals.[36] We suspect that there will be further interest in how intensification alters the composition of foods, particularly in respect of micronutrients such as vitamins and minerals, and how the production process determines their output and alters the food in unexpected ways. This was dramatically illustrated by the surprise discovery by Swedish researchers of high levels of acrylamide, a known mutagen, in a range of 'cooked', processed foods including chips, breakfast cereals and crispbreads. This led to an urgent WHO international symposium, once other countries also found acrylamides present.[37] More important than changes due to the agricultural processes might be health implications of what happens further down supply chain. For example, in standard milling of white flour, as much as 60 to 90 per cent of vitamins B6 and E, folate and other nutrients are lost.[38]

Food and Biodiversity

According to the UN Environment Programme (UNEP), global biodiversity is changing at an unprecedented rate, driven by a combination of land conservation, climate change, pollution, unsustainable harvesting of natural resources and the introduction of exotic species.[39, 40, 41] Meanwhile, evidence from evolutionary nutritionists is beginning to suggest that human physiology requires a rethink of human nutritional needs in terms of its diversity or variety of species consumed, not just their quantities and proportion. Two leading researchers in this area have argued that 'because of an unusually large repertoire of essential nutrients (macronutrients, micronutrients, vitamins and minerals, and certain phytochemicals and zoo-chemicals whose essentiality is yet to be defined), the human species is peculiarly destined to be at once highly ecologically dependent and with a requirement to locate various ecological niches for its survival.'[42] In fact, evidence for actual biodiversity loss is already considerable. Current agricultural production is based upon crop strains, and often uses an agricultural practice which, far from increasing biodiversity, is shrinking it.[43] In the US, over 75 per cent of potato production comes from four closely related varieties,[44] 76 per cent of snap beans from three strains, and 96 per cent of pea production from just two pea varieties;[45] 95 per cent of cabbage, 91 per cent of field maize, 94 per cent of pea, and 81 per cent of the tomato varieties have been lost; an estimated 80–90 per cent of vegetable and fruit varieties, strong in the 19th century, were lost by the end of the 20th century.[46] Of the 10,000 wheat varieties in use in China in 1949, only 1000 remained by the 1970s. In India, farmers grew more than 30,000 traditional varieties of rice half a century ago; now ten modern varieties account for more than 75 per cent of rice grown in that country.[47] Often such 'modern' varieties of plants are grown for their weight, or volume, or predictability, or responsiveness to fertilizers, with not enough thought given to either the biodiversity or nutrition diversity.

Potatoes are another crop where the concentration of varieties has accelerated. In the Peruvian highlands, a single farm may grow 30 to 40 distinct varieties of potato, each having slightly different optimal soil, water, light and temperature regimes which, given time, a farmer can manage. Diversity like this requires skill. By contrast, in The Netherlands, a different skill has been applied to grow just one potato. One variety now covers 80 per cent of all potato land.[48]

Overall, according to the FAO, around three-quarters of the world's agricultural diversity was lost in the 20th century,[49] a direct result of the pursuit of uniformity, control and predictability, characteristics sought by businesses and processors. They want routine, regular supplies. Fast-food firms want the same potato worldwide. Together, agriculture and food culture are shrinking biodiversity in the field.

This shrinkage has not gone unnoticed by some environmental guardians, who are increasingly arguing that more effective national and international policies are required. The UNEP's 'Global Environment Outlook' report underlines the case for maintaining the health and status of biological entities on the planet; we never know when we might need them, especially since ten of the top 25 retail drugs in the world in 1997 were derived from natural resources. There could be both health and commercial advantages in retaining biodiversity. It is not just plants that are narrowing in genetic source. More than 90 per cent of all the commercially produced turkeys in the world now come from three breeding flocks. The system is ripe for a new strain of avian flu to evolve for which these birds have no resistance.[50]

Human activity including agriculture now threatens huge numbers of species of fish, mammals and plants. The natural world is said to be going through the sixth great period of extinction of the last half billion years,[51] and agriculture plays a significant part in this tragedy by losing and degrading habitats; it is also destroying pollinators whose presence enhances crop yields. Animal grazing is also a significant threat to biodiversity, as is drainage by destroying wetlands,[52] and there are concerns that precipitate use of biotechnology could release organisms into the environment that are impossible to control or retrieve and that will have their impact on biodiversity (as recognized in the Convention on Biological Diversity).

In policy terms the choice is becoming clear. Agriculture and food production can continue to degrade biodiversity or they can be redesigned to enhance it. The viability of this latter policy approach is already being made manifest by the sustainable agriculture movement who are supporting, and learning from, traditional farming methods as well as a wide variety of new cropping techniques.[53, 54] An International Treaty for Plant Genetic Resources for Agriculture has been developed, but only 60 countries had signed by mid-2003, of which 30 had ratified the treaty. It was due to come into force in 2004, once 40 countries had ratified it.[55] Such protection measures are painfully slow to introduce as they threaten many interests, not least predators seeking to exert intellectual property rights.

WATER

Water is critical not just for direct consumption and agriculture but for hygiene.[56] Water covers 70 per cent of the planet, but 97.5 per cent is ocean, not usable in industry, agriculture or as drinking water. Of the 110,000 billion cubic metres of rainwater that falls on earth, only 12,500 billion are accessible and usable, yet this amount has been calculated to be sufficient for human use. Water for domestic use (ie drinking) accounts for only 8 per cent of the water available for human use, with agriculture using 70 per cent and industry 22 per cent.[57] Modern agriculture is, just like its consumer counterpart, a greedy consumer of water. The demand for water is expected to grow in all regions of the world over coming decades. A new water world order is emerging. Countries like the US and Canada have vast water resources; others such as Taiwan, Saudi Arabia and Germany are actually in water deficit.[58] The UNEP anticipates growing worldwide 'water stress' in coming years.[59]

Agriculture is both victim and perpetrator of our water wastefulness, already accounting for 70 per cent of the freshwater withdrawals in the world and acknowledged by the FAO as mainly responsible for freshwater scarcity:[60] lack of water increases the chance of cropland being degraded; yet agriculture is a major cause of this stress as it also pollutes drinking-water quality in less developed countries which cannot afford the filtration of contaminants.[61] UK water companies have had to spend £1 billion over a number of years to filter pesticide residues out of drinking water to get it to meet tougher new standards, illustrating how environmental costs can be externalized, and only the application of a 'polluter pays' principle both puts a monetary cost on this pollution and encourages re-internalization of the cost. The water used by agriculture, drawn from lakes, rivers and underground sources, is used mostly for irrigation which helps provide 40 per cent of world food production[62] (since 1961, the area of land being irrigated to increase yields by watering crops has risen from 139 million hectares worldwide to over 260 millions).[63] As a result, water tables are predicted to fall due to over-irrigation and intensive crop production,[64] and salination to increase as a direct consequence.

Access to safe water is denied to 1.4 billion people, with 2.3 billion lacking adequate sanitation; an estimated 7 million die each year due to water-borne diseases, including 2.2 million children under 5 years of age. While in the 20th century, human population tripled, water use grew by a factor of six,[65] and by 2020 the availability of water for humans is expected to drop by one-third. Water scarcity or stress – having less than 1700 cubic metres of water per person per year – is estimated to affect 40 per cent of humanity by 2050. The consequences are likely to be a rise in food prices and health threats, and poor countries are likely to be most heavily affected.[66]

Large-scale irrigation is probably reaching its limits. Projected crop irrigation requirements by 2050, according to the World-watch Institute, would require another 24 River Niles,[67] even with the use of better technology – using drips rather than constant water spraying. It would now be sensible to invest in enhancing the capacity of soil to retain moisture, which is a feature of sustainable agriculture systems. Again, both on and off the land, water security is set to become an even more serious issue.[68, 69, 70]

Pollution and Pesticides

Over the last 30 years, since the publication Rachel Carson's *Silent Spring*, a study of the environmental impact of pesticides,[71] methods of food production have gradually become major public and environmental health problems. Food is a vector by which pollution can enter human bodies and create biological damage. Table 6.1 summarizes how environmental factors can affect health and gives figures on the health impact of food pollution.

Some of the food supply chain's environmental pollution is clearly related to intensification. A medium-sized US feedlot (a mass cattle factory farm in the open air) can have an annual throughput of 20,000 cows, producing as much sewage as a town of 320,000 people[73] and creating a serious problem with waste disposal. (The Netherlands, as a small, intensively farmed country, has critical problems in this respect and is having to reduce its output of manure onto land.) Persistent organic pollutants (POPs) – toxic synthetic compounds – accumulate in the food chain, persist in the environment and travel by being bioaccumulated (as animals eat each other, so the POP is stored in fat and thus consumed and stored). Most humans have around

Table 6.1 *Environmental factors affecting health*[72]

	Polluted air	Poor sanitation and waste disposal	Polluted water or poor water management	Polluted food	Unhealthy housing	Global environmental change
Acute respiratory affections	✓				✓	
Diarrhoeal diseases		✓	✓	✓		✓
Other infections	✓	✓	✓	✓		
Malaria and other vector-borne diseases		✓	✓		✓	✓
Injuries and poisonings	✓		✓	✓	✓	✓
Mental health conditions					✓	
Cardiovascular diseases	✓					✓
Cancer	✓			✓		✓
Chronic respiratory diseases	✓					

Source: UNEP (2000) *Global Environment Outlook 2000*, London: Earthscan, based on WHO data

500 POPs stored in their body fat that have been created since the chemical revolution of the 1920s.[74] Pesticides are a key route for POPs, notably through aldrin, chlordane, DDT, dieldrin, endrin and heptachlor. POPs have a malign impact on humans, wildlife, land and water.

It is not unusual for daily US diets to contain food items contaminated by between three and seven POPs. The main POP-contaminated food items in the US have been found to be: butter, cantaloupe melons, cucumbers/pickles, meatloaf, peanuts, popcorn, radishes, spinach, summer squash and winter squash,[75] all containing levels of POPs which may individually be within safety limits, but which collectively pose a risk, according to health standards set by the Center for Disease Control's Agency for Toxic Substances and Disease Registry and the Environmental Protection Agency. Public health physicians argue that the only health strategy is prevention,[76] yet world pesticide usage grew 26-fold in the second half of the 20th century.[77] In California, the state which accounts for 25 per cent of all pesticides used in the USA, toxic pesticide use on fruit and vegetables actually increased between 1991 and 1995.

Although overall the quantity of pesticides applied in industrial agriculture has declined recently, the toxicity of what is used has increased by an estimated factor of 10- to 100-fold since 1975. The pesticides are packing a heavier punch. Despite this, resistance is spreading; POPs are becoming less effective: 1000 species of insects, plant diseases and weeds are now resistant, an environmental impact known as the 'treadmill effect'. However, so extensive is the reach of POPs that even crops grown without them may contain them, while crops grown using pesticides contain much higher levels.[78] And despite strong evidence of the negative impact of POPs and their connection with pesticides, governments only recently agreed the Stockholm Convention to phase out their use. If governments want raised controls on pollutants, such as from pesticides, regulations are essential, although encouraging increased fruit and vegetable intake to reduce rates of CHD and cancers tacitly encourages increased intake of POPs through that route. However most epidemiological evidence suggests that the relative risk of POPs is offset by the gain from the nutrient intake of fruit and vegetables.[79] From a policy point of view there need not be a trade-off of risks and benefits; why should this be an either/or when it is possible to aim for a win–win?

WASTE

One of the defining characteristics of modern food supply is that, as the chain has lengthened, the packaging that comes out at the end has increased. There is a strong correlation between economic growth and growth of waste,[80] and considerable correlation with urbanization. The scale of the problem is significant. In the EU, for example, waste generation per capita from household and commercial activities already exceeds, by 100kg, the target of only 200kg per capita per year, set by the EU Fifth Environmental Action Plan in 2001.[81] Much of this waste goes into landfill, creating more pollution and health hazards.

Whilst food always appears in most waste statistics, it is not in fact the prime source of waste at the household level. In the UK, for instance, of the 435 million tonnes of waste disposed of each year, the household dustbin accounts for only 6 per cent of the total, 8 per cent of which is sewage sludge; 36 per cent comes from commerce and industry, and half is produced by primary industries such as mining, dredging, quarrying and farming.[82] However, mountains of consumer garbage is associated with their food purchasing patterns – wrappers derived from a combination of steel, aluminium, glass, textile, paper, plastic, polystyrene, cardboard and other fibrous forms. Six billion glass containers are used annually in the UK, of which only 30 per cent are recycled.[83]

Where some of this waste is re-usable (such as bottles which can be both re-used and recycled), incidence of re-use is declining in affluent Western societies. Some of the waste is in putrescible form (ie compostable), and could be returned to refertilize the soil. In 1999, the US Environmental Protection Agency estimated that municipal solid waste consisted of 25.2 million tons of foodwaste,[84] of which containers and packaging were the largest percentage by volume at 76 million tons or 33.1 per cent of municipal product waste.

Much waste can only be recycled by industrial processes (the fate of steel, aluminium and plastic), but thereby expending double energy, first to be produced and then to be recycled. In addition, much packaging that may be ideal for marketing or microbiological safety is in fact unsuitable for recycling. Drink cans, for instance, may have bodies of aluminium, tops and caps of steel, linings of plastic and outers of paint or paper and even another layer of 'tamper-proof' plastic, making their recycling extremely difficult and energy-wasteful. And it is the local state which carries the can, literally, for household and street waste; even if waste-removal services are privatized, there will be a local authority offering the contracts. It is for this reason that public bodies are rethinking waste. They see an opportunity both to clean the streets and protect the environment.[85] In Canberra, Australia, the authority now recovers 66 per cent of its waste and Edmonton, Canada, achieves a target of 70 per cent.[86]

There is an incentive for the food industry to cut unnecessary waste. In the UK, compared to 30 years ago, yoghurt pots are 60 per cent lighter. Compared to ten years ago, milk bottles are 30 per cent lighter, food cans 40 per cent lighter and drink cartons 16 per cent lighter.[87] In Germany, since the introduction of tougher recycling regulations in 1991, which put the responsibility on manufacturers for their packaging waste and recycling, per capita consumption of packaging has declined from 94.7kg in 1991 to 82kg by 1998. Tough regulations work.[88]

Soil and Land

Barring a vast future investment in hydroponics, human food production will always rely upon soil. Yet reviews of the status of soil make sobering reading.[89] Soil structure is being damaged by desertification, pollution, water damage, clear-cutting of forests and overgrazing, leading to loss of humus. Once lost, soil is irreplaceable in the short term. The average loss of soil humus in recent decades has been around 30 times more than the rate throughout the ten millennia of settled agriculture.[90] Demand for animal protein has a knock-on effect on cropland and soil: if consumer demand for meat continues to rise, pressure to crop the land to feed animals will exploit the soil over and above the pressure from population. Meat consumption drives a vicious circle of energy and other fossil fuel inputs, chemicals, water and protein as feed, let alone labour.

In Africa, for instance, more and more population-pressured land has been taken into cultivation in response to commodity prices. UNEP argues that much of the growth of production is simply due to the use of new land. The paradox is that hunger remains; and, by turns, there is land degradation as people search for new sources.[91] In addition, soil erosion encourages farmers to use increasing amounts of fertilizers and agrochemicals, which in turn threaten the health of the land. In the Asia–Pacific region, an estimated 850 million hectares or 13 per cent of cultivatable land is formally designated as 'degraded' due to a variety of reasons ranging from salinity and poor nutrient balance to contamination. 'Desertification' is spreading: in Western Europe, the area of land being cultivated has fallen significantly for arable and croplands and for pasturelands over the last 30 years. Water damage from over-grazing and poor agricultural practices is damaging European soils generally, and housing is seen as more valuable than soil kept for food.

Climate Change

The Intergovernmental Panel on Climate Change recommends a reduction of between 60 and 80 per cent of greenhouse gases just in order to stabilize climate change, not even to reduce it; the implications for health have also been mapped.[92] World energy use will grow by two-thirds between 2000 and 2030 despite its known impact on atmospheric and climate change.[93] Its consequences (including flooding, temperature rise, volatility, storms and spread of infection and parasites) hit the socially marginalized and poor countries particularly hard because they cannot buy their way out of such crises. But even more affluent countries are affected by the spread of disease and crop failures: the US is under siege from West Nile fever carried by mosquitoes, and Europe is likely to be re-infected by malaria as temperatures rise. Equally crucially key cash crops such as coffee and tea in some of the major growing regions will, over coming decades, be vulnerable to global warming.[94, 95] The fear is that farmers will be forced into the higher, cooler, mountainous areas, intensifying pressure on sensitive forests and threatening wildlife and the quality and quantity of water supplies. These crops, central to the development agenda, not to mention the taste-buds of affluent consumers, could fall by as much as a third,[96, 97] and the impact on the economies of the countries affected could be serious. Agriculture earns Kenya an estimated $675 million a year in exports, and of this, $515 million comes from tea and coffee exports; for Uganda, annual agricultural exports are worth around $434 million with tea and coffee worth $422 million. This decline would coincide with the urgent need to raise yields to feed a growing global population. Rising temperatures, linked with emissions of greenhouse gases, can damage the ability of vital crops such as rice, maize and wheat, to flower and set seed.

Researchers at the International Rice Research Institute (IRRI) based in Manila in the Philippines estimate that average global temperatures in the tropics could climb by as much as three degrees centigrade by 2100. Their studies indicate that, for every one-degree centigrade rise in areas such as the tropics, yields could tumble by as much as 10 per cent. Such findings indicate that large numbers of people, many already on low incomes, will face acute food insecurity unless the world acts to address climate change by reducing emissions of carbon dioxide and other greenhouse gases. It is also why governments are beginning to take climate change seriously. Even in the US, which has resisted action on climate change, the Pentagon is now rightly concerned about the military, political and food security implications of climate change altering world food supplies.[98] Continual energy use – particularly oil – is driving environmental change.

Urban Drift

People all over the world migrate to the cities to look for employment and better economic prospects; usually they yearn for better access to amenities, services and food.[99] The result is increased strain on the country to feed growing cities. Cities and towns offer positive features such as access to education and health care. Caloric intake tends to rise with urbanization, and children can usually attain a better dietary status; but set against these positive features is evidence of how urbanization also brings the threat of marginalization: increased poverty, inequalities, unemployment and dependency. Urbanization alters nutritional status: intake of carbohydrates, meat, sugar and edible oils increases; populations are more exposed to junk foods high in salt and sugar and low in fibre; and intake of unprocessed foods drops.

Yet against these worrying features, the range of diet can actually improve in urban settings. Their populations have greater economic pulling power and usually offer more diversity of foods than are available in rural areas. In cities, the overwhelming preponderance of food is purchased, although there is a surprisingly high amount of urban agriculture 'hidden' in urban areas. An estimated $500 millions' worth of fruit and vegetables is produced by urban farmers worldwide.[100] Whilst traditionally urban areas have been fed by their hinterland, this is no longer the case in Western cities. Londoners, for instance, consume 2 million tonnes of food annually, sourced from all over the world,[101] and the FAO estimates that in a city of 10 million people, 6000 tonnes of food may need to be imported on a daily basis.[102]

Between 1950 and 1990, the world's towns and cities grew twice as fast as rural areas. Many cities in the world already have huge populations, such as Dhaka in Bangladesh with a population of 9 million growing at an annual rate of 5 per cent, an additional 1300 people per day.[103] Other Asian cities are now growing at a rate of 3 per cent per year, while African cities are growing at a rate of 4 per cent per year. In the Phillipines, for instance, 31 per cent of the population was urbanized in 1970; by 2001, 59 per cent lived in towns. Much of this growth occurred following the opening up of the economy in the late 1990s. The metropolitan area of Manila, which had a population of 200,000 in the early 1900s, grew rapidly to 9 million in 1999 and is set to top 16 million by 2016.[104] Even in a country such as Fiji, the demographic transition has been rapid. In 1966, 33 per cent lived in towns; by 1996 this had risen by half again to 46 per cent. Of this urbanized population, 93 per cent lives in just five centres, a concentration that is common in many countries.[105] In China, the number of cities with over 1 million inhabitants increased from 22 to 37 in just 12 years from 1985 to 1997. According to the UN Population Division, China's urban population will grow from 246 million in 1985 to an estimated 536 million by 2005 and 763 million by 2020.[106]

For the first time in human history, more than half of humanity is urban. In Latin America and the Caribbean, urbanization is 75 per cent; in Asia 38 per cent and in Africa 37 per cent.[107] In 2025, an estimated two out of three people in the world will live in urban areas: fed by whom and on what? The World Bank has estimated that there will be over 1 billion urban poor in the 21st century.[108] In 1950, the number of people living in cities was about the same in industrialized and in developing countries – about 300 million.[109] By 2000, some 2 billion people lived in cities in developing countries, more than twice the number of urban dwellers in industrialized countries. As populations in cities expand, so does the demand for food to feed all the people who are living there. Even allowing for urban food production – gardens, smallholdings, even window boxes – the majority of food in the city must be bought, and poor families often spend as much as 60 per cent to 80 per cent of their income on food, approximately a third more on food than their rural neighbours.

At its Habitat 2 conference in Turkey, the UN mapped out the urgency of the task, concluding that urban or peri-urban agriculture will have to make a comeback.[110] In Kathmandu, 37 per cent of urban gardeners already grow all the vegetables they consume, while in Hong Kong, 45 per cent of demand for vegetables is supplied from 5 to 6 per cent of the land mass. Across the world, there is a burgeoning movement of local authorities, small farmers and ecology-conscious consumers arguing for this modern urban agriculture sector: 'growing our own'. The WHO Regional Office for Europe Nutrition Programme believes too that local produce provides food security for otherwise marginalized populations.[111] This development is particularly urgent in Eastern Europe where some catastrophic collapses in currencies have been experienced; importing needs a strong currency. Projects such as community gardens and city farms have already sprung up in the industrial heartlands, showing that urban food production has a social as well as economic value.[112] Real market economies can offer valuable encouragement to local production to provide for more local need. If well planned, well located and well organized, urban agriculture initiatives can be a strong source of income for urban workers and help boost local economies.[113, 114] In Cuba, after the collapse of the Soviet Union, ecological urban agriculture and gardens became a direct necessity and has helped to feed and keep the population healthy, much as it helps any population in time of war.

Energy and Efficiency

Consumers expect their food to be safe and to look good but sometimes fail to recognize that the complex system of distribution and production that keeps their supermarket shelves so liberally stocked is in fact highly polluting. The UK food, drink and tobacco sector emits 4.5 million tonnes of carbon annually (compared to the chemical industry's 7 million and the iron and steel industry's 10 million), and creates nearly 6 million tonnes of waste, most of which is dumped in landfill sites.[115]

Environmental damage by food production is linked to energy use required by types of packaging (see Table 6.2); by

Table 6.2 *Energy used by product/packaging combinations for peas*[119]

	Steel can 420 cm^3	Alum. can 220 cm^3	Glass one-way 360 cm^3	Multi-layer pouch 600 cm^3	Frozen carton	Fresh carton	Fresh imported
Total energy used in MJ/kg	18.0	40.0	20.0	16.0	24.0	9.0	25.0

types of freight used (see Table 6.3); and by types of industry (see Table 6.4). A typical British household of four people annually emits 4.2 tonnes of carbon dioxide from their house, 4.4 tonnes from their car, and 8 tonnes from the production, processing,

Table 6.3 *Energy use and emissions for modes of freight transport*[120]

	Rail	Water	Road	Air
Primary energy consumption KJ/tonne-km	677	423	2890	15,839
Specific total emission g/tonne-km				
Carbon dioxide	41	30	207	1206
Hydrocarbons	0.06	0.04	0.3	2.0
VOC	0.08	0.1	1.1	3.0
Nitrogen oxides	0.2	0.4	3.6	5.5
Carbon monoxide	0.05	0.12	2.4	1.4

Table 6.4 *Energy used per year by various UK food industries*[121]

	Energy use (10J)	% of UK total energy use
Direct fuel to agriculture	121	
Fertilizers and agrochemicals	129	
Machinery, manufacture and repair	51	6.2
Transport to and from farm	16	
Imports of animal food	60	
Food processing	527	6.0
Food distribution	451	5.0
Imports of human food	208	2.3
Home cooking	728	
Waste disposal	26	8.5
Total		**28.0**

packaging and distribution of the food it eats.[116] Travelling to and from the food shops by car adds further environmental burden and reduced physical activity, contributing to obesity. The large rise in energy use on farms is illustrated in Table 6.5 which gives the growth of energy use in the production of US maize over time.[117] Intensification, by reducing human labour, has placed the energy focus on machinery and inputs. Between 1945 and 1985, US maize production changed in its energy mix.[118] For example, labour was reduced fivefold while energy input from machinery increased by a factor of 2.5. More strikingly, energy input from fertilizers and irrigation increased 15- and 18-fold respectively, yet over the 40-year period, the yield/ratio actually declined from 3.4 to 2.9.

Table 6.6 shows the impact on both emissions and direct energy use of different modes of transport used to freight food. From a sustainability perspective, the optimum policy is to consume food that is produced as locally as possible.

Food miles

The notion of 'food miles' has been developed for calculating the distance that food travels between primary producer and end consumer. Studies suggest that a major source of pollution from the food supply chain is its increasing dependency on transport,

Table 6.5 *Energy input in US maize production, 1945–1985, in MJ/ha*[122]

Activity	Energy input 1945 in MJ	1945 in %	1985 in MJ	1985 in %
Labour	130	1.2	25	0.1
Machinery	1701	16.3	4255	9.9
Draught animals	0	0	0	0
Fuel	5969	57.3	5342	12.4
Manure	–	–	–	–
Fertilizers	974	9.3	15,650	36.3
Lime	192	1.8	560	1.3
Seeds	673	6.5	2174	5.0
Insecticides	0	0	251	0.6
Herbicides	0	0	1463	3.4
Irrigation	522	5.0	9405	21.8
Drying	38	0.4	3177	7.4
Electricity	33	0.3	418	1.0
Transport	184	1.8	372	0.9
Total input	10,416	100	43,092	100
Yields	35,647		123,728	
Yield/input	3.4		2.9	

Table 6.6 *Emissions and energy use by modes of freight transport*[123]

Mode	Description	CO^2 Emissions (grammes CO^2/ tonne-kilometre)	Energy consumption (MJ/tonne-kilometre)
Air	Short-haul	1580	23.7
	Long-haul	570	8.5
Road	Transit van	97	1.7
	Medium lorry	85	1.5
	Large lorry	63	1.1
Ship	Roll-on/roll-off	40	0.55
	Bulk carrier	10	0.15

both in transporting the food within the food system – from farmer to depot to processor to retailer – and then to get the consumer to the retailer.[124] Some of this transportation appears almost comical, with a classic study of a West German yoghurt bought 'locally' using, in fact, ingredients – food and packaging – assembled from hundreds or thousands of kilometres away.[125] A US study of data on transport arriving at Chicago's terminal market from 1981–1999 found that produce arriving by truck from within the continental US (ie excluding externally sourced food) had risen from an average of 1245 miles to 1518 miles, a 22 per cent increase,[126] with the national food-supply system using 4 to 17 times more fuel than did the localized system.

An everyday meal is a minor piece of social history. Each ingredient will have travelled some distance, probably be based on a plant which has been grown thousands of miles from its biological origins, and the foods will probably have been trucked long distances. Table 6.7 disaggregates a meal eaten by one of the authors in his home in London in the summer of 2001. This tells a story of food miles that could be replicated throughout the world.

According to the UK Department of Transport, despite approximately the same tonnage of food being consumed annually within the UK, over the last two decades the amount of food being transported on roads has increased by 30 per cent;[128] and the average distance it has travelled has increased by nearly 60 per cent.[129] Between 1989 and 1999 there was a 90 per cent increase in agricultural and food products traded by road in the UK.[130] Worse still, total UK airfreight doubled over the same period and is predicted to increase at 7.5 per cent each year until 2010.[131] Not only is the same amount of food being transported further, but British consumers are travelling further to get it, and most often use cars to do so. The distance travelled for shopping in general rose by 60 per cent between 1975–1976 and 1989–1991; but the travel taken by car more than doubled.[132] Far from hypermarkets being 'convenient', they in fact generate more, not fewer, trips for food shopping (UK Government figures also indicating that the mileage of trips to town centre food shops is less than half that of trips taken to edge-of-town stores),[133] illustrating the economic dimension of externalities: the price of modern foods does not reflect the true price of food production.[134, 135]

Table 6.7 *A simple dinner at home*[127]

This meal for two was cooked at home and comprised:

- Pasta with herbs, dried tomatoes and olive oil
- Fruit salad with blackberry vinegar
- wine and water (LC), ginger beer and beer shandy (TL).

Food item	Botanical name of key ingredient(s)	Plant/cultural origin	Grown/made in	Bought in	Food miles (est)	Distance from point of purchase
DISH: PASTA						
Pasta shells	*Triticum durum*	Egypt	Italy	Clapham	900	1 mile
Dried tomatoes	*Lycopersicon lycopersicum*	South America	Italy	Westminster	1000	3 miles
Olive oil	*Olea europaea*	Eastern Mediterranean	Italy	Westminster	1000	3 miles
Salt	N/a	Sea	Sea (Maldon, Essex)	Clapham	40	1 mile
Pepper	*Peper nigrum*	Malacor/ Tranvancore Forests	India	Wandsworth Common	4700	½ mile
Mint	*Menta spicata*	Uncertain origin	Our garden	N/a	0	0
Basil	*Ocimum basilicum*	India/SE Asia/NE Africa	Our garden	N/a	0	0

Food item	Botanical name of key ingredient(s)	Plant/cultural origin	Grown/made in	Bought in	Food miles (est)	Distance from point of purchase
DISH: FRUIT SALAD						
Mulberry	*Morus nigra*	Central/Eastern China	Thames Embankment	Picked in a garden by the Thames	3	3 miles
Cherries	*Prunus avium*	Asia Minor/ Mediterranean	US (Washington)	Tooting	4600	1 mile
Nectarine	*Amygdalus persica*	Uncertain (China)	Italy	Tooting	1000	1 mile
Peach	*Amygdalus persica*	China	Italy	Tooting	1000	1 mile
Blackberry	*Rubus ulmifolius*	Europe	Our garden	N/a	0	0
Melon	*Cucumis melo*	Middle East	Spain	Tooting	600	1 mile
Loganberry	*Rubus loganobaccus*	US	Our garden	N/a	0	0
Blackberry vinegar	N/a	Common/England	Doncaster	Clapham	170	1 mile

Table 6.7 *(continued)*

Food item	Botanical name of key ingredient(s)	Plant/cultural origin	Grown/made in	Bought in	Food miles (est)	Distance from point of purchase
DRINKS:						
Beer	*Hordeum ditichon* + *Humulus lupulus*	Barley – Near East; Hops – Europe/W Asia	Wandsworth (brewery)	Wandsworth Common	2 miles	½ mile
Ginger beer	*Zingiber officinale* + *Beta vulgaris*	Ginger – Asia Sugar – SE Asia	Ginger from India, sugar from East Anglia, brewed in Newcastle	Wandsworth Common	4800 + + 100 + 200	½ mile
Wine	*Vitis vinifera*	Shiraz (France) & Cabernet Sauvignon (France)	Australia	Wandsworth Common	10000	½ mile
Water (bottled)	*aqua*	Aquifer	France	Wandsworth Common	450	½ mile

'Ghost Acres'

'Ghost acres' was a notion developed by food analyst George Borgstrom to refer to bought-in feedstuffs used in intensive agriculture.[136] (The term 'ecological footprint' is another term for this use.)[137] A study of The Netherlands, for example, found that while the average human on the planet has 0.28 hectares of arable land available to feed him or herself, a citizen in The Netherlands actually uses 0.45 hectares of arable land.[138] In other words, a Dutch citizen is also reliant on 'ghost acres' outside of The Netherlands to supply his or her food. Even the UK's putatively efficient food system sucks in the food products of other people's land and seas: especially the EU: soya, citrus, fishmeal, maize, manioc and many crops are grown in and for Europe in huge aggregate quantities, with the EU actually importing more fruit and vegetables than it exports.[139] One study has estimated the net import of 'hidden land' into the UK was 4.1 million hectares in 1995,[140] with much of the produce fed to animals, the fuel of intensive husbandry. This form of global sourcing, however, is not only energy-inefficient, but it is also doubtful whether it improves global 'equity', and helps local farmers to meet the goals of sustainable development.[141]

A study of London's resource flow found that food was a key source of the city's consumption and environmental impact.[142] Of the 6.9 million tonnes of food consumed in London in 2000, 81 per cent was imported from outside the Greater London area. Food accounted for 14 per cent of the city's total consumption (calculated as imports plus production, less exports) and for 2 per cent of the waste, but in addition, 94 million litres of bottled water were consumed in one year, creating an estimated 2260 tonnes of plastic waste. Londoners use 6.63 hectares of earth space each, with food accounting for 41 per cent of those ghost acres. For London to become sustainable, its overall consumption needs to reduce by 35 per cent by 2020 and by 80 per cent by 2050.

Like other so-called efficient food producing economies, the UK's population of farmers is declining[143] and ageing. Because people still have to eat, production is being drawn in from other countries not always blessed with such a fertile and benign climate as ours. Despite ideal conditions for top fruit production (especially soft fruits and berries), for instance, nearly four out of five pears consumed in the UK and two-thirds of its apples[144] are now imported,[145] imports coming from Chile, Australia, the US and South Africa as well as from throughout the EU. The policy question raised by the use of ghost acres in a rich growing country like the UK is why it should be fed by others when it can produce its own, has the land and climate to do so and when others are perhaps more needy, both within their own populations and in neighbouring ones.

The world of sustainable agriculture is beginning to grapple with the complexities raised by such environmental costs. The image of ecological production may appear localist and 'natural' but the reality is not so simple; 70 per cent of organic produce sold in the UK, for instance, is imported. The structure of the agricultural market has hindered profitable production from meeting consumer demand. If consumers want to be environmentally benign, the solution is not necessarily to purchase organic produce:[146] 'local-ness' of produce should be as important a factor as the system of production, and for organic produce to warrant high ecological favour, it ought be produced and consumed as locally as possible. Long-distance organic food is not environmentally benign, even if the demand for it sends ecologically oriented signals to the supply chain. As can be seen from Table 6.8, locally produced food, even if produced intensively, can be less environmentally damaging than organic food that has travelled intercontinentally.

Eating up the Fish?

Nowhere is the conflict between environmental and human needs more apparent than over fish. While nutritionists advise regular consumption of fish (particularly oily fish for its omega-3 essential fatty acids), rising concern is expressed by marine ecologists about the seas and fish farms reaching their ecological limits. As with organic foods, the problems are not just about production but also its environmental 'friendliness': the favoured Spanish dish of *baccalao* (salt cod), for example, may have been caught in the Baltic, but it lands in Norway and is processed in Scotland before being served as a 'Spanish' meal. Fish also strongly suggests the need for public policy to weave a sensible mix between human health, ecology and culture.

Table 6.8 *Weekly costs of food and drink in the UK (organic and non-organic)*

Modes of production and transport	Expenditure on food and drink (£ per person per week)	External cost from farm (£)	External cost from transport (£)	Total external costs (£)	Real cost of food (price + externalities)	Externalities as % of price paid by consumers
Conventional; local	16.94	1.563	0.004	1.57	18.51	9.3%
Conventional; national; road	16.94	1.563	0.096	1.66	18.60	9.8%
Conventional; national; rail and road	16.94	1.563	0.022	1.59	18.53	9.4%
Conventional; global–continental	16.94	1.563	1.190	2.75	19.69	16.3%
Organic; local	16.94	0.516	0.004	0.52	17.46	3.0%
Organic; national; road	16.94	0.516	0.096	0.61	17.55	3.6%
Organic; national; rail and road	16.94	0.516	0.022	0.54	17.48	3.1%
Organic; global–continental	16.94	0.516	1.190	1.71	18.65	10.1%

Source: Pretty, Hine et al (2001)[147]

One review of the fish situation has stated: 'Major stresses are evident in world food-producing systems, particularly land degradation, declining freshwater stores, and fisheries depletion.'[148] The FAO now accepts that three-quarters of the world's seas are 'maximally exploited'.[149] The myth of seas as endless sources of bounty is over, and independent reviews talk of scarcity in crisis proportions.[150] World fish catch (from the sea) has never reached the 100 million tonnes dream anticipated by policy-makers in the 1970s. While between 1985 and 2000, world fish catch was stuck at between 80 and 90 million tonnes per annum, by 1999, fisheries production had reached 125 million tonnes, with 92 from sea (or capture) fishing and 33 million tonnes from acquaculture; of these, 30 million tonnes were reduced to animal feed or oil. The rapid growth of aquaculture is a new pressure on fish stocks, and some analysts are concerned that rising demand for fishmeal (fish being caught wild to feed to farmed fish)[151] could place an even heavier pressure on fish stocks. To add to the problem, an estimated 20 million tonnes of fish harvested from the seas are discarded per year.

Poor people, in particular, are highly dependent on fish in developing countries.[152] Whereas supply to rich areas like the North Americas rose by 27 per cent between 1978 and 1990 and in Europe it rose by 23 per cent over the same period, supply to Africa declined by 2.9 per cent and to South America by 7.9 per cent. Regions vary considerably in their fish catch and culture.[153] Measured as fish caught per person, West Asia has experienced major drops in its fish catch since the 1970s, since the trend for developing countries to turn to fish and fish products to build exports. While the number of fishers, including fish farmers, has grown to feed this trade, the FAO and the World Bank estimate that 23 million Western Asians remain income-poor, (ie living on less than $1 a day) working either as fishers or in related jobs.[154]

The salmon zones off North-west America have also shown significant declines, with 24 sub-species on their danger list. In North America and Europe, where intensive agricultural farming is practised, run-off from fertilizers and manures into rivers and seas has led to a phenomenon known as 'nutrient-loading'. This can mean that coastal fisheries are seriously affected by nitrogen and phosphorus. One of the biggest recent ecological scandals has been the pollution from intensive shrimp farming in Asia to meet the West's demand,[155] and ancient coastal environments such as mangrove swamps have been destroyed in subsidized export strategies.[156]

Although the global fish catch (that is, sea catch plus aquaculture production) rose dramatically from 20 million tonnes in 1950 to 130 million tonnes in 2000,[157] much of the growth from the mid-1980s could be accounted for by China and much of that by aquaculture, with the direct catch from inland and sea sources (ie excluding aquaculture) reaching 94.8 million tonnes in 2000 and was valued at $81 billion (see Figures 6.2 and 6.3). There is now real evidence that the stocks are being depleted due to overfishing – enabled by technological advances such as 'factory fishing' (huge trawlers indiscriminately vacuuming the seas for fish) as well as wastage. The North American cod banks, for instance, once proverbially rich, are now empty, and in Canada in 1992, thousands of workers in Newfoundland and Labrador were affected by a complete ban on cod fishing.[158]

The seriousness of the fish stock situation was emphasized by the FAO in a 1999 submission to the WTO, a body designed to facilitate trade, encapsulating over-fishing problem simply as 'too many vessels or excessive harvesting power in a growing number of fisheries'.[159] Fleet sizes grew rapidly in the 1970s and 1980s worldwide and boats got larger, more 'efficient' and less

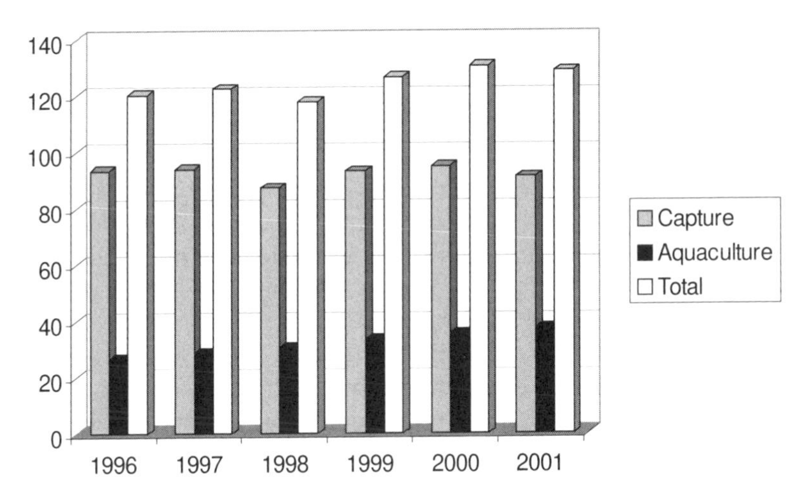

Source: State of the World's Fisheries and Aquaculture 2002, FAO, Rome, Table 1, p4

Figure 6.2 *World fisheries production, 1996–2001*

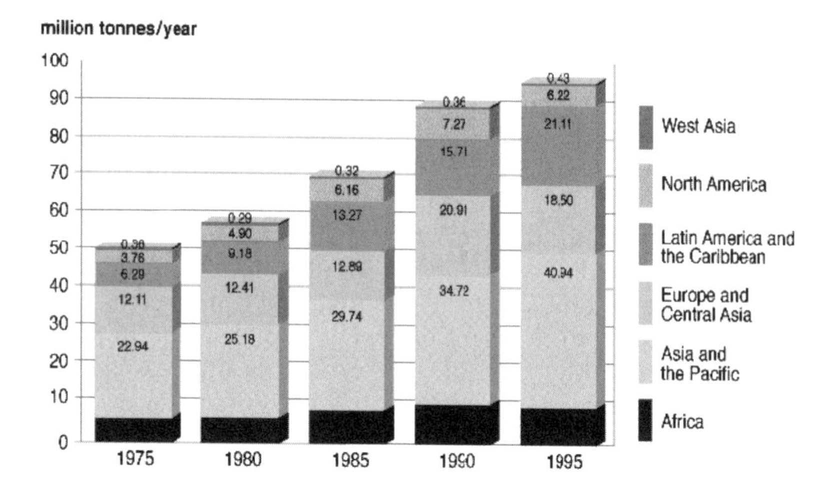

Source: FAO (1997c) cited in UNEP (2000) *Global Environment Outlook 2000*

Figure 6.3 *Global marine fish catch, by region, 1975–1995*

discriminating about what they trawled. Fish are not only being trawled out, but they are now also inadvertent concentrators of contaminants like POPs, especially PCBs. Even inland, the US Environmental Protection Agency has produced evidence of the widespread chemical contamination of rivers and lakes by POPs and, in 1999, PCBs were the single greatest risk cited in advice notices to US fish consumers.[160]

In the UK, where most fish is consumed not as whole or fresh fish but as fish products (such as the 'fish finger' in batter) despite consistent health advice to consumers to consume more fish, sales of both fish and fish products have steadily declined over the last half-century (see Figure 6.4). Our nation surrounded by sea is seeing an increased proportion of fish products imported from seas far away, and shellfisheries in decline: an integrated management designed to maintain shellfish production has been lacking; instead there have been arbitrary standards set, pollution from agriculture, inadequate powers being given to local authorities and poor marketing. This mix has not been helped by a decline in consumer demand.[161] Meanwhile across the English Channel, French shellfisheries thrive at the heart of a tourist industry and are prized in the French cuisine, while in the

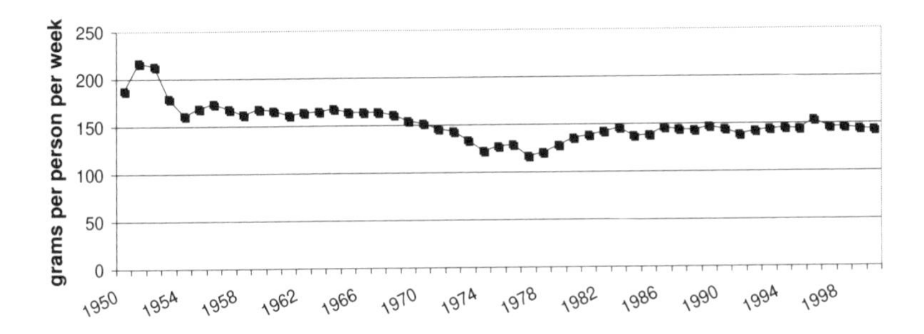

Source: DEFRA (2002) *National Food Survey*, http://www.defra.gov.uk/esg/work_htm/index/food.htm

Figure 6.4 *Decline of UK household fish consumption, 1950–2000*

UK, we have allowed a once-traditional and indigenous part of our food culture to flounder.

While many national nutrition guidelines worldwide stress the need for a significant intake of fish in the diet, now the environmentally literate advice is beginning to be the almost exact opposite.[162] The seas are in such poor shape that they ought to be protected in order to allow stocks to regenerate. Aquaculture accounts for much of recent decades' uptake of fish stocks, yet has an adverse environmental impact.[163] Three to four tonnes of marine fish, for example, can be used to produce one tonne of farmed fish;[164] Chinese aquaculture fishponds may yield up to 35 million tonnes of fish a year, but they are beset by infections and pollution; much aquaculture suffers from high stocking density, poor water quality, eutrophication and pollutants.[165] New investment in skills and facilities, and ecological controls on aquaculture are urgently needed. Aquaculture may rightly become a source of growth and supply of fish for the future, but not at any cost.

Fisheries policy is seriously under-discussed in food policy, compared to agricultural policy. In the European Union, for example, there are very few environmental or public health lobbyists on the side of the Common Fisheries Policy, but there are legions who pressurize decision-making on the Common Agricultural Policy. The corporate sector, including such companies as Unilever, has begun to take a lead, but its overall impact so far is slight and what remains to be done is enormous.

MEAT

Intensification of meat production is coming under increasing criticism for its adverse environmental impacts. Despite its social and cultural importance, meat is notoriously inefficient as a converter of energy. Vast quantities of grains are produced to feed animals. It takes 7kg of feed to produce 1kg of feedlot-produced meat, 2kg to produce 1kg of poultry meat and 4kg to produce 1kg of pig meat.[166] In addition, according to the International Food Policy Research Institute (the IFPRI), per capita demand for beef, poultry and pigmeat in China, for instance, is set to double by 2020, and the US grain trade is keen to exploit this anticipated increase in Chinese demand as Chinese incomes rise. Ecologically, of course, it makes bad sense to use oil to transport heavy grain to feed to animals the other side of the world and meeting the aspiration for meat as prosperity rises needs to be seen against the environmental impacts of using land for cattle rearing and grain production to feed them.[167]

A curtailment of feeding animals in rich countries would not automatically be translated into improved diets for the poor in developing countries. IFPRI has calculated that a dramatic fall in meat consumption by 50 per cent, for example, would only deliver approximately 1 or 2 per cent decline in child malnourishment.[168] It warns against oversimple solutions, such as mass vegetarianism, although there is good evidence that the vegetarian diet can be entirely satisfactory for health.[169]

This 'meat or plants?' debate illustrates a lifestyle conundrum in relation to environmental versus human health goals. Just as importantly, if an increase in the consumption of animal-based foods contributes to the incidence of diet-related diseases, surely an increase in meat and dairy production should not be encouraged.

ANTIBIOTICS

Intensive agriculture such as meat production would not have taken its current form without a whole panoply of veterinary inputs and support. In this respect, the widespread use of drugs such as antibiotics had knock-on health and environmental effects: antibiotic resistance is building up in both humans and animals, making it difficult to cure new infectious diseases, and making new antibiotics both more elusive and more expensive. 'Wonder drugs' have generated 'super-bugs',[170] and in the US, for instance, antimicrobials are typically purchased and used without a veterinary prescription.[171] Pathogens thrive, according to Professor Tony McMichael, an epidemiologist, because the globalization of economic activities and culture, the escalation of travel and trade, and our increasing use of intensified food production and processing, other technologies, antibiotics, and various medical procedures are all reshaping the world of microbial relations. Pathogens live today in a world of changing, mostly increasing, opportunity.[172]

In 20 years, some forms of *Salmonella* have developed multiple drug resistance, the number increasing from 5 per cent to 95 per cent today. *Methicillen-resistant staphylococcus aureus* (MRSA) has grown from 2 per cent to 40 per cent in just one decade.[173]

The US General Accounting Office first raised its concerns about the cost to human health of excessive use of antibiotics in intensive animal rearing in 1977.[174] Two decades later, the US Federal bodies had still failed to produce an integrated health policy.[175] In fact, the situation is now alarming. The effectiveness of antibiotics which have saved tens of millions of lives in the last half century is being radically undermined by a number of features:

- excessive prescription by doctors;
- poor use by patients (eg not completing a course of treatment);
- routine use of antibiotics as growth promoters in intensive animal husbandry and in veterinary practice;
- the capacity of bacteria to adapt and produce new antibiotic-resistant strains.

There is now an ample scientific evidence for the gradual erosion of the effectiveness of antibiotics.[176, 177] Antibiotic-resistant tuberculosis, for instance, rose in New York from 1–2 per cent of cases in 1950 to over 30 per cent in the 1990s.[178] In 1998, the UK House of Lords Committee on Science and Technology concluded that imprudent use of antibacterial drugs had made many new drugs worthless;[179] it concluded that the only way to foster their effectiveness is to restrict their use. A 1997 WHO conference also recommended the termination of the use of antibiotics as growth promoters in farm animals if they are also used for human health.[180]

In 1986, Sweden banned the use of growth promoters; the UK had banned the use of penicillin and tetracycline for growth promotion as early as the 1970s; Denmark banned virginiamycin in January 1998 and reduced its use of antimicrobials in food by 54 per cent by 2001. This caused a loss, it has to be said of €1.04 per pig; but there were no economic losses for poultry. The entire strategy was judged a success when the World Health Organization conducted a review in 2002.[181] Canada has called for voluntary reduction. This piecemeal situation will probably change gradually, following the EU's ban on the use of four antibiotics as growth promoters – bacitracin zinc, spramycin, virginiamycin and tylosin phosphate – which took effect in 1999.

Antibiotics are central to the production of cheap meat. It is estimated, for example, that 70 per cent of all antibiotics in the US are used on healthy pigs, poultry and beef cattle.[182] With controls on antibiotic use and other animal welfare protection strategies being enacted into law, particularly in the EU, there appears to be a geographical relocation of poultry production to areas of the world where land and labour are so cheap, and regulation and inspection less than tough.[183]

There is a growing conviction among medical and veterinary science, as well as amongst some food producers, that this situation is unacceptable. In 2002, a US review commended the prudent use of antibiotics and stressed ecological as well as the human-health implications, warning that resistant bacteria can outcompete, and propagate faster than, non-resistant bacteria.[184] Similarly, the European Commission has proposed the phasing out of antibiotics used as growth-promoting feed additives (with some coccidiostat exceptions) by 2006. Some larger companies have pre-empted this and already taken unilateral action: McDonald's decided in 2003, for instance, to curtail its use of meat from livestock treated with antibiotics.[185]

Keep Eating the Fruit: A UK Case Study

Health education advice is for the population to eat lots of fruit and vegetables: they are rich in vitamins and other micronutrients that help prevent some of the degenerative diseases we reviewed in Chapter 2; in particular, there is strong evidence for their protective effect against chronic diseases such as CHD and cancer.[186] Since 1990, the WHO has recommended that people should eat at least 400g (approximately 5 portions) of fruit and vegetables per day,[187] which could reduce overall deaths from such diseases by up to 20 per cent.[188] This is a positive message forming the core of much health education and policy. But in a country such as the UK, the production of fruit and vegetables is declining and the deficit is made up by huge increases in imports. According to national agricultural statistics, over the period 1989/91 to 2000, the UK area given over to fruit production declined from 46,700 hectares to 34,200 hectares (see Figure 6.5),[189] and total production in this period declined from 527,000 tonnes in 1989/91 to 305,000 tonnes in 2000, although imports shot up. The result has been a large deficit in the national food trade gap which has grown by value over recent decades.

Nonetheless, while the UK has experienced a long-term decline in per capita consumption of fresh vegetables, there has been an increase in consumption of fresh fruit: often in processed form, either pre-cooked or as juice, but also as fresh produce which still accounts for 61.6 per cent of all fruit and vegetable sales.[190]

From a nutritional perspective, progress is painfully slow, but over time and overall the UK population is very gradually increasing its intake of fruit and vegetables. At this rate, according to researchers at Oxford University, the British will meet the WHO guideline of consuming 400g per person per day only by 2047 (Figure 6.6).[191]

At present levels, UK consumers consume less fresh fruit per head than many of their European counterparts. British children – particularly those growing up in poverty – are eating considerably less than the recommended five portions of fruit and vegetables a day; consumption of fruit and vegetables by children fell over the last 20 years of the 20th century.[192] In fact, fewer than 20 per cent of 2 to 15-year-olds eat fruit and vegetables more than once per day and the typical diet of children and adolescents is rich in fat, sugar and salt.

A study of 2635 schoolchildren aged 11–16 years in schools in England and Wales in 2001 found that on average they consumed only one-third of the recommended 35 portions of fruit and vegetables per week. Five per cent of the sample reported that

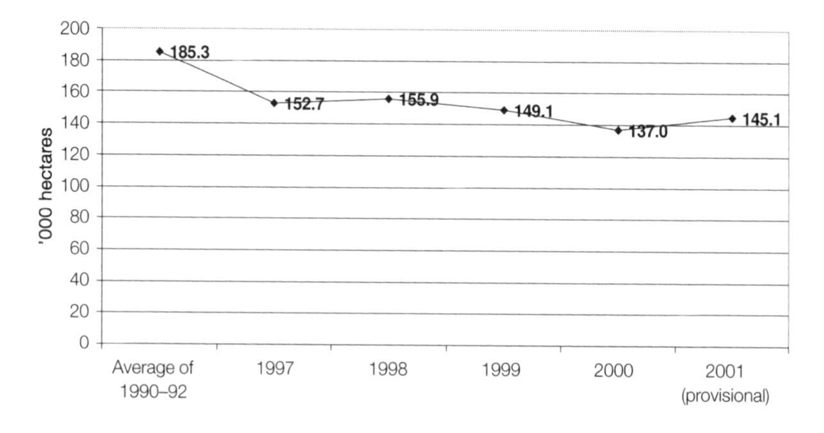

Source: DEFRA/Scottish Executive Environment and Rural Affairs Department/ Department of Agriculture and Rural Development (Northern Ireland)/National Assembly for Wales Agriculture Department (2001) *Agriculture in the United Kingdom*, p42

Figure 6.5 *UK production area of fruit and vegetables, 1990–2001*

they had eaten no vegetables at all in the previous seven days, and 6 per cent reported eating no fruit at all in that period.[193] The study confirmed theories that lower-income households consume less fruit and vegetables, but proportionately more sweet foods, soft drinks, crisps and chips than their richer counterparts. Health education must be speedily implemented to alter such figures for the better.

The irony of this picture from a business perspective is that fresh produce has become one of the most value-added sectors within food retailing, with supermarkets in the UK making good profits from sales of fruit and vegetables proportionate to their shelf space. In the US, the 20th century saw a decline in the state-based sourcing of fruit and its replacement by a concentration of production in California and to a lesser extent in Florida;[194] at the same time, there was a search for all-year-round fruit: instead of capitalizing on the seasons, traders and retailers, of strawberries for instance, pursued locations which offered compliant labour, land and national regulatory regimes.[195]

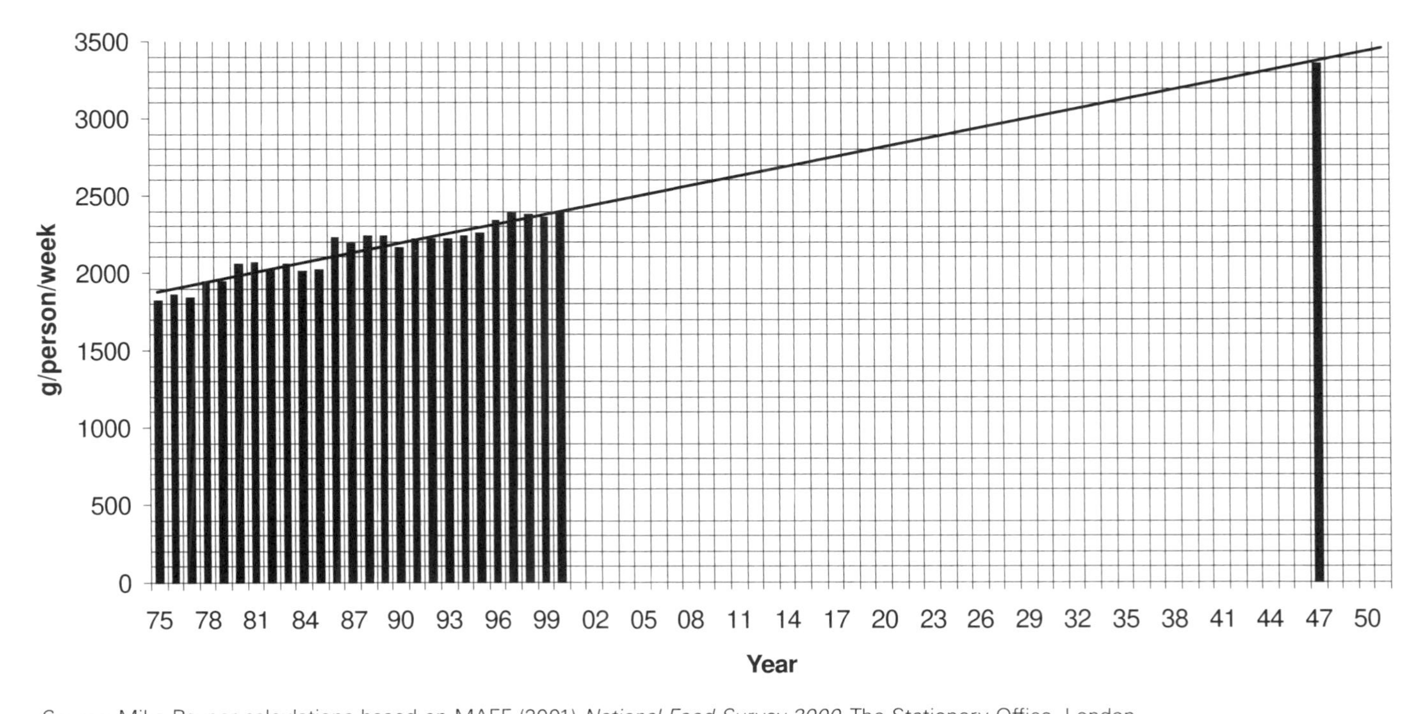

Source: Mike Rayner calculations based on MAFF (2001) *National Food Survey 2000*, The Stationery Office, London

Figure 6.6 *UK fruit and vegetable consumption, 1975–2000, with COMA targets to 2045*

The Clash of Farming and Biology: Have Humans Got the Wrong Bodies?

An examination of the relationship between environmental and human health raises an important question as to how the well-being and disease patterns of humans are related to the environment in which they find themselves. The new science of eco-nutrition and our greater understanding of biological systems suggest a close and symbiotic relationship between the two, while, in the period of the Productionist paradigm, there has been a disastrous disconnection of production, ecology and human health.[196] The future lies in re-creating the connections and in developing policies, business solutions and a food culture that respect these connections.

The recognition that diet and health *are* connected has moved human biology centre-stage in the war of paradigms. What are humans biologically programmed to require, thrive on and get ill from? Could genetic screening find out which genotypes are disposed to particular diseases? Within the Life Sciences Integrated paradigm, the response to these questions is clear: if modern lifestyles are not burning up enough of our energy in the form of food, and if we decline to eat less because food is pleasure and because culture is built around it, then food must be altered to fit our new circumstances. In other words, the problem is not the diet but human biology, and a whole new way of conceiving both public policy and food supply chains opens up. Suddenly, commercial and state policy can focus responsibility for diet and health care onto the consumer, and this position – that there is nothing wrong with the food; the problems are all in the genes – is the sophisticated health spin powering the future of much food and health policy.

According to palaeontologists, the earliest experimentation with systems of agriculture only became possible once settled, rather than hunter–gathering, societal forms emerged during the so-called 'Neolithic Farming Revolution', eight and ten millennia ago. Whether the Neolithic change to agriculture from hunter-gathering was as clean or clear cut as has been traditionally thought is now the subject of some argument among researchers into the origins of the species.[197] As settled agriculture spread, and particularly as new seeds and then domestic animals were developed, a new civilization became possible: people could live in a different way; diet could become a matter of choice of culture – literally agriculture, cultivation of the land – or plantation rather than subject to necessity or availability.[198]

Agriculture enabled *Homo sapiens* eventually to prosper. People were now able to eat a diet different from the one for which human evolution had prepared them: culture, as well as raw necessity and biological predispositions and determinants, began to shape diet. As Professor Michael Crawford puts it, the foods humans 'ate throughout 99.8 per cent of [their] history and those [they] eat today are different in many ways'.[199] Agriculture pressed back wild culture. Crawford again: 'Up to the invention of agriculture, the species that was destined to be wildly successful, perform brilliantly (in evolutionary terms) and to dominate all living things [ie humans], lived on food that was wild.'[200] And this enabled a new eating experience. The Neolithic Revolution domesticated animals and plants but also humans.

Some ecologists are now questioning whether this 'humanization' of progress threatens the planet's capacity to carry civilization itself, let alone whether it can be healthy.[201, 202, 203] The 'deep ecology' position claims that the notion of progress associated with Western or Anglo-Saxon capitalism as the solution for a better life is hindered by technology. This view rejects the argument that commerce needs to be and can be humanized to get it back on the right track,[204] while others argue that this same insight should send urgent signals to commerce, including in particular the food industry, to cut energy use, reduce waste and change production methods, for instance by reintroducing biodiversity into farming and thence to diets.[205]

The environmentally friendly rule for the 21st century probably ought to be to eat a diet that is both local and that maximizes biodiversity. But to be able to eat a variety of foods, soundly produced, is 'contingent on biodiversity' according to Professors Wahlquvist and Specht, two Australians who have taken a special interest in the relationship between ecology, biology and nutrition, asserting that, 'with a week as a time frame, at least 20, and probably as many as 30 biologically distinct types of food, with the emphasis on plant food, are required'.[206] There are a number of potential far-reaching implications for food supply and farming in particular from this sort of perspective. Biodiversity needs to be on the consumer's plate, routinely.

There is a current vibrant debate raging about the appropriateness of an evolutionary perspective on diet and health.[207] An eco-epidemiology paradigm has been proposed to examine individuals and how they connect with their context.[208] Some argue that policy needs better to understand the Paeleolithic origins of human biology and food preferences,[209] while others argue that it is difficult to apply any historical suppositions into a modern time frame.[210] The core argument is whether humans have evolved to eat certain foods, and whether our present physiology is determined by our evolutionary past; the policy question is whether today's agriculture can provide us with foods that we need physiologically and socio-psychologically as befits our evolutionary past.

Throughout this chapter, the understanding mapped on the diet–health–environment connection suggests the need for a radically more integrated approach to public health. Too often, one has the feeling that while the proponents of the Life Sciences Integrated paradigm are planning the future of the food supply, the ecologists are bemoaning its present, and public health is studying the past. Counting bodies as they fall off the cliff may be good epidemiology but it is poor public policy. In the next chapter we argue that a new ecological public health movement needs to learn stringent lessons about structures and policy, the current lack of which goes to the heart of the paradigm wars.

CHAPTER 7

FOOD DEMOCRACY OR FOOD CONTROL?

'A day will come when the only fields of battle will be markets opening up to trade and minds opening up to ideas. A day will come when the bullets and the bombs will be replaced by votes, by the universal suffrage of the peoples.'

Victor Hugo (1802–1885) in a speech to the Congrès des amis de la paix universelle, Paris, 22 August 1849

CORE ARGUMENTS

Many institutions of food governance are out of date and struggling to maintain consumer trust. They also have failed fully to address the changing nature of the food economy and the challenges raised by public health on the one hand, and by corporate influence on the other hand. Institutions need to rethink the relationship of the global to the local, while rebuilding public trust by engaging with social concerns. Food and health policy needs to be a central concern for the state since it is our sole mediator between increasingly powerful interests and the consumer. All too often, the institutions of food governance fail to integrate food with health; the many existing policy commitments to which governments lend their name are not being actively enough pursued; state support for Productionism is now out of date. A central tension for the future of food governance will be negotiating the (im)-balance between 'food democracy' and 'food control': food policy can only be legitimized if created through a process that is democratic rather than 'top-down'. Leadership will be needed to achieve a new vision for food governance.

Why is Governance an Issue?

In this chapter we argue that there is a crisis of institutions and of governance – that curious English word that refers to the science and practice of government – over what to do about food and health. Governance, according to Richards and Smith, 'is generally a descriptive label that is used to highlight the changing nature of the policy process in recent decades. In particular, it sensitizes us to the ever-increasing variety of terrain and actors involved in the making of public policy. Thus, governance demands that we consider all the actors and locations beyond the "core executive" involved in the policy-making process.'[1] This chapter sets out to detail the wider terrain and actors involved in food governance, not least where the 'push' and 'pull' of food policy-making is located. In this respect we argue that there needs to be more room for what we call 'food democracy' within the more centralized food governance environment currently in place.[2]

One of the major battles in terms of governance, from our point of view, is to get health, in an integrated sense, taken more seriously in the institutions of food governance. Too often in local, national, regional and global institutions, while many divergent views proliferate, they fail to integrate food and health. Today, more than at any time in human history, institutions central to food and health governance are multi-level: they can be placed in at least five levels of governance, from the global to the community (see Table 7.1). There is tension between these levels, with political debate raging about whether power is sliding up to the global institutions, whether more local institutions have lost their power and influence, or whether the global bodies are ultimately having to bow down to national and more accountable local levels.

The tussle between globalization and localization, then, is being fought out in food and health. Food enforcement has to happen locally, yet legislation is increasingly being set, not at national but at international level; through such bodies as the European Union, the Asia-Pacific Economic Cooperation or the North American Free Trade Agreement, and with the creation of newer bodies such as the WTO.

At the global level, there are many bodies which have either a pre-eminent role in relation to national ministries and agencies

Table 7.1 *Multi-level governance in relation to food and health*

Level of governance	Example of institutions	Food and health role
Global	UN, WTO, Codex Alimentarius Commission	Intergovernmental negotiations; coordination of expert consultations; setting and sharing policy agenda and standards
Regional	North American Free Trade Agreement, European Union, ASEAN	Set trade rules between Member States; develop regulation; cross-border food safety issues
National	200+ nation states	Legislation and regulation; health care, policy covering food supply chain; dietary guidelines
Sub-national	Regional health bodies, elected regional assemblies	Coordination of local initiatives; regional voice and policy
Local/community	Town or village council; health authorities; community centres	Delivery of local services such as food law enforcement, primary health care, dietetic advice

or which have a coordinating role; others such as the World Bank, the WTO, and UN bodies such as the WHO act as initiators and think-tanks for global food and health approaches and they have considerable influence on the panoply of national government as well as in the subtle framing of the food supply chain within and between nation states.[3] Table 7.2 lists some examples of these institutions.

Such institutions frame and tussle over existing policy commitments; their role is significant, both in why the state has supported Productionism and why the costs can no longer be borne. A constant theme of debates about governance is responsibility and control over the food supply chain. The creation of mechanisms for an integrated government overview of food policy is now urgently required. There is an immense body of

Table 7.2 *Global institutions involved in food and health*

Remit	Examples of organizations/bodies
Public health	WHO, FAO
Children and health	UNICEF, UNESCO
Global economic bodies with health impact	World Bank, International Monetary Fund, UN Conference on Trade and Development (UNCTAD), WTO, World Intellectual Property Organization (WIPO), Organisation for Economic Co-operation and Development (OECD)
Intergovernmental agreements with a health impact	Bio-safety Convention, International Conference on Nutrition, Basel Convention on Hazardous Waste
Emergency aid	World Food Programme, International Committee of the Red Cross/Crescent
Environmental health	Global Panel on Climate Change, UN Conference on Environment and Development (UNCED), International Maritime Organization
Commercial interests	International Chamber of Commerce, transnational corporations, International Federation of Pharmaceutical Manufacturers Associations
Regional bodies with health role	European Union, Regional Offices of the WHO and FAO
Trade associations	International Hospitals Federation
Networks to promote public health	Healthy Cities Network (WHO), International Baby Food Action Network (IBFAN), Local Agenda 21 Network, Pesticides Action Network, Tobacco Free Initiative (WHO)
Professional associations	International Union of Health Education
Non-governmental organizations	Greenpeace, Friends of the Earth, Oxfam, Médecins sans Frontières, Médecins du Monde, World Federation of Public Health Associations

knowledge waiting to inform food governance and to encourage a change of policy direction but there are few institutions or mechanisms by which this can happen. The results of three decades of work in nutrition, epidemiology, medical and ecological sciences today in effect enables much of society to recognize the vital need for healthy eating and a life-time diet for optimum health. And tens of thousands of nutritionists and health professionals in both the developing and developed world are, through their professional activities or through policy involvement, trying to help others to achieve or understand better health through the food they eat. But progress in implementing population-wide change has been frustratingly slow. Yet the dynamics of the food supply chain are moving far faster than policy-making to promote health issues. Policy is reactive too often. Food governance, over the last half-century, has been rendered even more problematic by:

- the revolution in food supply and distribution: the lengthening of the supply chain;
- advances in nutritional understanding;
- changes to the political architecture: the emergence of new trading blocs and international institutions;
- changing consumer lifestyles: less physical activity, more women working in the waged labour force;
- population demographics: ageing populations and smaller households;
- political ideology: the collapse of Communism and the triumph of individualism and free-market politics.

This new 'architecture' makes it more challenging to make the links between different levels of government and the different food sectors and it is rare too to encounter anyone in government with an overall vision for food policy, let alone responsibility for delivering it. By default, an industry-driven vision of the food supply chain has taken centre-stage. The food supply chain is so huge and so important, in commercial terms, that it cannot operate in a policy or paradigm vacuum.

Mediation on the Food Wars is urgently required from the state; but the state finds it problematic to connect with democratic tendencies within food cultures.

Civil Society Emerges

Part of the challenge for governance is how the state and institutions can engage with the people – hence the emergence of 'stakeholders' or civil society. These new processes underline the importance of NGOs in all walks of life. For example, the US has 1–2 million NGOs, and by the mid-1990s, about 1 million NGOs were operating in India, 210,000 in Brazil, 96,000 in the Philippines, 27,000 in Chile, 20,000 in Egypt and 11,000 in Thailand.[4] A key priority of governance is to understand and define the role of NGOs in the policy process. While some NGOs have established a loud voice and a strong presence in food policy in recent years, they have yet to convert these into a firm position in terms of delivery and actively shaping overall policy integration.

Historically, until a quarter of a century ago, the representation of civil society in food and health policy was relatively weak, but, towards the end of the 20th century, activism grew dramatically over a whole series of food and health issues such as baby foods, labelling, contaminants, genetic modification, animal welfare, farming practices, labour rights, social justice and hunger. Despite this flowering, however, there has not been the large organized food equivalents of such environmental groups as Greenpeace, Friends of the Earth and the Worldwide Fund for Nature (WWF). NGOs have led representations, for example, on hunger and on the inappropriate marketing of infant formulae and breastmilk substitutes – which engendered a formidable global coalition in the form of the International Baby Food Action Network (IBFAN), but these are exceptions in public health. The public health world is, however, disproportionately dominated by more powerful organizations at the professional end of the lobbying spectrum, and industry relatively highly funded and well represented in the corridors of food governance.[5, 6, 7] Now governance must address the changing influence and emergence of new coalitions of interest.

Food and health discourse within government has been strongly led by scientific interests focused on the health benefits of the different components of foodstuffs. The epidemiological and public health arguments outlined in Chapter 2 have tended to be downplayed. The arrival of better coordinated NGO activism into the global food and health world would be surely welcome.[8] As the evidence of the nutrition transition has emerged, there are clearly good grounds to campaign about the negative health effects of Western foods, for example, particularly with the opening up of national markets following the liberalization of global trade. There is a strong need for sharing information about the international marketing practices of giant corporations across the globe. For example, the publicity engendered by the McDonald's legal case against two British activists in the early 1990s spawned many support groups which continued to put the spotlight on corporate activity in food and health long after the trial was over.[9]

Often, in the face of emerging activism, the food industry's strategy is to deny links between consumption and ill health, while agreeing to fund scientists to study their product's risks and to adopt new softer tactics such as corporate social responsibility when confronted with the evidence. (Such was the tobacco industry's defence of its interests.) But often such defensive tactics are designed merely to limit the ingress of more radical thinking. If corporate accountability is to be real, it has to be more than skin-deep; otherwise it can be dismissed as 'greenwash'.[10] The reductions to conflict between civil society and certain commercial interests at least require the state to take a more pro-active role in food and health governance, specifically.[11]

Building on Existing Policy Commitments

Besides their national commitments and policy objectives, governments are often already signed up to existing commitments which signal the directions food and health policy could take. Table 7.3 gives a list of some key global commitments to nutrition, food safety and wider sustainable development. These go back to the founding of the UN system in the 1940s, with some governments placing a stronger emphasis on aspects of food and health than others. The point is that there is along history of formal concern; but these attempts are far from being consistently integrated.

Such global commitments are reminders that there is not a policy vacuum: health is well-trodden territory. In theory, the

Table 7.3 *List of global commitments*

Occasion	Date	Nutrition	Safety	Sustainable food supply
Universal Declaration of Human Rights	1948	+	+	
UN Covenant on Economic, Social and Cultural Rights	1966	+		
Stockholm Environment and Development Conference	1972			+
World Food Conference (Universal Declaration on the Eradication of Hunger and Malnutrition)	1974	+	+	
Convention on the Rights of the Child	1989	+	+	
Innocenti Declaration on Breastfeeding	1991	+	+	
UN Conference on Environment and Development and Rio Declaration, UN Framework Convention on Climate Change and on Biological Diversity	1992			+
International Conference on Nutrition	1992	+	+	+
World Conference on Human Rights, Vienna, and Vienna Declaration and Programme of Action	1993	+		+
UN Fourth World Conference on Women and Beijing Declaration and Platform for Action	1995	+		+
World Food Summit	1996	+	+	+
UN Habitat 2 and Istanbul Declaration	1996			+
UN General Comment on the Right to Adequate Food[12]	1999	+	+	+
World Health Assembly (Resolutions 53.15, 51.17, 53.18)	2000	+	+	+
World Food Summit (Rome)	2002	+	+	+
World Summit on Sustainable Development (Johannesburg)	2002	+	+	+

nutrition transition and the environmental degradation from intensive food production systems ought to be addressed with such policy frameworks in place, but the resolutions, binding agreements and commitments are often not, in reality, followed up at the national level or supported by adequate funding. In the real world of politics, sustainable development takes a lower priority than conventional approaches to trade and economic growth; just as 'health' is perceived as the budget line that can be raised only if the economy can afford it, so 'sustainable development' is defined as something that can be bolted on rather than transform what is meant by economics.[13] Nutrition is deemed a matter for health ministries generally, not something that should concern the Ministry of Culture or Trade specifically. Food safety is the problem of food agencies or ministries of agriculture or health, not the concern of the Finance Ministry unless or until the bill becomes too high. This compartmentalization must be tackled.

How Global Institutions Frame Food and Health

Trade has long been the totem pole around which much agricultural and food politics dances.[14, 15] At the great agriculture and food conferences of the 1940s at which Productionism was put in place – in 1942 at Hot Springs, 1945 in Quebec, 1947 in Washington DC and 1948 in Cairo[16] – a keen conviction was expressed that the optimum way to deliver health was by raising the quantity of food and selling and transporting it to the people. In our modern food world, the focus has shifted from the national to the global and regional. Yet institutions of governance are by and large local or national; the EU is exceptional. Attempts to internationalize decision-making for the common good are locked into regional or sectional politics framed by rich nations. UN bodies such as the FAO, WHO, UNICEF, UNCTAD and UNEP were sidelined by the creation of the WTO in 1994, and by the political and economic policy supremacy of financial institutions (also set up in the 1940s) such as the World Bank and the International Monetary Fund.

The General Agreement on Tariffs and Trade (GATT) is a set of trading rules now overseen by the WTO. Like the UN, the

GATT came into existence in the 1940s and was signed in 1948 by two dozen countries committed to reducing tariffs and barriers to trade and to trade as the means to development and wealth-creation. Food issues under the provisions of the GATT were slow to be incorporated because of resistance from the developed countries, but finally, at the so-called Uruguay Round, the GATT was revised in 1994 to include a huge set of trade-related rules for food and agriculture. This is now policed and developed by the WTO secretariat.

Besides its general commitment to trade liberalization, the GATT contains a number of subsidiary agreements, such as the Agreement on Agriculture (AoA), on Technical Barriers to Trade (TBT) and Sanitary and Phytosanitary Standards (SPS), where food and health issues are encoded. The US and the EU fought hard in the Uruguay Round and thereafter to position their agricultures to advantage and to protect the interests of their food industries, in particular their positions on biotechnology. This saw the rich world protecting its trade interests against weaker and less-subsidized agricultural exporting countries, known as the Cairns Group (17 agriculture-exporting countries committed to a market-oriented trading system) and the informal 'Like Minded Group' of developing countries. Non-exporting countries have been much weaker, yet are dependent upon world prices that are the outcome.[17]

This balance of forces was dramatically altered at the meeting in September 2003 at Cancun, Mexico. Previously, campaigners had sought to inject into the WTO's Agreement on Agriculture special conditions such as a 'development box', to protect developing country's interests, but the confrontation in Cancun side-stepped such strategies. There, to EU and US negotiators' astonishment, a group of 21 countries from the South led by Brazil, China and India emerged and confronted the food hyperpowers. The new G-21 Group took a hard line, demanding that the US and the EU reduce their barriers, cut subsidies and open up markets to their cheaper commodities,[18, 19] without which conditions it would not entertain any kind of agreement. It presented evidence of the poor being penalized, indicating that, in rich countries, agriculture represents less than 2 per cent of total national income and employment, whereas in middle-income countries, agriculture accounts for 17 per cent of GDP, rising to 35 per cent in the poorest countries; yet the interests of 'rich country' agribusiness has predominated. G-21 said that enough was enough. As a result of this pressure, the entire Cancun negotiations went into abeyance, with the challenge from civil society groups for the West to deliver a more equitable food economy ringing loud. Pressure on the EU and US to reform their highly subsidized agricultures intensified; but their response was to threaten a return to bilateral trade agreements. The US 2002 Farm Bill had donated $52 billion to crop and dairy subsidies, increasing them by 67 per cent.[20] Small farmers and growers in the developing world simply cannot compete with these deep pockets. It should be noted that even within rich regions, subsidies can be highly discriminatory. For instance, in the US, 60 per cent of farmers receive no support at all, while the biggest 7 per cent receive 50 per cent of government pay-outs.[21]

Global Standards

The SPS (Sanitary and Phytosanitary Standards) agreement within the GATT is particularly important for food and health. In 2000, the WTO and the WHO began a process of serious negotiation, aiming to better understand and monitor the linkages between food safety, environment, food security, nutrition and biotechnology.[22] Food safety in particular has been a persisting problem. The 1994 GATT had catapulted a previously low-key world body, the Codex Alimentarius Commission ('Codex'), into the hot seat of setting world food-safety standards as benchmarks for other nations. While Codex is a UN body, set up in 1965 with the FAO and WHO as its secretariat, it was to be available to national governments. The 1994 GATT move, however, meant that Codex set the world standards of arbitration of food trade disputes.

Codex membership is by national delegations, but these often contain corporate interests.[23] When it was realized that Codex would have this new powerful arbitration role under the proposed Uruguay Round GATT, a decade-long fight to clean up Codex procedures ensued.[24, 25] After some pressure, there has been some improvement in governance and procedures: for example, some NGOs (but only international ones) are now accorded observer status.[26, 27, 28] The working culture of Codex is one of risk management, taking foods and sectors of food, on a one-by-one basis. While the emphasis in Codex is on food safety and standards in order to facilitate trade, developing countries stand at limited advantage, lacking the budgets required to improve their own measures and standards. When country total health expenditure averages just $4 per capita,[29] attendance at costly meetings such as Codex is problematic. In 2002 there was a proposal to create a trust fund to aid poorer countries to attend, but there were worries that this could be funded by the food industry and compromise objectivity.

Notwithstanding its worthy brief, Codex lacks an overall vision of how to integrate food and health, and there is little sign that it is going to address crises such as the worldwide epidemic of degenerative disease. In addition, Codex is a technocratic body, dealing with only single issues and unlikely to provide the necessary leadership in integrating food and public health policy.

Health policy is also suffering as a result of tension between leading world economic institutions such as the World Bank and the WTO on the one hand, and the 'social' institutions of the UN. The World Bank's own experience of investigating national economies has shown health to be both a barrier to economic progress and a potential vehicle for development and, while it has been in the vanguard of promoting the monetarization of social capital,[30] it now realizes the limits of this approach and is espousing instead the importance of the social bonds and relationships that hold peoples together and give support in communities.[31] In moving to guide the framing of conditions for public health, the World Bank has engineered reactions, and the WTO has responded by trying to alter its image and be more inclusive. The key issue is accountability and openness in food and health governance. The WHO has a World Health Assembly but there is no parallel citizen's voice for the WTO. Economically, there is no contest, which is why countries have clamoured to join the WTO and companies expend considerable effort monitoring its activities and lobbying their national governments to ensure their views are pursued.

This heightened political atmosphere about global governance is both promising and threatening for public health. On the positive side, as far as improving democracy and transparency about world negotiations are concerned, NGOs and developing countries feel much more capable of confronting entrenched Northern power blocs such as the US and the EU (after such successes as managing to delay the Multilateral Agreement on Investment [MAI], an attempt through the OECD to introduce a new regime for investment and finance).[32, 33] On the negative side, however, there is still much to be done to educate the (potential) food and health lobby about the roles and powers of the GATT and the WTO. A revitalized WHO, under Dr Gro Harland Brundtland as Director-General during the late 1990s, began to position itself to assume public health leadership[34] and to embrace a wider role for public health in all aspects of food and nutrition policy.[35] But to do this successfully, the WHO needs to take a stronger line in the dynamics of the global food system and to play a fuller part in confronting the compartmentalization of health.

Indeed, the WHO recognized its need for a new role:[36, 37] outlining a broader base for international relations and collaborative strategies that will place greater emphasis on international health security to include the health consequences of trade in commodities harmful to health. The argument is that major transnational health issues require:

- global intersectoral action through transnational cooperation and partnerships;
- an enhanced role for international legal instruments, standard setting and global norms;
- more comprehensive forms of global vigilance, research, monitoring and assessment;
- global research programmes that concentrate on developing cost-effective technologies to improve the status of the poor;
- human resource development; and
- ongoing comparative assessments and cross-fertilization of experiences regarding health system reform.

Injecting Health into Regional Institutions: The EU Case

The European Union is an interesting battleground and illustration of how the new ecological health approach we have argued for can be helped or hindered. On the one hand, the EU, through its Common Agricultural Policy (CAP), is a bastion of

Productionism and a bizarre mix of protectionism, internal liberalization and external mercantilism. On the other hand, due to public pressure and food scandals, the EU has experienced a crisis of confidence and a period of introspection which has improved its chances for openness and new ideas. While its lack of a strong health lobby weakens its capacity for the needed integration of policy, CAP reform is thus a key symbolic battleground in the midst of our three core paradigms.

The foundation of the CAP was a classic expression of the Productionist vision for health. With Europe wracked by World War II and experience of famines (such as that in The Netherlands in 1944) and food shortages, its priority was to rebuild agriculture as a home-grown policy for the reconstruction of post-War Europe. Enshrined in 1957, it had grown by 2000 to account for half of all the EU budget, representing 15 Member States.

Public health by contrast only formally entered the EU remit with the Amsterdam and Maastricht Treaties of the 1990s. The revised Treaty of Rome, Article 152, states that the European Commission will take a strong health line in all its policies, but this is not helped by a small staff in the public health group (while the Agriculture Directorate has hundreds). To date, the only health audits of Commission food policies in general and the CAP in particular have been externally conducted and have not filtered into mainstream policy-making.[39,40] It was, in fact, in 2003, the Swedish government which conducted the most comprehensive health audit yet of the CAP, producing a radical, evidence-based set of policy recommendations to:[41]

- phase out all consumption aid to the manufacturers of dairy products with a high-fat content
- limit the School Milk Measure to include only milk products with a low-fat content;
- introduce a school measure for fruit and vegetables;
- develop a plan to phase out tobacco subsidies within a reasonable time frame;
- redistribute agricultural support so that it favours the fruit and vegetable sector and encourages increased consumption.

CAP spent €32.6 billion in production subsidies in 2000. For that not to be health audited weakens the commitment to audit policies for health. A 2002 European Council decision finally gave the European Commission some decent funds, €312 million over five years. But much of that was to improve health information.[42] The EU's policies, like its institutions, grow incrementally rather than through revolutions; the CAP has always been in a constant state of change, however, giving it some chance for a health dimension to be injected into it over time. The internal contradictions of CAP have meant that its Productionist orientation has been under immense internal pressure, and its bill for subsidies is substantial. In 2002, the EC announced radical plans to move away from subsidizing commodities and farmers' production directly and to start funding rural development in general and environmental protection in particular.[43, 44] It remains to be seen whether the fiscal burden of the CAP will drop but the policy shift is significant, even though most focus is on food safety rather than wider public health. Only time and politics will tell if the demise of the Productionist paradigm in the CAP will yield in the direction of the Ecologically Integrated paradigm or the Life Sciences Integrated paradigm.[45]

At an individual country level, there are already interesting integrated policy initiatives underway. For example in Finland, where almost 90 per cent of the food consumed is produced nationally, the food industry is working in cooperation with the entire food production chain to unify activities into a national quality strategy, developed at the end of the 1990s, aimed at addressing people's concerns about the source of their food products, how they have been produced and who are involved in the various phases of the production chain. Watchwords for this strategy are safety, transparency and openness.[46] We would argue that the fourth watchword and strategy goal – currently missing – should be health. The UK government's Curry Commission might be another interesting case study to watch for the level of policy integration. There is pressure for this to accelerate.[47, 48, 49, 50]

Agriculture, Subsidies and Health

At the heart of both the CAP reform process and of global food governance in general is an international conflict over subsidies. Currently, subsidies are disproportionately the policy instrument of rich nations. The total cost of agricultural subsidies in all

OECD countries in 1998 was some $362 billion dollars, two and a half times the combined GDP of all the least-developed countries.[51] In 2003 the UN's world economic survey put world-wide farm subsidies at over $300 billion,[52] an immense fiscal sum alongside the few million dollars that national governments give to proactive health education programmes. In the US, too, the agricultural industry is seen as one of the more powerful lobby groups: in a three-year period, the food sector contributed around $25 million to President George W Bush's and other politicians' campaigns, and five of the top people running the US Department of Agriculture in 2002 were linked to the meat industry.[53] The penetration of the corridors of power by such interests raises questions about the influences on food governance.

The extent of producer subsidies to the farm sector in some countries is summarized in Table 7.4, using an indicator developed by the OECD known as the producer support estimate (PSE). It shows how countries in the so-called Cairns Group of less subsidized food producers (including Australia, Brazil and New Zealand) have dramatically lower PSEs than do high subsidizers such as the EU, the US or Switzerland. Table 7.5 gives the list of US federal subsidies to agriculture at the end of the 20th century: fewer than the EU, but seriously dented by President Bush's Farm Bill 2002 which announced increased subsidies of $190 billion over the following decade.

Much international food trade debate and practice in the last two decades was dominated by USA–EU spats and jockeying for position: they each accuse the other of unfairly subsidizing their own agriculture. In fact, both subsidize heavily compared, say, to a de-regulated farm economy such as New Zealand's. In 1999, the estimated producer subsidy equivalent for an EU farmer was 49 per cent, (nearly half his income), whereas for a US farmer it was 24 per cent and for a New Zealander it was 2 per cent.[56]

The policy debate about subsidies is often pitched as a battle between developed and developing countries. In fact, subsidies are distorted within countries too.[57, 58] In the early 1990s the UK House of Lords estimated that 20 per cent of UK farmers received 80 per cent of subsidies, a figure often quoted, while a more recent estimate found that 16 per cent of farm holdings received 69 per cent of subsidies, and a study by Oxford University for the UK government in 2003 showed that producers in six eastern British counties – the 'grain barons' – received more

Table 7.4 *Farm subsidies, by country; OECD Producer Support Estimate, 1986–2001 (US$ millions)*[54]

		1986–88	1999–2001	1999	2000	2001p
Australia	USD mn	1285	947	1135	878	827
	Percentage PSE	9	5	6	4	4
Canada	USD mn	5667	3930	3709	4153	3928
	Percentage PSE	34	18	18	19	17
Czech Republic (1)	USD mn	1670	655	849	532	585
	Percentage PSE	38	19	24	16	17
European Union	USD mn	93719	99343	115330	89617	93083
	Percentage PSE	42	36	39	34	35
Hungary (1)	USD mn	891	881	1151	912	580
	Percentage PSE	17	18	23	20	12
Iceland	USD mn	193	136	161	139	108
	Percentage PSE	74	63	67	62	59
Japan	USD mn	49498	51980	53809	54888	47242
	Percentage PSE	62	60	61	61	59
Korea	USD mn	12120	18170	18335	19337	16838
	Percentage PSE	70	66	66	67	64
Mexico	USD mn	-266	5694	4515	6032	6537
	Percentage PSE	-1	18	15	19	19
New Zealand	USD mn	476	67	77	71	52
	Percentage PSE	11	1	1	1	1
Norway	USD mn	2628	2274	2511	2138	2173
	Percentage PSE	66	66	67	64	67
Poland (1)	USD mn	528	1676	2584	997	1447
	Percentage PSE	4	12	19	7	10
Slovak Republic (1)	USD mn	675	292	389	335	151
	Percentage PSE	35	20	25	23	11
Switzerland	USD mn	5063	4480	4869	4356	4214
	Percentage PSE	73	70	72	70	69
Turkey	USD mn	2779	6522	7707	7882	3978
	Percentage PSE	14	21	23	24	15
United States	USD mn	41839	51256	55433	49333	49001
	Percentage PSE	25	23	25	22	21
OECD*	USD mn	238936	248302	272563	241599	230744
	Percentage PSE	38	33	35	32	31

Notes: p: provisional. NPC: Nominal Protection Coefficient
NAC: Nominal Assistance Coefficient. EU-12 for 1986-94, EU-15 from 1995, EU includes ex-GDR from 1990.
(1) For Czech Republic, Hungary, Poland and Slovak Republic: the figure in the first column refers to 1991-93.
*Austria, Finland and Sweden are included in the OECD totals for all years and in the EU from 1995.

Table 7.5 *US federal subsidies to agriculture, 1997–2001*[55]

US Federal subsidies $m	1997	1998	1999	2000	2001
Income subsidies	6120	6001	5046	5049	4046
Price supports	–	–	–	1127	1056
Deficiency payments	–	1792	5895	6425	5000
Conservation programmes	1693	1441	1493	1615	1667
Disaster assistance	–	2841	7804	8493	3632
Others	257	138	356	190	278
Total	8070	12,213	22,899	22,899	15,679

than £540 million out of the more than £2 billion in CAP aid which the UK receives annually.[59] East Britain's grain farmers take £121 from each consumer, whereas other farmers get £41 from each taxpayer, and three out of England's nine regions take half the total CAP subsidy received by England.

Even though subsidy figures are inequitable, and economists argue even over their calculation, supporters of the public health should look at them carefully. The current Productionist paradigm is awash with public money in the form of subsidies, which could be directed towards different ends; there is also an opportunity to approach finance ministers and treasuries for cost savings. At present, current food policies and the supply chain are delivering huge externalized costs. Health-care bills throughout the world are rising, as we saw in Chapter 2. Taxpayers and consumers are taking the externalized financial consequences for current food supply under a number of headings:

- When they buy the food.
- When they pay for health care due to diet-related disease.
- When they pay to subsidize farmers, if their government does so.
- When no one pays for environmental damage.
- When no one pays for the social and family care of looking after sick relatives.

Health proponents can link arms with proponents of world and social justice. Subsidies damage the developing world.[60] The subsidy system constrains choice by allowing the state to dump over-production and diet-related health impacts on the rest of the world. Current subsidies have outlived their historical health value. What is needed now is not a CAP but a Common Food Policy, incorporating human and environmental health.

Injecting the New Health into National Institutions

There is wide experience of policy interventions on food and health from which to draw and learn. The US, often derided as the home of junk food, has also initiated some strong health actions: it was the first country to introduce mandatory nutrition labelling on food products with the 1990 Nutrition Labeling and Education Act; it also introduced alcohol labelling, and developed a 'food pyramid' to explain simply nutritional advice to consumers;[61, 62] it introduced guidelines on public school meals. Many countries have learned well from such national food and health education initiatives, even though they may not be strong enough to compete with, for example, the deep pockets of food business' advertising and marketing efforts. Progress is painfully slow; new strategies and new policy innovation are needed.

A number of Nordic countries too – Finland, Sweden and Norway – are much celebrated in the public health policy literature for their efforts at injecting a human health dimension into their food supply to reduce the incidence of diet-related diseases and illness. In part, these schemes were medically driven but they were also inspired by the Scandinavian corporatist, consensual tradition of social democracy: this attributes to the state a potentially benign and socially responsible role vis à vis the individual citizen, working with industry, unions and public interest groups to establish the public good. These countries have evolved workable public health policies to address all aspects of food supply.[63]

More recently, these Scandinavian countries have begun to extend their policy framework by creating better coordinated Nutrition and Physical Activity Councils and strategies to tackle obesity. Norway's official Nutrition and Food Policy of 1975 was designed to combat the incidence of cardiovascular disease, which accounted for around half of all that country's deaths,[64] and to reduce the proportion of fat in the national diet from 40 per cent to 35 per cent, a goal first achieved in 1991.[65, 66] The farm lobby saw the value of adapting to the emerging diet–health paradigm and helped introduce an integrated national food policy, linking agriculture, food processing, consumers and health and rural affairs.[67, 68, 69, 70]

In the early 1970s, Finland had the highest rate of coronary mortality in the world.[71] Through a project in North Karelia (a region with the highest internal rate), the Finnish government and health services set out to tackle this toll,[72] and targeted smoking, blood-pressure control and diet, with preventive campaigns throughout the country. The proportion of saturated fats in total fat consumption in Finland declined, while fish and vegetable consumption rose,[73] and a 55 per cent fall in the male mortality rate from coronary heart disease was recorded in the period 1972–1992 as a result. And all this health improvement occurred 'without the need for extra resource allocation',[74] with close collaboration between health agencies and the food industry. The strategy was systematic, with a clear overall vision.

Sweden's moves towards an integrated food and health policy was fired first by a food-safety crisis, following a horrific outbreak of *Salmonellosis* in the early 1950s which killed 100 people. This led to the setting up of the country's National Food Administration which aimed to link food production with high health standards. Painful though it was, the Swedish farmers accepted that it was in its long-term interests to meet tougher health criteria than were internationally stipulated.[75]

Later, in the 1990s, Sweden reinforced its public and environmental health policy with employment and food quality objectives, following heavy criticism about monoculture in forestry and farming;[76] further, both the agriculture and the environment ministries are developing programmes to reduce fossil fuel and energy use[77] by a factor of four, through increasing technological sophistication.[78] Sweden is also exploring how to achieve tough targets on reducing greenhouse gases emitted from the entire food supply chain[79] – one of the climate change recommendations of the WHO, the World Meteorological Organization and the UN Environment Programme.[80]

These countries' efforts to integrate food and agricultural policies with health goals have entailed energetic professional and personal commitment from advocates. Interestingly, Sweden, like Norway, recognizes the importance of the cultural dimension of food policy. They both permit no TV food advertisements targeted at children under 12 years of age,[81] a policy much admired elsewhere for its protection of children from junk food advertising but hated by the world advertising industry. This child protection régime is now under attack (led by the British advertising industry) in the name of creating a common EU-wide framework on advertising.

Such efforts to integrate food and health are under some strain, partly due to EU membership and also due to pressure from globalization and the need to meet the dictates of the GATT. Respectful as we are about the Nordic experience, it does have its limitations. The North Karelia initiative, for instance, would be harder to implement today. In the 1970s, for example, US culture was less of a force on youth culture. Finland was outside the EU. There was no multi-channel, satellite TV beaming in commercial messages.

For all their limitations, the Nordic experiments are enormously important. They also show the value of thinking about policy-making in terms of context, culture and process.[82] The Nordic experience is living proof that policy battles can be won by health interests, that public and environmental health can be fused with food and agricultural policy, and that improvements in health can go hand in hand with sound economies.

Agencies: another response to the crisis of governance?

One response to difficulties in governance has been the setting up of agencies – government bodies wholly funded out of taxation, yet which can be presented as 'independent' and science-based. Science policy experience suggests that the key factor in determining the impact of evidence and views on policy outcomes may be the framing assumptions which are in place to start with.[83, 84] In respect of food agencies, a key goal has been to set out to restore public trust in governance. The plethora of agencies covering environmental protection, medicines and food safety suggests that the crisis in trust has been extensive. The creation of such agencies creates another problem, however: how to integrate multi-sectoral as well as multi-level interests.

In the UK the Food Standards Agency was created in 2000 after the greatest period of food crisis since World War II, with a brief of remaining out of environmental affairs and of focusing mainly on food safety, but also on nutrition. Similarly, the European Food Safety Authority (EFSA) came into being in 2003 after years of loss of public confidence about standards.[85] In the southern hemisphere too, building on the National Food Authority of Australia, a food safety body set up in the late 1980s merged in the early 1990s with New Zealand's regulatory bodies to create the Australia–New Zealand Food Authority (ANZFA) and it was relaunched in 2002 as Food Standards Australia New Zealand.[86] Denmark's food agency began as an environmental agency, set up in the 1970s in the wake of concerns about pesticides; it evolved into a multi-sectoral body, first answering to the Environment Ministry, then to Health, and it is currently re-integrated into a newly created Ministry of Food, from the ashes of the old agriculture ministry.[87]

Another solution to governance problems could lie in advisory bodies. The post-War generation of food thinkers in the UK created a Committee on Medical Aspects of Food Policy (COMA) which, when it was abolished in 2000, was replaced by a Standing Advisory Committee on Nutrition (SACN) which had a much narrower remit and lost COMA's policy-making function. Specialist advice is clearly needed if policy is to be informed by evidence, but there is a danger of a real imbalance and a lack of integration. There needs to be a mechanism built in that gives governance bodies informed, evidence-led direction. Food Policy Councils may provide the key: these were proposed at the 1992 International Conference on Nutrition as vehicles through which the diversity of advice and evidence could be channelled to policy-makers. When Norway's Inter-ministerial Council to coordinate policy implementation was deemed passive, an advisory body, the National Nutrition Council, filled the vacuum and, using information as a key weapon, became the institutional force for injecting health into wider food and social policies. Former opponents of policy, such as the food and agriculture industries, found themselves more willing to listen to arguments from nutritionists and health authorities. The Nutrition Council's functions have now been re-integrated into the state machinery, locating it within the Directorate for Health and Social Welfare and integrating it with agricultural, fishery, price, consumer and trade policies as well as educational and research policy.

Food agencies are an opportunity, at least in theory, to build creative tension into food governance between the setters of standards and the measures that they implement.[88] Only time will tell if they will have a truly positive effect. If their role is to bolster the survival chances of the existing Productionist paradigm or to pursue one particular paradigm, they will become objects of contention themselves rather than an agency for conflict resolution.

THE EMERGING BATTLE LINES: FOOD DEMOCRACY VERSUS FOOD CONTROL

The history of food governance can usefully be understood as a long struggle between two conflicting forces: 'food democracy' and 'food control': the latter suggests relatively few people exerting power to shape the food supply; the policy framework is *dirigiste*; decisions are 'top-down'; the views and interests of others are mediated through the controllers' eyes; there is limited dialogue; and few resources are allocated to investigate ranges of policy options. 'Food control' presupposes people, animals, plants and the environment being controlled in order to maintain order, authority and predictability.

'Food democracy', on the other hand, gives scope for a more inclusive approach to food policy. Its ethos is 'bottom-up', considering the diversity of views and interests in the mass of the population and food supply chain; the needs of the many are favoured over the few; mutuality and symbiosis are pursued. There is genuinely open debate between alternative and opposing views. The core assumption of 'food democracy' is that the public good – in this case the ecological and public health – will be improved by the democratic process.

Some of the clearest examples of 'food control' occur in times of war or under authoritarian political regimes. Forced farm collectivization and imposition of tightly enforced rationing are extreme examples. Under wartime circumstances, governments frequently impose tough regulations and rules, ordering the food supply chain and rationing access to food. In today's architecture of food governance, bodies such as the WTO or EU have considered powers of control, themselves fought over.[89, 90] Whilst developing nations battle for land, consumers in developed countries, more reliant on the market to meet needs, battle for 'food democracy' in respect of access to unadulterated food at affordable prices and of regaining trust in food manufacturing processes and in the supply chain.

A challenge for 'food democracy' is to decide where the public interest lies and to set food priorities on behalf of and consulting the wider citizenry – not just consumer organizations but voices representing all other interests. The emergence of NGO coalitions across sectors in this respect prefigures what governments ought to consider too – the integration of previously discrete policy 'boxes'. But another clear position, coming from the political right, is to argue that modernization of institutions of governance is unnecessary and undesirable because the quickest and most democratic form of governance is the free market and direct accountability through consumer sovereignty. But there are problems with this 'consumer votes' approach when anti-health food forces have at their disposal deep pockets to influence the consumer vote to their ends.

Human Liberty and Consumer Choice

Central to any discussion of 'food democracy' or 'food control' is the question of human liberty. Is health an individual responsibility and decision? Is food a personal affair? If people want to eat out-of-season foods sourced from the four corners of the world and if the market will meet their needs, should they not be allowed to do so? This liberty argument may be superficially attractive but is ultimately flawed, and the rallying cry of individual freedom could be a smokescreen for food sectors maintaining the right to sell sub-standard foods. Central, too, is the issue of the state's role in allocating priorities, rights and responsibilities for whole populations and societies which ensure individual as well as collective liberties. One argument consistently raised, not least by food companies, is that modern consumers have market power through 'voting' with their daily purchases of food and consumption patterns. They luxuriate in a historically unparalleled range of choice. No one forces them to consume as they do. It is true that a visit to a hypermarket with its 20,000 or more items on sale is truly astonishing to those who have never experienced it. But, from a health perspective, these modern cathedrals of choice are in fact the forums in which the public does its food penance; it consumes now but 'pays' later.

This so-called 'consumer votes' theory works mainly for the affluent in both the developed and developing worlds. As we have seen, consumer prices do not reflect the worldwide picture of environmental or health costs caused by the nature of the production and distribution which gave them this extraordinary range. In addition, despite the apparently wide choice available to consumers, they often do not have full information as to what it is that they are choosing between. Labels do not always convey comprehensive contents or the origins and nature of production. Food labels cannot really tell the full story of what went into the food or its health attributes.[93]

Beneath the patina of consumer choice there often lurks a creeping homogenization. For instance, there is a shrinkage in the number of food shops that consumers can patronize: about half of all the food and drink purchased in the UK now comes from just 1000 hypermarket outlets controlled by just four dominant retail players, and in an increasing number of countries levels of market concentration are not dissimilar. And on supermarket shelves, it is highly likely that consumers will find only leading brands in particular product categories. There may have been 10,000 new food items launched annually in Europe in the 1990s, but only hundreds have survived the brutalities of the competitive retail environment for more than a few months, let alone years.[94]

The 'consumer votes' theory assumes that consumers frame for themselves the context within which they choose goods. In practice, however, rules of trade are made by highly political and powerful international forces which frame trade policies to suit their own policy and commercial realities. Even supporters of the Productionist paradigm agree with civic movements that the outcome of this process distorts choices within local food economies.[95, 96, 97]

Conclusion

For food governance to meet the challenges that we have outlined in this book, it must engage with the real complexity of multi-level policy and control, and food policies, institutions and instruments need to be improved so as to achieve:

- better coordination between the different levels of governance in relation to food policy;
- synchronicity of action;
- lack of domination by commercial interests;
- full stakeholder participation giving priority to including voices previously excluded;
- more imaginative use of participative democracy techniques;
- good research, evaluation and monitoring so as to enable policy-making to learn, adapt and incorporate solid evidence;
- better mechanisms for getting policy to respond to evidence and evidence to feed into policy;
- balancing 'global' with 'local' food policy objectives.

Improving food and health nationally and worldwide is a process which requires that policy take public health evidence more seriously than commercial rhetoric about leaving markets unimpeded by state rules and intervention; leadership to deliver the right mix of policy measures; that there is a group of food-committed citizens who will not take 'no' for an answer; and that there is a commitment to constant improvement in the policy- and food-delivering process. At the core of the new food governance must be an iterative process of regular policy review and an ability to experiment. At its simplest, this process will be one in which government sets up a policy review leading to collection of evidence, leading to audit of policy, leading to a decision to maintain or modify policy; leading to new ideas and policies being sought, leading to scoping and experimentation, testing and consideration for policy, leading to a political decision to change, which then is monitored and upon which evidence is sought and evaluated.

Much food and health policy suffers an evidence deficit but there are encouraging signs of change. There remains the even more important need to link human and ecological health with culture and science in order to deliver an authentic health-centred food supply, which in turn requires commitments from research foundations and funders as well as from governments. Ultimately, however, only government can set policy frameworks that have real legitimacy, are truly in the public interests and are 'food democratic'.

CHAPTER 8
THE FUTURE

'Lay then the axe to the root and teach governments humanity.'

Thomas Paine, 1737–1809

CORE ARGUMENTS

We have seen how there are clear tensions in the policy framework of food and health. The Productionist paradigm is fast eroding, not just on human and environmental health grounds, but also on hard economic grounds. It is not yet clear how the food supply chain will finally adapt and respond to the health crisis or which of the other two paradigms will replace it. The terrain is fissured by the competing areas for policy, business and science and what their roles should be. As ever, the relationship between evidence and policy formulation is problematic, and some elements of policy are being pursued without or despite evidence; others are strong in evidence but not strongly supported by policy consensus and actions. A period of experimentation is underway in which new solutions to this crisis are being proposed by the corporate sector, agriculture and NGOs, and social movements vie with advocates of corporate responsibility to win the battle for consumer culture and state attention.

A battle is underway about what the public interest is. The paradigmatic approach offers a tool to explore various scenarios for the future and helps to clarify differences and perspectives between public (as opposed to private) and corporate policy. Although food policy is highly contested space, it is now possible to outline distinct policy options, where health is interpreted and pursued differently. Which course is ultimately (or temporarily) followed will be decided by the strength of forces and ideologies in and beyond the state and will be framed by the degree of organization and coordination of the particular paradigm. Currently, the likely 'winner' in the paradigmatic struggle to replace productionism is the Life Sciences Integrated paradigm; but this dominance is vulnerable to external shocks, which may alter the preparedness of consumers to act, not just think, like citizens with a long-term commitment to ecological sustainability, which is the core of a new ecological public health. Consumer behaviour is not the only key factor. The relationships between science, technology and innovation is being fought over by adherents of both paradigms.

Introduction

At the conclusion of this book, we return to its starting point: the Food Wars are battles over food policy. They occur within and between states, corporate sectors and civil society; they frame food's relationship to health and well-being, on an individual and a domestic, as well as a population, level. We have shown that the shape of food and health is determined by such issues as:

- the cost and affordability of a good diet;
- equitable food distribution within and between countries;
- access to food;
- local, regional or global food sourcing;
- how food is grown, processed and marketed;
- where and how food is bought and consumed;
- food's impact on the environment;
- food's role in cultural life;
- the role of science and technology;
- the distribution of power and governance across the supply chain.

Whether policy-makers make such issues *explicit* or leave them *implicit*, they require policy directives. The challenge, as we suggested in the last chapter, is for a political grip to be taken of these policy choices in order to optimize population and ecological health, and to assert the public interest.

What are the viable alternatives to the Productionist paradigm? As we have shown, evidence of the social, economic, environmental and health problems associated with food supply is mounting fast. Policy should feed the search for evidence and evidence should inform policy: there should be mutual feedback. Ideally, policy is made on best evidence, with data synthesized on a systematic basis. As one paper has argued: 'Evidence-based nutrition is the application of the best available systematically assembled evidence in setting nutrition policy and practice.'[1] In practice, however, evidence often fails to inform policy: policy-makers are often just not interested in the evidence because it embarrasses a politically driven agenda. As we saw in the case of evidence-based nutrition guidelines, there can be a tortuous relationship between how such population-oriented advice ought to be applied and what actual populations eat. The science of policy-making is inevitably political and experience suggests that we should not be holding our breath for health to be put at the heart of food policy processes.

Policy-making is not precise: it is a product of politics and the (im)balance of forces. It is a matter of timing, mechanized by champions, vision and imagination – both private and popular. It is framed by 'grand' contexts such as wars and the long sweep of historical forces.[2] Policy-making about food and health is therefore a social process, informed by different understanding, as Table 8.1 illustrates.

Which Paradigm Will Triumph?

Although Productionism is weakening in power and public appeal, it is uncertain which of the other two paradigms will replace it for the long term. It is possible, as Thomas Kuhn originally argued when setting out his view of paradigms in scientific thought, for the two paradigms to coexist. One danger with both the Ecologically Integrated and Life Sciences Integrated paradigms, while they offer divergent scientific viewpoints on the worlds of food and health, is the polarization within 'healthy' food production between 'rich' and 'poor' consumers – those who can afford such foods and those who cannot. We see

Table 8.1 *Different approaches to food and health policy, by paradigm*

Policy focus	Productionist paradigm	Life Sciences Integrated paradigm	Ecologically Integrated paradigm
Relationship to general economy	'Trickle-down' theory; primacy of market solutions; inequality inevitable	Supremacy of individualized consumption; Corporation-led due to need for large private sector science budgets	Population approach via genuine stakeholder consultation; health as economic determinant; inequalities require societal action
Direction for health policy	Individual risk; reliance on charity; safety is prime concern	Public–private partnerships; personal insurance; risk management and hazards control	Social insurance including primary care, welfare and public health services; sustainability and public health are integrated
Approach to diet, disease and health	Implicit acceptance of societal burden of disease; inability to act on problems of over- and under-nutrition	The right to be 'unhealthy'; a medical problem; individual choice is key driver; demand will affect supply; niche markets	The right to be well; entire food supply geared to deliver health
Food business	Commodity focus; industrial-scale ingredients and processing; costs of ill health not included in price of goods	Commodity focus with niches; underpinned by public costs but subject to pressure to shift costs from public to private	Costs internalized where possible; need to develop more robust mass production controls; emphasis on 'natural' products and processing

Policy focus	Productionist paradigm	Life Sciences Integrated paradigm	Ecologically Integrated paradigm
Environment	Tendency towards monoculture; limited consideration of costs; pressure on resources to produce food; *ad hoc* adjustment; industrial–chemical dependency	Reinforces monocultural tendencies but some rhetorical concern about diversity; gradualist; acceptance of importance; hi-tech industrial approach to problems; tries to reduce industrial–chemical dependency	Biodiversity at heart of thinking; ecological assumptions; development of robust ecological systems; minimized industrial–chemical use
Consumer culture	Individual responsibility; self-protection; consumerism dependent on willingness to pay as consumer	Access and benefits according to capacity to pay	Societal responsibility based on a citizenship model; defined rights as citizenship; authentic stakeholder involvement
Role of the state	Minimal involvement; avoidance of 'nanny state' action; resources best left to market forces but use of public subsidy to implement policy	Balance of public and private sector; rhetoric of minimal state accompanied by strong state action in some sectors; enabling regulation	Sets common framework; provider of resources; corrective lever on the imbalance between individual and social forces

such polarization in production as socially unsustainable as well as unjust: a pool of ill health and inequality in communities acts as a drag upon society as a whole.[3]

The Food Wars' ultimate challenge will, therefore, be to clarify the organizational and practical aspects of the food supply chain and to chart its new direction. Childhood obesity, for instance, is an issue demanding answers about whether to control marketing and advertising and about moving food culture in a more health-enhancing direction. Equally, whether or not to adopt GM crops and whether individualized/personalized nutrition will improve population health profiles are very real issues which will help determine which paradigm becomes pre-eminent. Much will also depend on whether research into cutting-edge issues for each paradigm receive equal resources and funds. There is real concern that commercial research is overwhelmingly dominated by investment in the Life Sciences paradigm, with little going into clarifying evidence on different interpretations of the biological sciences. Likewise, funding and resources are helplessly skewed when it comes to public information and education. For every $1 spent by the World Health Organization on trying to improve the nutrition of the world's population, $500 are spent by the food industry on promoting processed foods.[4]

The key deciding factors as to which paradigm triumphs overall, and which becomes pre-eminent in the food supply chain or whether there is a 'truce' include:

- the costs of ill health and health care as fiscal burdens;
- environmental pressures such as climate change, energy use and water shortage and their impact on people;
- the perception, by key opinion-formers, and social groups with the capacity to pressurize the state, of the risks;
- the role of scientists and other specialists in articulating the urgency of the health case in public debate;
- the tolerance for levels of food inequalities within and between societies;
- whether powerful, coherent international alliances promoting 'food democracy' emerge;
- the level of public confidence and trust in the food supply chain and in science and technology;
- the funding available to competing notions of science and technology in each paradigm;

- whether commerce implements social responsibility and sustainable business practices;
- any unforeseen new crises arising from current food supply practices.

Much depends, of course, on the balance of forces and resources, but one thing is certain: the solution to the food supply chain is too important to leave to chance. Social pressure will win progressive change. History suggests that, if threatened by contagion, and presented with evidence that prevention is possible but requires new policies or infrastructure, people will accept taxes and discomfort necessary for effective drains or good water systems or improved housing or transport. Even vested interests resistant to change will, when they accept the inevitable, support the flow of funds. Such an era may be upon us in the Food Wars.

Clausewitz, the 19th-century Prussian military tactician, famously argued that war is politics conducted by other means.[5] Food, we have argued here, is an illustration of that dictum. The battles, the war fronts, and the marshalling of forces all are signs of powerful interests pursuing their goals. For example, two of the early cases for the WTO's trade disputes mechanism were based around a 'health' issue – namely the issue of recombinant BST and beef hormones.[6] This led to huge tariffs ($117 million per year) being imposed by the WTO on the EU for its ban on hormones, and the US retaliated with a case in 1999 against the EU on the grounds that its hormone ban was discriminatory and spurious public health.[7] This, and other cases, showed for the first time how the EU and US giants of world food trade could be brought to court and be adjudicated against. Further, the emergence of the G-21 group of countries allied against the EU and US at the WTO revision talks at Cancun in 2003 was another illustration of shifting global food politics, as a result of which even the food power elites have been sensitized to the new world order,[8] and to the delicacy of public opinion.

The Paradigmatic Analysis

We have made much of the value of exploring ideas through the paradigms and frameworks surrounding food in public policy. Food is a material and biological reality so deeply entrenched that culture as well as social movements are required to help set new food frameworks and paradigms. Food marketing already taps into deep cultural mores and beliefs when moulding consumer behaviour in order to sell food products. It is only people *and* organizations *and* social forces which can take this on and shape the food economy differently.

If the paradigmatic framework we have outlined is apt, it has profound implications for the future of the entire food supply chain. It is not that one paradigm is hostile to industry or to science and technology, and that the other favours it; on the contrary, the paradigms offer *different* approaches to industry, to consumers, to health, to science and technology, to managing the environment, and so on. Terms such as 'lifestyle' and 'choice' that current policy-makers hold dear will vary in meaning according to which paradigm holds sway.

Within the Productionist paradigm, the policy emphasis has been focused on unleashing the productive capacity in the food supply chain (particularly land and labour) and to aim for quantity and efficiency of output, defined in terms of yield, throughput and profitability. Its assumption is that the public good requires sufficiency and, in turn, that sufficiency will deliver the public good. In the pre-World War II and post-War worlds scarred by hunger and maldistribution, this policy made both short-term and long-term sense. Productionism has in its time been immensely successful. Output rose dramatically and more mouths were fed.

So why is the Productionist paradigm under such strain? First, evidence has mounted about its externalized health costs: those associated with diet are rising dramatically, expenditure on health care is now enormous and there is now a double burden of disease, due to hunger, degenerative disease, underconsumption and over- or malconsumption.

Second, there has been a persistent thread of criticism as to Productionism's emphasis on quantity before quality. Despite Productionism's astonishing production capacity, it has possessed a tendency to over-production, distorting food processing and marketing practices without enough consideration for health outcomes. In addition, the environmental impacts of intensive agricultural systems, such as residues from pesticides and damage to wildlife, biodiversity (both on and off the farm) and diet itself surfaced first, to be followed by more drastic pressures such as water scarcity, falling fish stocks, soil loss and climate change.

Third, the policy focus of Productionism upon primary production is no longer in synch with the contemporary food economy where the shift has been away from the agricultural/rural sectors to what happens *off* the farm: that is, away from production to the new cultures of consumption. Food processors, retailers and catering/food service sectors have in effect changed the food economy, with retailers being pre-eminent among these. The labour process, as well as agricultural policy, has had to adapt to this competitive change and is marked by low wages and poor working conditions in many countries.

Fourth, wide-ranging societal changes such as urbanization, globalization, supermarketization and media-based food culture now shape rapid dietary shift. The rise in popularity of fats, meats, soft drinks, sugary and processed foods has accompanied the decline of indigenous, fresh foods. Although facilitated by trade liberalization, this emerging 'globo-food' culture is not wholly attributable to the 'pushing' or manipulation of consumer tastes, although huge sums are expended to bend tastes to suit the products of powerful company sales; consumer aspirations also 'pull' these changes: that is, there are consumer aspirations that Productionism has serviced. But the maximization of output has not necessarily been good for consumer health, and policy critics condemn the over-supply of the Productionist era.[9, 10]

Fifth, there has been a clear failure of policy to address the health, social and environmental problems associated with Productionism in an integrated way. Policy has often been conceived in a 'sectoral' manner and what is needed now is the integration of sectoral interests in our policy framework. What is the point, for instance, of increasing food production by using irrigation if this depletes water tables? Why give advice to consumers to eat more fish when stocks are in peril or fishing practices are unsustainable? Too many policy actions that make sense in one area undermine actions elsewhere, and the overall policy coherence lacking under Productionism is now required. We need to juxtapose farming not only with sustainability, but also with health and tackle food safety alongside all other aspects of food-related health. Instead, too often policy is addressed in discrete boxes. Partly this is a problem of institutional architecture, but it is also due to a lack of vision.

Table 8.2 *Features of the Productionist paradigm under contest in the Food Wars*

Policy area	Examples of features being contested
Technical	Transformation of nature (GM, seeds, land); impact of intensification; motives for investment and research (private/corporate financial returns or public goods?); the use of food processing aids (eg additives)
Commercial	Managing corporate power; reliance on fossil fuels; trade liberalization benefiting some rather than all; market concentration and competition policy; coincidence of mass and niche markets' reliance on 'cheap' labour and low-cost production centres
Health	Individualization; emphasis on safety rather than on degenerative diseases; provision of health care services; externalized costs borne by society and families; rise of obesity and diseases such as diabetes
Food culture	Globalization, branding and marketing framing choice; food mores affected by consciousness industries and rapidity of change; artificial creation of choice and need; the role of the consumer
Food in society	Food justice and the role of trade; inequalities; simultaneous under- and over-consumption; problems of access; neuroses associated with food (phobias, anorexia, etc)
Environment	Unsustainability; externalized costs borne by nature; dramatic fall in biodiversity; pollution, contamination, residues
Governance	Degree of regulation; delineation of the state's role; problems of multi-level governance; rise of NGOs and 'civil' society; restoration of public trust

These are fundamental, not peripheral, difficulties and explain why the Productionist paradigm has increasingly become contested. Table 8.2 summarizes some keys areas where the policy framework, in place from the late 1940s even to the present, needs an overhaul. A new systemic approach is required. Clear demarcations have emerged in health, as we have seen, between an individualized medical model of food and health and a population or societal model, both offering the prevention of disease and illness – one targeting 'at risk' individuals and the other developing population-based strategies.

QUESTIONS EMERGING FROM CIVIL SOCIETY

After decades in which big government and big business have dominated food policy discourse, the last three decades of the 20th century witnessed a re-emergence of political and public involvement in food policy in the form of social movements. These often emerged as 'single-issue' campaigns ranging across such issues as famine, hunger, pesticides, animal welfare, organics, socially responsible investment, fair trade, labelling, marketing practices, cooking skills and genetic modification, as well as disease-specific health conditions such as heart disease or diabetes. Collectively, food movements have emerged, which question the status quo and try to bring food, diet, health, the environment and food culture into wider public consciousness. These groups questioned the new consumerism and viewed the technological outpourings of the food supply chain with some scepticism. Thus, over the last two decades of the century, at first hesitantly and then with increasing coherence and confidence, a new worldview emerged. This food counter-culture articulated that there was an alternative to the mechanistic approach to food pursued within the Productionist paradigm that was 'denaturing' and de-valuing food.[11] It was driven by full awareness of the commercial forces controlling the means of food production;[12, 13] and so, by the time of the WTO meeting in Seattle in December 1999, a defining moment, the Productionist paradigm was clearly in trouble, not least due to new social awareness of how trade rules and competition policy worked in the interests of the already powerful.[14, 15]

This crisis of legitimacy was brought to a head by massive popular and NGO activity against the new global world order, with mass picketing and demonstrations following the world's political elite wherever it gathered – at meetings of the G7, at the World Economic Forum at Davos in 2001 or at Cancun, Mexico, in 2003 – aiming to direct political attention towards such issues as access to food markets and the distortions of international farm trade and subsidies. Further issues also meriting public and political attention must be: competition in food markets; curative health care, marketing and advertising; and the role of diet in the increase in rates of degenerative disease.

Civic and popular movements have been both vocal and effective in articulating elements of the Ecologically Integrated paradigm. NGOs have been courageous in confrontations over issues such as contaminants and chemicals in food, GM foods, world hunger and famine, the impact of trade rules on developing countries, the advertising of foods, labour conditions and ethical investment, environmental damage and animal welfare. Scientists working within the food system have sometimes dismissed these social movements as either scientifically illiterate and ungrateful for all the efforts of the food industry in delivering progress and improvements, and public critics have often been relegated as pursuing naïve views of health in preferring the 'natural' over the 'artificial'. Nonetheless the public confrontations over Productionism have done their work in highlighting scientific uncertainties and the need to shape policy on food and health beyond simply the traditional medical model of health progress.

A food movement which has posed a genuine alternative to the Productionist paradigm is the sustainable agriculture movement. With gradual momentum, it has begun to provide scientific and social scientific evidence for the validity of the Ecologically Integrated paradigm. (Ironically, the Life Sciences Integrated paradigm is also now attempting to claim the sustainability agenda for its own legitimacy; but this reinforces the point.) The agroecology movement,[16] for instance, found itself responding to sustainable agriculture in creating a food agriculture that was not punitive to ecology.[17] It drew upon both science and cultural experience in testing its own practices and experimenting with new ways of combining cropping and skills. Sustainable food systems can both raise output and reduce inputs, while being more socially equitable, feeding people in need, providing jobs and being environmentally beneficial (not just benign).[18]

A country like Cuba has shown what is possible. Faced with the collapse of its long-term colonial mentor, the Soviet Union, and in the face of a continued trade blockade from the US, its neighbour, Cuba had rapidly to transform a large proportion of its agriculture from sugar to real food; it had to produce food directly for its people without the costly inputs of fertilizers and pesticides used in the past and paid for by trade in sugar with the USSR. In the space of a few years, Cuba had developed a whole new system of agroecology.[19] The Cuban approach was multi-level: central government encouraged community action and participation but also provided research support and infrastructure. Cuba provides some evidence that only when countries accept they are in a crisis will solutions outside the dominant paradigm be entertained.

Sadly, such dramatic changes on a country-wide basis towards an ecological goal are few and far between. While the questioning of Productionism and the ambition of an ecological public health generally remain embryonic, popular movements – from farmers' movements such as Via Campesina (the international farm and peasant coalition)[20], the actions of the French farmers in response to McDonald's, the Food Justice movements in the UK[21], the work on food justice in Brazil,[22] to the farmers' resistance in India[23] and the seikatsu clubs (people's co-ops) in Japan[24] – express the aspiration to a safe, affordable and accessible food supply. 'Food democratic' pressures are building up and take many forms: in one country, attempts to rebuild or retain local foodways and the reduction in food miles; in another country, attempts to reconnect land with health and to meet basic human needs. On some fronts, what was 'fringe' is now moving centre-stage: organic agriculture, for example, and some community supported agriculture (CSA) and urban agriculture, movements with worldwide voices and recognized at official levels, receiving varying degrees of political support.

While there is less coherence in urban areas, one development is worthy of comment. In both developing and developed worlds, there is a rebirth of cooperatives and attempts to create 'alternative' distribution systems. (The original 'mutuals' were the first organized attempt to create options for the working poor in the early days of industrialization.) In the UK, for instance, the old Co-operative Movement, born in the 1840s as the 'working person's friend', first introduced the supermarket format in the 1950s, only to be overtaken by smaller independent chains adopting its own modern methods and marginalized by rapid retail concentration from the 1970s. It is interesting, then, to see the 'Co-op' with its emphasis on sustainability and health, begin to reconnect with the modern food movement.[25]

Another important challenge to the cultural tenets of productionism is the 'Slow Food' Movement, founded in Italy but now developing around the world.[26] As its name implies, this group believes that food should be about pleasure and social integration: it celebrates local cuisines and produce; it promotes skills and artisanal food; and it presents a counter-culture to the ubiquitous fast-food culture born from the restlessness of modernity and travel on the road.[27] Whilst it remains to be seen whether this movement gains genuine mass appeal, it is certainly successful in questioning the ethic of cheap and bland 'globo-food' and in celebrating diversity, regionalism and skill.

Localism is another potentially significant cultural movement. In the wake of a growth of global sourcing from around the world by rich country retailers and traders, new global giants already exist and are poised to expand dramatically. Yet in practice, present food supply chains are still overwhelmingly regional rather than global: factories and workers are more likely to be regionally located where the food is to be sold. Even some of the global giants espouse 'glocalization' (sic), the ownership of 'local' plants, and brands or fascias which give the illusion of national branding. It is not just food campaigners who are interested in the (re)localization of food: major companies are also looking to target 'local' consumers. In the words of Niall Fitzgerald, the former Co-chair and Chief Executive of Unilever: 'Every consumer is local. They are driven by who they are, what they believe in and what the mores and traditions of that society are.'[28] But this trend by large companies to reposition themselves as 'authentic' can also be seen as a sign of some desperation and a corporate response to consumers' revolt against the mass-produced.[30]

Can reborn localism pose a threat to transnational supply chains? While there is cheap oil, the search for sources of supply where land and labour are cheap is likely to encourage the trend to all-year-round a-seasonality in value chains, while at the same time, the 'proximity principle' – maximizing the production of those foods that can suitably be grown and produced as locally as possible – remains popular.[30] If consumers have tasted diversity and got used to the ready availability of a far wider range of foods, can they be persuaded to return to more restricted diets? It would be short-sighted of any localist movement to prevent citizens of, for example, Eastern European countries with dire winter climates from having access to fruit and vegetables all year round. And, it would make little sense in Finland or the UK, for example, to try to grow mangoes to be able to meet local demand or to localize production for its own sake. The issues of cost and of environmental externalities policy-makers will have to address include: persuading consumers to pay the full price and take the environmental consequences of their new choices; or delivering nutritional diversity while skilling growers into farm practices that ensure biodiversity. How much choice do consumers really want? Balancing of all environmental, health and economic considerations is vital to new food policy.

Many food companies see the need for change in how food is produced and marketed. Yet, while some companies are actively pursuing only the Life Sciences route, others are hedging their bets and are considering the Ecologically Integrated paradigm. There is acute awareness in the corporate sector worldwide that the loosening of borders which brought such unprecedented growth and marketing frenzy is now facing some reaction: that there is unease about quality issues and about the social, environmental and health costs associated with the food revolution of the last half-century. Companies recognize that they have not been passive participants in the cultural transition, and they are also acutely aware of the risks of litigation – particularly in the US – if health can be shown to have been affected by their actions. Retailers, in particular, cannot be seen to be unresponsive to consumers' concerns, however irrational or rational their views might seem to the food technologist. Obesity, in particular, is forcing major corporations to revise their product ranges and adjust their marketing practices – whether this is cosmetic or fundamental remains to be seen.

As a direct result of these recognitions, icons of global branding such as Starbucks and McDonald's have attempted to integrate 'corporate social responsibility' into their everyday business practices. In 2002, for instance, Starbucks introduced Fair Trade coffee as part of its product offering and also started sourcing some milk from cows not injected with the milk-yield-enhancing Bovine Somatotrophin (BST), a growth-promoting hormone. In 2001, in the US, McDonald's implemented 'animal welfare' measures for its suppliers of eggs, requiring them to a commitment to end de-beaking and increasing ranging space for each animal; in 2003, the company refined its policy to include phasing out of animal growth-promotion antibiotics in the 2.5 billion lb of chicken, beef and pork it buys annually.[31] In 2004, it refined its nutrition policy to begin to address or neutralize its nutrition critics by reducing portion sizes.

Unilever, too, has become similarly 'responsible': in response to evidence about the dramatic global loss of fish stocks, it helped set up the Marine Stewardship Council which campaigns for the responsible management and protection of fish stocks. (Unilever has also set up an internal agricultural sustainability programme.) Such developments show that business can engage with the ecological discourse and can fund action. To 'fundamentalist' critics, these changes may seem small and mere 'window-dressing', but they are surely to be welcomed, if only for the messages they send widely throughout the food value chain. Ultimately, only external auditing and time will determine whether such moves are mere tokens of corporate social accountability (or 'greenwash') or real transitions.

Some of these initiatives emanate from within companies themselves, while others have emerged from civil society demanding corporate change: boards of directors and chief executives are having to face increasingly literate, well-informed and even angry shareholders. The Fair Trade movement, for example, has spawned many initiatives, codes and shareholder actions in the last two decades. Reminiscent of the reborn Cooperative Movement, the Fair Trade movement first expressed itself two centuries ago as a reaction against slavery and colonialism (many early companies were religion-based, with Quakers establishing such 19th-century 'pure food' companies as Rowntree and Cadbury), but re-emerged in the 1980s as evidence about poor working conditions and returns for producers of primary products was uncovered by NGOs and researchers.[32, 33] Today, it is an important force in 17 affluent food countries (particularly for product sectors such as tea, confectionery and coffee) linking 350 commodity producers and representing 4.5 million farmers in developing countries.[34] Such initiatives deliver health to the primary producers, whose livelihoods in producing such commodities has sometimes been one of near immiseration.[35]

Part of the appeal of movements like Fair Trade and organics is their espousal of openness and transparency and their resolve to put food companies increasingly under scrutiny for their health impact. Many companies will want to be seen as contributing to public health, or at least to not contributing to ill health, and this requires new composite systems of auditing which link different aspects of food (such as ethics, health and environmental impact) into one format. Companies are already familiar with producing environmental or social audits, but this is likely to widen. In the UK, an experiment in developing a wider auditing of food companies, in this case of leading food retailers, was launched in 2002 but it wound down in late-2003 after resistance from the giant food retailers: the 'Race to the Top' project, funded by government and foundations, set out to develop a composite audit to provide verifiable evidence of social aspects such as worker conditions, environmental impact and human and nutritional health,[36] but the biggest UK supermarkets ultimately refused to participate, suggesting that they were more prepared to present themselves as corporate socially responsible 'citizens' on a single-issue basis. The Race to the Top project with its focus on the multiple dimensions of consumer concern was entirely consistent with the principles of the Ecologically Integrated paradigm.

The emergence of new food alliances is evidence that the hegemony of Productionism in food policy is cracking. NGOs and businesses increasingly make strategic decisions to engage with each other and to put joint heat on governments to deliver market reform. It is possible that out of this process will come new policy convergences and configurations. Health through food, however, has yet to achieve centrality in food governance, and government policy has yet to deliver the 'sustainability' it promises. If governments continue to allow massive targeting of food advertisements at children, for instance, is there any wonder that the political pressure builds up? Governments must deliver action, from taking a tough stance on the advertising industry to giving more resources to public education about food skills and addressing the role of food in affecting long-term health.

A particularly tough issue is competition policy; yet the rapid concentration of corporate power suggests a rethink of what is meant by a market is overdue. Is a market local, national, regional or international? Market share varies according to which geography is chosen. Huge food companies will argue that they only have 2 per cent of global sales, so competitiveness is ensured; but this may disguise awesome power within regions or particular market categories. Critics of supermarkets, for instance, have argued that current definitions of what a market is need to be more local and consumer friendly and that competition authorities need to be beefed up to protect the consumer interest, including how much time or effort they have to spend getting to and from the shop.[37] The International Federation of Agricultural Producers, a coalition of farmers, has initiated a global data collection on international concentration in the agricultural sector.[38] This is to be welcomed but the issues need to be thought through at intergovernmental level. As we showed in Chapter 4, food power spreads not just in agriculture itself (land ownership), but in farm inputs (chemicals, seeds), logistics (transport, shipping, aviation), retail distribution and food service. From the farm crisis and retail crisis, demand for a new world competition policy seems to be emerging. In the face of high concentration levels, recourse to voluntary codes of conduct is too weak a policy response. Voluntarism is no way to address market power. Cross-border mergers and acquisitions and foreign direct investment have health implications. They pose challenges to government and are key drivers of inappropriate food and health cultures for they do not pay the externalized costs.

What of the Future?

The emerging scientific consensus is that there has to be a real shift if we are to effect behavioural change in serious food issues, such as obesity: labelling and 'consumer advice' constitute a health policy that is too limited. Can obesity be effectively tackled by technical fixes such as nutraceuticals or low-fat products, or by the search for genetic markers which can be manipulated or responded to? Or is what is needed a combination of dietary restraint and environmental alterations such as improved transport policies and pricing policies and other measures marked by the Ecologically Integrated approach? What should be the emphasis on exercise being built safely into lifestyles? Is a medicalized, a cultural or an environmental approach the most effective? Obesity is a *leitmotif* for the modern food age, a symbol of surplus amidst hunger, and of more pragmatic ways of seeing food and health issues. A new agenda for food policy is emerging which requires that all sectors of the food world and governments:

- integrate public policy *across* sectors (health, environment, trade, transport, regulation, welfare and education) and *between* levels of governance (global, regional, national and sub-national/local);
- curtail currently externalized costs and encourage their re-internalization so that consumer prices are fair and realistic and operate properly as a key instrument for market efficiencies;
- take a long-term approach that can be delivered in incremental short-term and medium-term actions and via reasonable policy instruments;
- deliver a rational, sustainable food supply that links ecological and population health with citizens' rights and in an equitable manner;
- harness existing beneficial trends while side-lining obstacles and parochialism.

Looking For a Political Lead

Central to our analysis of the policy challenge is the need to address the extent of change in the food supply chain. The impact of supermarket concentration, the collapse of rural employment accompanying urbanization, the environmental externalities, the difficulty of regulating rapidly changing food markets and supply chains; all these surely add to the complexity of food governance. It can be argued that in a transnationalizing world, 'old' *dirigiste* models of food governance may not be appropriate any more, and a dual system of food standards and 'rules-setting' has emerged, one created by the state, the other by companies setting regimes (such as the Eurepgap system[39] which began as an attempt by some European food companies to agree standards on pesticide residues, but has become worldwide and broadened its remit to other food aspirations, and to deliver standardization, consistency and supply-chain coherence).

What companies are tentatively beginning to do – namely re-invent food governance – needs to be led by the democratic process and to be fully accountable. Coherence at and between all levels of governance is essential: policy needs to move beyond the current era of appeals, recommendations, commitments, and resolutions – all voluntarist – to a tougher era of binding agreements with legal gravitas. The enormity of the nutrition transition will not be diminished with mere, bland nutritional advice. Food policy is a political challenge: it requires leadership, ambition and vision to give coherence to the hitherto uncoordinated. We challenge ministers and ministries of agriculture, food, fisheries and health to give this vital process far higher priority. To date, no country's government has fully addressed the full range of problems outlined in this Food Wars thesis.

Data is another vital tool in the Food Wars: too much data that ought to be in the public domain is commercially confidential; major companies spend large budgets on consumer information; the public sector globally has patchy information. Despite admirable efforts by the UN bodies, data on some countries is limited. Without information and best evidence, policy-making can too easily return to whim and political pragmatism. On the other hand, we are wary of delaying policy action pending better information. Often calls for research can be an elegant form of procrastination. The real deficiencies are, however, the political will and leadership to confront powerful vested interests.[40] Sometimes policy has to be made and actions have to be taken on the basis of seriously restricted information; but enough is known about the Food Wars for policy-makers to act on key issues, and pressure on politicians to give that leadership should surely mount.

The nearest most countries get to visionary plans of action are their National Food and Nutrition Action Plans, as recommended by the 1992 International Conference on Nutrition. Many countries have produced such plans of actions. The problem is that they do not or cannot confront the combined effects of the cultural transition and the food supply chain restructuring that we have described here. The FAO conference to reconvene its 1996 World Food Summit in 2002, the so-called 'WFS+5' conference, was a missed opportunity. It failed to achieve political momentum and wide enough support. An immediate task is for countries to collaborate on new official guidelines that link nutrition with ecology. None exist at present. Food culture in an era with abundant choice requires new cultural mores, new rules for everyday eating and living (see Table 8.3).

We have argued that ultimately only the state (or collectivities of states) have the leverage to re-frame the direction taken by the

Table 8.3 *Some tentative rules for food and ecological health (adults)*

- Eat simply.
- Eat no more than you expend in energy.
- Encourage supply to deliver to each according to need.
- Eat a plant-based diet with flesh more sparingly, if at all.
- Celebrate variety; eat in a manner that encourages biodiversity in the field and thence to the plate.
- Eat for nutrition and the environment; think fossil fuels; energy = oil.
- Eat a range of foods of between 20 and 30 species per week.
- Eat seasonally, where possible.
- Eat according to the proximity principle, as locally as you can.
- Support local suppliers; food is work.
- Be prepared to pay the full externalized costs; if you do not, others will.
- Drink water, not soft drinks.
- Without being neurotic, be aware of the hidden ingredients.
- Enjoy food in the short term, but think about its impact long term.

current powerful forces within the modern food supply chain and to reclaim the public interest. What used to be termed 'public' policy is now a trilateral bargaining of state(s), corporations and civil society. But only the state can facilitate that process and exert legal leverage.

It is time to re-invent food and health public policy and to re-invigorate what is meant by 'public' and which strategic goals, instruments and measures ought to be deployed. The old 'top-down' model of public policy had States deciding what they wanted to do, and which are the best, most efficient means for delivering what they set out to do. Taxes and fiscal measures? Advice, education, information? Regulations and laws? The patrician state would decide. We do not reject these instruments but are concerned about the ends to which they are used. Regulations, advice and fiscal measures are still core weapons in the armoury of the Food Wars. It has been more fashionable and politically expedient since the 1980s to see mechanisms such as codes of conduct, the issuing of advice, the publicizing of league tables of performance and 'naming and shaming' and the creation of voluntarist partnerships rather than the use of legislation and sanctions. These have their place, but they depend upon consensus among partners to work. We do not believe that consensus yet exists. The threat of sanction of legal action has to be there, and used appropriately. The European food safety crises of the 1990s led to legislative change and such 'hard' policy measures are more likely to be deployed when state credibility is at stake.

In the modern era, however, the responsibility of consumers is also considerable. Too often, governments and companies put the emphasis on consumers to look after their own health. We have questioned that policy approach, but modern consumerism now suggests that consumers will have to become 'food citizens'; to think about and eat their food against longer time horizons; to judge food not just by price but by quality and long-term impact on health and environment. New cultural mores are required and need to be integrated. What good is a fair-traded chocolate if it is poor for health, or is long on food miles? And yet the fair-traded chocolate gives a decent living – and therefore health – to the primary producer. A welcome debate about what these social criteria for food could be needs to start, linking social considerations with human health and environmental health, as well as economic factors.

These are dynamic times in food policy and in health. The Food Wars thesis proposes that new approaches to food policy are emerging,[41] with new policy choices and alliances and business opportunities. The new era has more complexity, but also more opportunity for strong alliances to tackle together the forces of ill health and inequity. For any policy problem, four broad policy options emerge:

- to do nothing and allow 'market forces' to run their course;
- to look to corporate solutions;
- to frame market conditions;
- to empower civil society to demand and consume differently.

These broad policy options are expanded in Table 8.4 which gives a typology of some general policy directions and options.

We are optimistic. We see signs of pressure to re-invigorate food and health policy: to deliver human and ecological health combining efficiency with authenticity and integrity. Some of the key variables likely to determine the future of food include:

- the relative strength of social forces and ideologies in and beyond the state;

Table 8.4 *Some broad policy options for tackling food and health*

Policy option	Key player	Strategic focus	Examples of actions and instruments
1 Laisser-faire (ie do nothing/give up)	The 'hands off' state	Allow market dynamics to run their course	'It's all too late'; 'leave it to Wal-Mart'
	The 'monitoring' state	Collect more data	'We do not know enough to act' → data collection; monitoring
2 Corporate action	Corporate sector	Product	New product development (eg functional foods); standards-setting
		Technology	New processes
		Prices	Reduction of transaction costs*
		Corporate social responsibility	Voluntary labelling; industry-wide partnerships; consumer information; voluntary auditing and data
		Insurance	Reward positive health behaviour through premiums
3 Frame market conditions	State–corporate–civil society partnerships, politically led processes (ie state)	Investment	R&D support; company audits; use of pension funds to invest in healthy foods
		Fiscal measures	Tax (fat, energy, marketing etc)
		Competition policy	Set travel-to-shop definitions; encourage new enterprises
		Regulations	Marketing controls
		Alter production	Investment; targets; R&D/science and technology; use of public procurement budgets

Table 8.4 *(continued)*

Policy option	Key player	Strategic focus	Examples of actions and instruments
		Infrastructure	Investment in human capital and physical infrastructure; update markets;
		Standards-setting	Guidelines; codes of practice; mandatory regulations; law enforcement
		Institutional architecture	Food policy councils; 'independent' agencies; reform of ministries; improved cross-departmental coordination
		Education	Training; education in schools; consumer advice
4 Civil Society action	NGOs, civil society	Civil action	Campaigns; media coverage
		Re-energizing the state	Actions on pension investment
		Consumer education	Parents' juries; self-help; skills classes in school etc
		Law suits	Class actions against advertisers/manufacturers
		Capacity building	Skills training; school-based education on marketing, nutrition etc
		Improved supply chain	Develop 'best practice' to deliver integrity and authenticity; alternative supply chains (eg Co-ops; a civil vision for public procurement)

Note: *Transaction costs are the costs that occur in a product or production chain from one economic agent to the next.

- which paradigm dominates, which itself depends on political commitment and external pressure;
- how health is conceived – as an individual or a population issue;
- whether the environment is seen as an infrastructure for health or as a separate policy 'box';
- public pressure and the preparedness of consumers to act, not just think, like citizens with a long-term commitment to ecological sustainability;
- the degree of organization and coordination among forces adhering to particular paradigms;
- food insecurity taking hold in the developed world as well as the developing world: if drought or other ecological crisis occurred.

Food is such an enormous sector of the economy, so vital to people's needs and contexts and so powerfully fought over by companies wishing to capture markets that conflict is inevitable. The Productionist paradigm is on the way out. Currently the Life Sciences Integrated paradigm has most of the big commercial backing; far more resources need to be put into policies from within the Ecologically Integrated paradigm. This paradigm, while seriously under-funded, holds great promise, not least in meeting the environmental determinants of human survival. Even if the Life Sciences paradigm reduces its reliance on GM and exploitation of biology, we think it would still be folly for governments to rely on it; it is the Ecological paradigm which deserves stronger backing. As we have sought to show, there is an emerging and remarkable concern for both environmental and human health interests from within the Ecologically Integrated paradigm: a diet that is good for biodiversity is also good for human health, providing a rich variety of nutrients. In this very practical sense, as well as in theory, the goal of food and health policy surely must be for humanity to be at one with nature. But culture, as ever, might turn out to be the defining arbiter.

Notes and References

Introduction

1 Boyd Orr, J (1943) *Food and the People*, London: Pilot Press, p5
2 US Surgeon General (2001) Call to Action to Prevent and Decrease Overweight and Obesity, Washington, DC, Office of the Surgeon General, http://www.surgeongeneral.gov/topics/
3 Hamel, G (2002) 'Innovation Now!', *Fast Company*, December, pp114–124
4 Brabeck, P (2003) quoted in Food Navigator, *Nutrition and Health*, accessed 27 May 2003, http://foodnavigator.com/news.news.asp?id=7700

Chapter 1

1 FAO (2003) *State of Food Insecurity*, Rome: Food and Agriculture Organization
2 CNS Media, personal communication (2003), http://www.win.com
3 Tansey, G and Worsley, T (1995) *The Food System*, London: Earthscan
4 Kuhn, TS (1970) *The Structure of Scientific Revolutions*, Chicago: University of Chicago Press, pviii
5 Masterman, M 'The nature of a paradigm' in: Lakatos, I and Musgrave, A (eds) (1970) *Criticism and the Growth of Knowledge*, Cambridge: Cambridge University Press, pp59–89
6 Mintz, SW (1985) *Sweetness and Power*, New York: Viking
7 Seddon, Q (1989) *The Silent Revolution*, London: BBC Books
8 Clunies Ross, T and Hildyard, N (1993) *The Politics of Industrial Agriculture*, London: Earthscan
9 Boyd Orr, J (1966) *As I recall: The 1880s to the 1960s*, London: MacGibbon & Kee
10 Brandt, K (1944) *The Reconstruction of World Agriculture*, London: Allen & Unwin
11 Pretty, J (1998) *The Living Land*, London: Earthscan
12 Tansey, G (2002) *TRIPS with Everything: Intellectual Property and the Farming World*, Nottingham: Food Ethics Council

13 Rifkin, J (1999) *The Biotech Century*, London: Orion
14 Barling, D, de Vriend, H, Cornelese, JA, Ekstrand, B, et al (1999) 'The social aspects of food biotechnology: a European view', *Environmental Toxicology and Pharmacology*, 7, pp85–93
15 Conway, G (1998) *The Doubly Green Revolution*, London: Penguin
16 Company figures quoted in Vidal, J (2002) 'GM at the crossroads', *The Guardian*, 2 July
17 See the Symposium (1999) 'Interactions of diet and nutrition with genetic susceptibility in cancer', *Journal of Nutrition*, 129, 2, pp550S–551S
18 Müller, M and Kersten, S (2003) 'Nutrigenomics: goals and strategies', *Nature Reviews Genetics*, 4, pp315–322
19 For example, in the US see http://nutrigenomics.ucdavis.edu/
20 For example, in Europe see http://www.nutrigenomics.nl/
21 Fogg-Johnson, M and Merolli, A (2004) 'Nutrigenomics: the next wave in nutrition research', *Neutriceuticals World*, http://www.neutraceuticalsworld.com/marapr001.htm, accessed 29 February 2004
22 Singer, P, Daar, A and Castle, D (2003) *Nutrition and Genes: Science, Society and the Supermarket*, Toronto: University of Toronto, Center for Bioethics, November, http://www.utoronto.ca/jcb/main.html, accessed 9 March 2004
23 Nuffield Trust (2004) Seminar on Nutrigenomics, London: Nuffield Trust, 5 February
24 Shapiro B (1998) 'A new era of value creation', letter to Monsanto shareowners, 1 March
25 Gardner, B (2003) http://www.foodpolicy.info
26 Lheureux, K, Libeau-Dulos, M, et al (2003) *Review of GMOs under Research and Development and in the Pipeline in Europe*, Seville: European Science Technology Observatory / Ispra: Joint Research Centre of the European Commission
27 Conford, J (2001) *The Origins of the Organic Movement*, Edinburgh: Floris
28 Robertson, J (1979) *The Sane Alternative*, St Paul, MN: River Basiu Publishing
29 Gussow, JD (1991) *Chicken Little: Tomato Sauce and Agriculture*, New York: Bootstrap Press
30 Pretty, J (2002) *Agriculture*, London: Earthscan
31 Mannion, AM (1995) *Agriculture and Environmental Change*, Chichester: Wiley
32 Altieri, M (1996) *Agroecology: The Science of Sustainable Agriculture*, Boulder: Westview Press
33 McMichael, AJ (2001) *Human Frontiers, Environment and Disease*, Cambridge: Cambridge University Press

34 Pretty, J (1995) *Regenerating Agriculture*, London: Earthscan
35 Conford, P (2001) *The Origins of the Organic Movement*, Edinburgh: Floris Books
36 Stapledon, RG (1935) *The Land: Now and Tomorrow*, London: Faber and Faber
37 Boyd Orr, J (1943) *Food and the People*, Targets for Tomorrow 3. London: Pilot Press
38 McMichael, AJ (2001) *Human Frontiers, Environment and Disease*, Cambridge: Cambridge University Press
39 McMichael, AJ (2001) *Human Frontiers, Environment and Disease*, Cambridge: Cambridge University Press
40 Keys, A (1970) *Coronary Heart Disease in Seven Countries*, New York: American Heart Association
41 Walker, C and Cannon, G (1984) *The Food Scandal*, London: Century
42 Trusswell, AS (1987) 'The Evolution of Dietary Recommendations, Goals and Guidelines', *American Journal of Clinical Nutrition*, 45, pp1060–1072
43 Marmot, M and Wilkinson, RG (eds) (1999) *Social Determinants of Health*, Oxford: Oxford University Press
44 Smil, V (2000) *Feeding the World: A Challenge for the 21st Century*, Cambridge MA: MIT Press
45 Dyson, T (1996) *Population and Food: Global Trends and Future Prospects*, London: Routledge
46 Dealer, S (1996) *Lethal Legacy: BSE – The Search for the Truth*, London: Bloomsbury
47 McKee, M, Lang, T and Roberts, J (1996) 'Deregulating health: policy lessons from the BSE affair', *Journal of the Royal Society of Medicine*, 1996, 89, pp424–426
48 UK CJD Surveillance Unit, Edinburgh, http://www.cjd.ed.ac.uk
49 Trichopoulou, A, Millstone, E, Lang, T, Eames, M et al (2000) *European Policy on Food Safety*, Luxembourg: European Parliament Directorate General for Research Office for Science and Technology Options Assessment (STOA). September 2000. PE292.026/Fin.St.
50 Lang, T and Rayner, G (eds) (2002) *Why Health is the key to Farming and Food*, London: UK Public Health Association, Chartered Institute of Environmental Health, Faculty of Public Health Medicine, National Heart Forum and Health Development Agency, pp32–33
51 Goodman, D and Redclift, M (1991) *Refashioning Nature: Food, Ecology and Culture*, London: Routledge
52 Norberg-Hodge, H (1991) *Ancient Futures: Learning from Ladakh*, San Francisco: Sierra Books
53 Cannon, G (1987) *The Politics of Food*, London: Century
54 Nestle, M (2002) *The Politics of Food*, San Francisco: University of California Books

55 Critser, G (2003) *Fat Land*, London: Penguin Books. Schlosser, E (2001) *Fast Food Nation*, Boston MA: Houghton Mifflin
56 ACC/SCN (2000) '4th Report on the World Nutrition Situation: Nutrition through the life cycle'. Geneva: United Nations Administrative Committee on Co-ordination Sub-Committee on Nutrition (ACC/SCN), p2
57 World Cancer Research Fund/American Institute for Cancer Research (1997) *Food, Nutrition and the Prevention of Cancer: A Global Perspective*, Washington DC: AICR
58 WHO (2003) *Global Burden of Disease Estimates*, Geneva: World Health Organization
59 Murray, CJL and Lopez, A D (1996) *The Global Burden of Disease*, Boston: Harvard School of Public Health
60 US Surgeon-General figures, quoted in Griffith, V and Buckley, N (2002) 'The scary fat end of the wedge', *Financial Times*, 12 July, p11
61 WHO (2002) 'World Health Report 2002: reducing risks, promoting healthy life', Geneva: World Health Organization, p60
62 WHO (2002) 'World Health Report 2002: reducing risks, promoting healthy life', Geneva: World Health Organization
63 Maniadakis, N and Rayner, M (1998) 'Coronary Heart Disease Statistics: Economics Supplement', London: British Heart Foundation (http://www.heartstats.org)
64 WHO (2000) 'Obesity: Preventing and Managing the Global Epidemic: Report of a WHO Consultation on Obesity', Geneva: World Health Organization WHO TRS 894
65 Powell, KE, Thompson, PD, Casperson, CJ and Kendrick, JS (1987) 'Physical activity and the incidence of coronary heart disease', *Annual Review of Public Health*, 8, pp253–287
66 Berlin, JA and Colditz, GA (1990) 'A meta-analysis of physical activity in the prevention of coronary heart disease', *American Journal of Epidemiology*, 132, pp612–628
67 UNEP (2002) *Global Environmental Outlook 3*, London: UN Environment Programme/Earthscan
68 World Watch Institute, http://www.earth-policy.org
69 Cochrane Collaboration, http://www.cochrane.org
70 Brunner, E, Rayner, M, Thorogood, M, Margetts, B et al (2001) 'Making public health nutrition relevant to evidence-based action', *Public Health Nutrition*, 4 (6), pp1297–1299
71 WHO (2002) *World Health Report 2002*, Geneva: World Health Organization. WHO (2003) *World Global Strategy on Diet, Physical Activity and Health*, Geneva: World Health Organization, 4 December, http://who.int/hpr/gs.strategy.document.shtml. WHO and FAO (2003) *Diet, Nutrition and the Prevention of Chronic Disease*, Technical Report Series 916, Geneva: World Health Organization

CHAPTER 2

1. Cited in Hill, C (1975) *The World Turned Upside Down: Radical Ideas During the English Revolution*, Harmondsworth: Penguin, p391
2. WHO & FAO (2003) 'Diet, Nutrition and the Prevention of Chronic Diseases', Technical Report Series 916, Geneva: World Health Organization and Rome: Food and Agriculture Organization
3. WHO (2003) *Draft Global Strategy on Diet, Physical Activity and Health*, Geneva: World Health Organization, http.//www.who.int/hpr/gs.strategy.document
4. WHO (2002) *World Health Report 2002*, Geneva: World Health Organization
5. WHO (2003) *Word Cancer Report*, Geneva: World Health Organization/International Agency for Research on Cancer
6. IOTF (2003) 'Call for international obesity review as overweight numbers reach 1.7 billion'; press release, London: International Obesity Task Force/International Association for the Study of Obesity
7. WHO (1998) 'Obesity: Preventing and Managing the Global Epidemic': Report of a WHO Consultation on Obesity. Geneva: World Health Organization
8. Commission on the Nutrition Challenges of the twenty-first Century (2000). 'Ending Malnutrition by 2020: An Agenda for Change in the Millennium. Final Report to the ACC/SCN', *Food and Nutrition Bulletin*, 21, 3, September. New York: United Nations University Press p19ff
9. Ewin, J (2001) *Fine Wines and Fish Oils: the Life of Hugh Macdonald Sinclair*, Oxford: Oxford University Press
10. Alexandratos, N (ed) (1995) *World Agriculture: Towards 2010: a FAO study*, Chichester: John Wiley & Son
11. Burinsma, J (ed) (2003) *World Agriculture: Towards 2015/2030*, Rome: Food and Agriculture Organization/London: Earthscan, p5
12. Murray, CJL and Lopez, AD (1996) *Global Burden of Disease*, Geneva: World Health Organization
13. Popkin, BM (1999) 'Urbanization, lifestyle changes and the nutrition transition', *World Development*, 27, 11, pp1905–1916
14. Cabellero, B and Popkin, B (eds) (2002) *The Nutrition Transition*, New York: Elsevier
15. Popkin, BM (2001) 'An overview on the nutrition transition and its health implication: the Bellagio Meeting', *Public Health Nutrition*, 5 (1A), pp93–103
16. Popkin, BM (1994) 'The nutrition transition in low-income countries: an emerging crisis', *Nutrition Reviews*, 52, pp285–298

17 Drewnoski, A and Popkin, B (1997) 'The Nutrition Transition: New Trends in the Global Diet', *Nutrition Reviews*, 55, pp31–43
18 Pena, M and Bacallao, J (eds) (2000) *Obesity and Poverty: a new public health challenge*, Washington DC: Pan American Health Organization (WHO)
19 Popkin, BM (1997) 'The Nutrition Transition in Low-Income Countries: An Emerging Crisis', *Nutrition Reviews*, 52, 9, pp285–298
20 WHO (2002) *The World Health Report 2002: Reducing Risks, Promoting Healthy Life*, Geneva: WHO
21 Lenfant, C (2001) 'Can we prevent cardiovascular diseases in low- and middle-income countries?', *Bulletin of the World Health Organization*, 79, 10, pp980–982
22 Robson, J (1981) 'Foreword' in Trowell, H and Burkitt, D (eds) *Western Diseases: their emergence and prevention*, London: Edward Arnold
23 Trowell, H and Burkitt, D (eds) (1981) *Western Diseases: their emergence and prevention*, London: Edward Arnold
24 Popkin, BM (1998) 'The Nutrition Transition and its Health Implications in Lower Income Countries', *Public Health Nutrition*, 1, pp5–21
25 Cabellero, B and Popkin, B (eds) (2002) *The Nutrition Transition*, New York: Elsevier
26 IFPRI (2002) *Living in the City: Challenges and Options for the Urban Poor*, Washington DC: International Food Policy Research Institute
27 Verster, A (1996) 'Nutrition in transition: the case of the Eastern Mediterranean Region' in Pietinen, P, Nishida, C and Khaltaev, N (eds) *Nutrition and Quality of Life: Health Issues for the 21st century*. Geneva: World Health Organization, pp57–65
28 Chen, J, Campbell, TC, Li, J and Peto, R (1990) *Diet, Lifestyle and Mortality in China: a Study of the Characteristics of 65 Counties*, Oxford: Oxford University Press
29 Geissler, C (1999) 'China: the soybean-pork dilemma', *Proceedings of the Nutrition Society*, 58, pp345–353
30 Dowler, E and Pryer, J (1998) 'Relationship of diet and nutritional status' in *Encyclopaedia of Nutrition*, New York: Academic Press
31 Lang, T (1997) 'The public health impact of globalisation of food trade', in Shetty, P and McPherson, K (eds) *Diet, Nutrition and Chronic Disease: Lessons from Contrasting Worlds*. Chichester: John Wiley and Sons
32 Heasman, M and Mellentin, J (1999) 'Responding to the functional food revolution', *Consumer Policy Review*, 9, 4, pp152–159
33 Schlosser, E (2001) *Fast Food Nation: the Dark Side of the All-American Meal*, New York: HarperCollins
34 Vidal, J (1997) *McLibel: Burger Culture on Trial*, London: Pan
35 Lang, T (2001) 'Trade, public health and food' in: McKee, M, Garner, P and Stott, R (eds), *International Co-operation in Health*, Oxford: Oxford University Press

36 Gardner, G and Halweil, B (2000) 'Underfed and Overfed: The Global Epidemic of Malnutrition', *Worldwatch paper* 150, Washington DC: Worldwatch Institute
37 Barker, DJP (ed) (1992) *Fetal and Infant Origins of Adult Disease*, London: British Medical Journal
38 Barker, DJP (2001) 'Cutting Edge', *THES*, 1 June, p22
39 Commission on the Nutrition Challenges of the 21st Century (2000). 'Ending Malnutrition by 2020: An Agenda for Change in the Millennium. Final Report to the ACC/SCN', *Food and Nutrition Bulletin*, 21, 3, Supplement, September. New York: United Nations University Press, p19
40 Pinstrup-Anderson, P (2001) *Achieving Sustainable Food Security for All: required policy action*. Washington DC: International Food Policy Research Institute
41 See the Projections in FAO (2000) *The State of Food Insecurity in the World*, Rome: Food and Agriculture Organization ppv, 6
42 Smil, V (2000) *Feeding the World: a challenge for the 21st century*, Cambridge MA: MIT Press
43 Dyson, T (1996) *Population and Food: Global Trends and Future Prospects*, London: Routledge
44 ACC/SCN (2000) *Nutrition through the Life Cycle: 4th Report on The World Nutrition Situation*. New York: United Nations Administrative Committee on Co-ordination Sub-Committee on Nutrition (ACC/SCN), p8
45 International Association for the Study of Obesity (2003) *Obesity Newsletter*, October, London: IASO
46 Royal College of Physicians of London (1983) 'Obesity. A report of the Royal College of Physicians', *Journal of the Royal College of Physicians of London*, 17, 1, pp5–65
47 Stearns, P (1997) *Fat History: Bodies and Beauty in the Modern West*, New York: New York University Press
48 WHO (1998) *Obesity: Preventing and Managing the Global Epidemic: Report of a WHO Consultation on Obesity*, Geneva: World Health Organization WHO/NUT/NCD/98.1, p1
49 Peña, M and Bacallao, J (eds) (2000) *Obesity and Poverty: a new public health challenge*, Washington DC: Pan American Health Organization (WHO)
50 WHO (2000) *Nutrition for Health and Development: A Global Agenda For Combating Malnutrition*. Geneva: World Health Organization, http://www.who.int/nut/db_bmi.htm
51 IOTF (2003) 'Call for international obesity review as overweight numbers reach 1.7 billion', press release, London: International Obesity Task Force/International Association for the Study of Obesity

52 Centre for Disease Control (2002), http://www.cdc.gov/nccdphp/dnpa/obesity/basics.htm
53 National Institutes of Health (1998) *Clinical Guidelines on the Identification, Evaluation, and Treatment of Overweight and Obesity in Adults*, Bethesda, Maryland: Department of Health and Human Services; National Institutes of Health; National Heart, Lung, and Blood Institute http://www.nhlbi.nih.gov/guidelines/obesity/ob_home.htm
54 OECD Health data, http://www.oecd.org/pdf/M00031000/M00031130.pdfpg. 5
55 Peña, M and Bacallao, J (eds) (2000) *Obesity and Poverty: A New Public Health Challenge*, Washington DC: PAHO Scientific Publication, No. 576
56 WHO (1998) *Obesity: Preventing and Managing the Global Epidemic: Report of a WHO Consultation on Obesity*, Geneva: World Health Organization WHO/NUT/NCD/98.1
57 Vepa, S et al (2002) *Food Insecurity Atlas of India*, Chennai: MS Swaminathan Research Foundation and UN World Food Programme
58 Verster, A (1996) 'Nutrition in transition: the case of the Eastern Mediterranean Region' in Pietinen, P, Nishida, C and Khaltaev, N (eds) *Nutrition and Quality of Life: Health Issues for the 21st century*, Geneva: World Health Organization pp57–65
59 Murata, M 'Secular trends in growth and changes in eating patterns of Japanese children', *Am J Clin Nutr* 2000; 72 (suppl): pp1379S–1383S.
60 Ebbeling, CB, Pawlak, DB and Ludwig, DS (2002) 'Childhood obesity: public health crisis, common sense cure', *The Lancet*, 360, 10 August, pp473–482
61 Hodge, AM, Dowse, GK, Gareeboo, H, Tuomilehto, J et al (1996) 'Incidence, increasing prevalence and predictors of change in obesity and fat distribution over 5 years in the rapidly developing population of Mauritius', *InternationalJournal of Obesity & Related Metabolic Disorders*, 20, 2, pp137–146
62 Chitson, P (1995) 'Integrated intervention programmes for combating diet-related chronic diseases' in Pietinen, P, Nishida, C and Khaltaev, N (eds) *Proceedings of the 2nd WHO Symposium on Health Issues for the 21st Century: Nutrition and Quality of Life, Kobe Japan, 24-26 November 1993*, Geneva: World Health Organization pp269–287
63 Kuczmarski, R et al (1994) 'Increasing Prevalence of Overweight Among US Adults', *JAMA*, 272, 3, pp205–211
64 CDC (2002) *Physical Activity and Good Nutrition: Essential Elements to Prevent Chronic Diseases and Obesity 2002*, Atlanta: Department of Health and Human Services, Centers for Disease Control and Prevention

65 Centre for Disease Control (2002), http://www.cdc.gov/nccdphp/dnpa/obesity/basics.htm

66 Barboza, D (2000) 'Rampant Obesity, A Debilitating Reality for the Urban Poor' *New York Times*, December 26, D5

67 Nestle, M (2002) *Food Politics*, Berkeley CA: University of California Press

68 Stunkard, AJ and Wadden, TA (eds) (1993) *Obesity: Theory and Therapy*, New York: Raven Press

69 Egger, G and Swinburn, B (1997) 'An "ecological" approach to the obesity pandemic', *British Medical Journal*, 315, pp477–480

70 Swinburn, B, Egger, G and Raza, F (1999) 'Dissecting Obesogenic Environments: The Development and Application of a Framework for Identifying and Prioritizing Environmental Interventions for Obesity', *Preventive Medicine*, 29, pp563–570

71 Swinburn, B and Egger, G (2002) 'Preventive strategies against weight gain and obesity', *Obesity Reviews*, 3, pp289–301

72 US Dept of Health and Human Services (1988) The Surgeon-General's Report on Nutrition and Health, Report 88–50210 Washington DC: DHHS. pp2–6

73 Murray, CJL and Lopez, AD (1997) 'Mortality by cause for eight regions of the world: Global Burden of Disease Study', *The Lancet*, 349, 3 May, pp1269–1276, 1347–1352, 1436–1442, 1498–1504. Murray, CJL and Lopez, AD (eds) (1996) *The Global Burden of Disease: A Comprehensive Assessment of Mortality and Disability from Diseases, Injuries and Risk Factors in 1990 and Projected to 2020*, Cambridge, MA: Harvard School of Public Health on behalf of the World Health Organization and the World Bank

74 National Institute of Public Health, Stockholm (1997) *Determinants of the burden of disease in the European Union*, Stockholm: NIPH

75 Mathers, E, Vos, T and Stevenson, C (1999) *The Burden of Disease and Injury in Australia*, Canberra: AIHW, 1999 (Catalogue No. PHE-17)

76 WHO and FAO (2003) *Diet, Nutrition and the Prevention of Chronic Diseases. Technical Report Series 916*, Geneva: World Health Organization and Rome: Food and Agriculture Organization

77 Commission on Macroeconomics and Health (2001) 'Macroeconomics and Health: Investing in Health for Economic Development', Geneva: World Health Organization

78 Cited in Kenkel, DS and Manning, W (1999) 'Economic Evaluation of Nutrition Policy Or There's No Such Thing As a Free Lunch', *Food Policy* 24, pp145–162, p148

79 Wanless, D (2002) *Securing Our Future Health: Taking a Long-Term View. Final Report*, London: H M Treasury, April

80 Maniadakis, N and Rayner, M (1998) *Coronary Heart Disease Statistics: Economics Supplement*, London: British Heart Foundation, http://www.heartstats.org

81 Wanless, D (2002) *Securing Our Future Health: Taking a Long-Term View. Final Report*, London: H M Treasury, April
82 Wanless, D (2004) *Securing Good Health for the Whole Population*, Final Report, London: HM Treasury, 25 February
83 Heart Protection Study Collaborative Group (2002) MRC/BHF Heart Protection Study of cholesterol lowering with simvastatin in 20, 536 high-risk individuals: a randomised placebo-controlled trial. *The Lancet*. 360, pp7–22
84 WHO (1999) *World Health Report 1999*, Geneva: World Health Organization
85 WHO (2002) *World Health Report 2002*, Geneva: World Health Organization
86 WHO (2003) *Diet, Nutrition and the Prevention of Chronic Diseases*, Technical Series 916, Geneva: World Health Organization
87 Dobson, AJ, Evans, A, Ferrario, M, Kuulasmaa, KA et al (1998) 'Changes in estimated coronary risk in the 1980s: data from 38 populations in the WHO MONICA Project. World Health Organization. Monitoring trends and determinants in cardiovascular diseases.' *Annals of Medicine*, 30, 2, pp199–205
88 Rayner, M (2000) 'Impact of Nutrition on Health', in Sussex, J (ed) *Improving Population Health in Industrialised Nations*, London: Office of Health Economics pp24–40
89 Trowell, H (1981) 'Hypertension, obesity, diabetes mellitus and coronary heart disease', in Trowell, H, Burkitt, D (eds) *Western Diseases: their emergence and prevention*, London: Edward Arnold
90 Zhou, B (1998) 'Diet and Cardiovascular disease in China in Diet', in Shetty, P and Gopalan, C (eds) *Nutrition and Chronic Disease: an Asian perspective*, London: Smith-Gordon
91 Doll, R and Peto, R (1981) 'The causes of cancer: quantitative estimates of avoidable risks of cancer in the United States today', *J National Cancer Institute*, 66, pp1191–1308
92 Miller, AB (2001) *Diet in cancer prevention*, Geneva: World Health Organization http://www.who.int/ncd/
93 WHO (1999) *World Health Report 1999*, Geneva: World Health Organization
94 Zheng, W, Sellers, TA, Doyle, TJ, Kushi, LH, Potter, JD and Folsom, AR (1995) 'Retinol, antioxidant vitamins, cancers of the upper digestive tract in a prospective cohort study of postmenopausal women', *American Journal of Epidemiology*, 142, pp955–960
95 World Cancer Research Fund/American Institute for Cancer Research (1997) *Food, nutrition and the prevention of cancer: a global perspective*, Washington DC: AICR
96 WHO (2003) *World Cancer Report*, Geneva: World Health Organization/International Agency for Research on Cancer pp62–67

97 World Cancer Research Fund/American Institute for Cancer Research (1997) *Food, nutrition and the prevention of cancer: a global perspective*, Washington DC: AICR
98 WHO (2002) *Diabetes Mellitus Factsheet 138*, April update. Geneva: World Health Organization
99 WHO and FAO (2003) *Diet, Nutrition and the Prevention of Chronic Diseases, Technical Report Series 916*, Geneva: World Health Organization and Rome: Food and Agriculture Organization, p72
100 International Diabetes Federation (2000) *Diabetes Atlas 2000*, Brussels: International Diabetes Federation
101 International Diabetes Federation. http://www.idf.org
102 Yajnik, CS (1998) 'Diabetes in Indians: small at birth or big as adults or both?', in Shetty, P and Gopalan, C (eds) *Nutrition and Chronic Disease: an Asian perspective*, London: Smith-Gordon
103 Ramachandran, A (1998) 'Epidemiology of non-insulin-dependent diabetes mellitus in India' in Shetty, P and Gopalan, C (eds) *Diet, Nutrition and Chronic Disease: an Asian perspective*, London: Smith-Gordon
104 Vannasaeng, S (1998) 'Current status and measures of control for diabetes mellitus in Thailand', in Shetty, P and Gopalan, C (eds) *Diet, Nutrition and Chronic Disease: an Asian perspective*, London: Smith-Gordon
105 Barker, DJP (ed) (1992) *Fetal and Infant Origins of Adult Disease*, London: British Medical Journal
106 Yajnik, CS (1998) 'Diabetes in Indians: small at birth or big as adults or both?', in Shetty, P and Gopalan, C (eds) *Nutrition and Chronic Disease: an Asian perspective*, London: Smith-Gordon
107 Buzby, J (2001) 'Effects of food safety perceptions on and food demand and global trade' in: Regmi, A (ed) *Changing structure of global food consumption and trade*, Washington DC: US Department of Agriculture, Agriculture and Trade Report WRS-01-1
108 Nestle, M (2003) *Safe Food: bacteria, biotechnology and bioterrorism*, Berkeley CA: University of California Press
109 WHO (2002) *World Health Report 2002*, Geneva: World Health Organization
110 Barling, D and Lang, T (2003) *Codex, the European Union and Developing Countries: an analysis of developments in international food standards setting*. Report to the Department for International Development, London: City University Institute of Health Sciences
111 WHO (2002) *Food safety and foodborne illness*, WHO Information Fact Sheet 237, Geneva: World Health Organization, January
112 WHO (2002) *Food safety – a worldwide public health issue*, Geneva: WHO, http://www.who.int/fsf/fctshtfs.htm
113 Buzby, J (2001) 'Effects of food safety perceptions on food demand and global trade', in: Regmi, A (ed) *Changing structure of global food*

consumption and trade, Washington DC: US Department of Agriculture, Agriculture and Trade Report WRS-01-1

114 Brundtland, GH (2001) Speech to 24th Session of Codex Alimentarius Commission, Geneva 2 July, Geneva: WHO

115 WHO (2002) *Food safety and foodborne illness*, WHO Information Fact Sheet 237, Geneva: World Health Organization, January

116 WHO (2002) *Food Safety*: Agenda item 12.3, 53rd World Health Assembly, WHO 53.15, 20 May

117 Brundtland, GH (2001) Speech to 24th Session of Codex Alimentarius Commission, Geneva 2 July, Geneva: WHO

118 Phillips, Lord, Bridgeman, J and Ferguson-Smith, M (2000) *The BSE Inquiry: Report: evidence and supporting papers of the Inquiry into the emergence and identification of Bovine Spongiform Encephalopathy (BSE) and variant Creutzfeldt-Jakob Disease (vCJD) and the action taken in response to it up to 20 March 1996*, 16 vols, London: The Stationery Office

119 WHO (2002) *Food safety – a worldwide public health issue*, Geneva: WHO http://www.who.int/fsf/fctshtfs.htm

120 DG SANCO (2000) *Trends and sources of zoonotic agents in animals, feedstuffs, food and man in the European Union in 1998*, Brussels: European Commision Part 1. Prepared by the Community Reference Laboratory on the Epidemiology of Zoonoses, BgVV, Berlin, Germany

121 Rayner, M and Peterson, S (2000) *European Cardiovascular Disease Statistics 2000*, Oxford: BHF Health Promotion Research Group, University of Oxford

122 Data from Murray and Lopez, National Institute of Public Health (Sweden) and WHO, compiled in Rayner, M and Peterson, S (2000) *European Cardiovascular Disease Statistics 2000*, Oxford: BHF Health Promotion Research Group University of Oxford

123 WHO (2002) *Food safety and foodborne illness*, WHO Information Fact Sheet 237, Geneva: World Health Organization, January

124 Scholte, JA (2000) *Globalization: a critical introduction*, London: HarperCollins

125 Vepa, S et al (2002) *Food Insecurity Atlas of India*, Chennai: MS Swaminathan Research Foundation and UN World Food Programme

126 World Bank (1999) *World Development Report 1999–2000*, New York: World Bank

127 Beaglehole, R and Bonita, R (1998) 'Public health at the crossroads: which way forward?', *The Lancet*, 351, 21 February, pp590–592

128 Howson, C, Fineberg, H and Bloom, B (1998) 'The pursuit of global health: the relevance of engagement for developed countries', *The Lancet*, 351, 21 February, pp586–590

129 United Nations Development Programme, *Human Development Report 1999*, New York: Oxford University Press and UNDP, 1999, p2.
130 Navarro, V (1998) 'Whose Globalization?' *American Journal of Public Health*, 88, 5, pp742–743
131 Barrientos, S, McClenaghan, S and Orton, L (1999) *Gender and Codes of Conduct: a case study from horticulture in South Africa*, London: Christian Aid
132 UNDP (2003) *Human Development Report 2003*, New York: UN Development Programme/Oxford University Press
133 UN (2000) Millennium Development Goals, New York: United Nations. http://www.developmentgoals.org/
134 Townsend, P and Gordon, D (eds) (2002) *World Poverty: new policies to defeat an old enemy*, Bristol: Policy Press
135 UNICEF (1998) *State of the World's Children 1998*, New York: UN Children's Fund/Oxford University Press
136 UNICEF (2002) *State of the World's Children 2002*, New York: UN Children's Fund/Oxford University Press
137 UNCEF (2002) *State of the World's Children*, New York: UN Children's Fund/Oxford University Press
138 This section draws on unpublished work by Dr Liz Dowler of Warwick University and Tim Lang
139 World Food Summit (1996) Rome Declaration, http://www.fao.org/wfs/index_en.htm
140 Maxwell, S and Smith, M (1995) 'Household Food Security: a Conceptual Review', in Maxwell, S and Frankenberger, T (eds) *Household Food Security: Concepts, Indicators, Measurements: A Technical Review*, New York and Rome: UNICEF and IFAD, p4
141 Maxwell, S (1988) 'National food security planning: first thoughts from Sudan', Paper presented to Workshop on Food Security in the Sudan, 3–5 October, Falmer: University of Sussex Institute of Development Studies.
142 Quoted in Killeen, D (2000) 'Food Security: A Challenge for Human Development. The Food We Eat', Seymour, J (ed) *Poverty in Plenty: a Human Development Report for the UK*, Earthscan: London
143 Riches, G (ed) (1997) *First World Hunger: Food Security and Welfare Politics*, Basingstoke: Macmillan Press
144 Boyd Orr, J (1966) *As I Recall The 1880s to the 1960s*, London: Macgibbon and Kee
145 Boyd Orr, J (1943) *Food and the People, Target for Tomorrow No 3*, London: Pilot Press
146 Lang, T (1999) 'Food and nutrition: the relationship between nutrition and public health' in Weil, O, McKee, M, Brodin, M and Oberlé, D (eds) *Priorities for Public Health Action in the European*

Union, Paris, Vandoeuvre-Les-Nancy & London: Société Française de Santé Publique

147 Acheson, D (1999) *Independent Inquiry into Inequalities in Health: Report*, London: The Stationery Office

148 James, WPT, Nelson, M, Ralph, A and Leather, S (1997) 'Socioeconomic determinants of health: The contribution of nutrition to inequalities in health', *British Medical Journal*, 314, 7093, pp1545–1549

149 Leather, S (1996) *The Making of Modern Malnutrition*, London: Caroline Walker Trust

150 LIPT (1996) *Low income, food, nutrition and health: strategies for improvement*. A report by the Low Income Project Team for the Nutrition Task Force, London: Department of Health

151 Eisinger, PK (1998) *Towards an End to Hunger in America*, Washington DC: Brookings Institute Press

152 Eisinger, PK (1998) *Towards an End to Hunger in America*, Washington DC: Brookings Institute Press

Chapter 3

1 Mintz, S (1996) *Tasting Food, Tasting Freedom*, Beacon Press, Boston MA

2 Cannon, G (1992) *Food and Health: The Experts Agree*, London: Consumers' Association

3 Hobsbawn, E (1988) *The Age of Capital*, London: Abacus

4 Hamlyn, C (1998) *Public Health and Social Justice in the Age of Chadwick: Britain 1800/1854*, Cambridge: Cambridge University Press

5 Finer, SE (1952) *The life and times of Sir Edwin Chadwick*, London: Methuen

6 Friedman, M and Friedman, R (1980) *Free to choose*, London: Secker and Warburg

7 Cockett, R (1995) *Thinking the Unthinkable: think-tanks and the economic counter-revolution, 1931–1983*, London: Fontana

8 Gabriel, Y and Lang, T (1995) *The Unmanageable Consumer*, London: Sage

9 Acheson, D (1988) *Public Health in England*, London: The Stationery Office.

10 Baum, F (1998) *The New Public Health: An Australian Perspective*, Melbourne: Oxford University Press

11 Ashton, J and Seymour, J (eds) (1988) *The New Public Health*, Milton Keynes: Open University Press

12 Draper, P (ed) (1991) *Health through public policy: the greening of public health*, London: Green Print
13 Lang, T, Barling, D and Caraher, M (2001) 'Food, social policy and the environment: towards a new model', *Social Policy and Administration*, 35, 5, 538–558
14 United Nations Environment Programme (2002) *Global Environment Outlook 3*, London: Earthscan
15 McMurray, C and Smith, R (2001) *Diseases of Globalization: socio-economic transitions and health*, London: Earthscan
16 Nutbeam, D (1998) *Health Promotion Glossary*, Geneva: World Health Organization
17 Nutbeam, D (1998) *Health Promotion Glossary*, Geneva: World Health Organization. WHO/HEP/98.1, p3
18 Ashton, J and Seymour, H (1988) *The New Public Health*, Buckingham: Open University Press
19 Lang, T and Heasman, H (2004) 'Diet and Nutrition Policy: A clash of ideas or an investment?', *Development*, 47, 2, pp64–74
20 Drummond, JC and Wilbraham, A (1958) *The Englishman's Food*, London: Jonathan Cape
21 Mann, J and Trusswell, S (eds) (2002) *Essentials of Human Nutrition*, Oxford: Oxford University Press
22 Atwater, WO (nd) *Investigations on the Chemistry and Economy of Food*, US Dept Agriculture, Bulletin 21
23 Atwater, WO (nd) *Foods, Nutritive Value and Cost*, US Dept Agriculture, Bulletin 23
24 Rowntree, BS (1902) *Poverty: a study of Town Life*, London: Macmillan
25 Rowntree, BS (1941) *Poverty and Progress: A Second Social Survey of York*, London: Longmans, Green & Co
26 Rowntree, BS (1913) *How the Labourer Lives*, London: Thomas Nelson & Sons
27 Rowntree, BS (1937) *The Human Needs of Labour*, London: Longman
28 Wood, TB and Gowland Hopkins, F (1915) *Food Economy in War Time*, Cambridge: Cambridge University Press
29 Quoted in Rowntree B S (1937) *The Human Needs of Labour*, London: Longman
30 Burnett, J (1979) *Plenty and Want: a social history of diet in England from 1815 to the present day*, London: Scolar Press
31 Drummond, JC and Wilbraham, A (1958) *The Englishman's Food*, London: Jonathan Cape p404ff
32 McCarrison, R (1936) The Cantor Lectures. Reproduced in McCarrison, R and Sinclair, H (1953) *Nutrition and Health*, London: Faber & Faber
33 Boyd Orr, J (1936) *Food, Health and Income: a report on A Survey of Adequacy of Diet in Relation to Income*, London: Macmillan

34 Boon, T (1997) 'Agreement and disagreement in the making of "World of Plenty"', in Smith, DF (ed) *Nutrition in Britain: Science, Scientists and Politics in the Twentieth Century*, London: Routledge
35 Keys, A (1952) 'Human atherosclerosis', *Circulation*, 5, pp115–118
36 Keys, A (ed) (1970) 'Coronary heart disease in seven countries', *Circulation*, 41 (suppl. 1), pp1–211
37 Nestle, M (1995) 'Mediterranean diets: historical and research overview', *American Journal of Clinical Nutrition*, 61, pp1313s–1320s
38 Ewin, J (2001) *Fine Wines and Fish Oils: the life of Hugh Macdonald Sinclair*, Oxford: Oxford University Press
39 Sinclair, H (1963) Preface to McCarrison R, Sinclair H (1953) *Nutrition and Health*, London: Faber & Faber, 3rd edition
40 Trowell, H and Burkitt, D (eds) (1981) *Western Diseases: their emergence and prevention*, London: Edward Arnold
41 Cannon, G (1992) *Food and health: the experts agree: an analysis of one hundred authoritative scientific reports on food, nutrition and public health published throughout the world in thirty years, between 1961 and 1991*, London: Consumers' Association
42 Lennon, D and Fieldhouse, P (1982) *Social Nutrition*, London: Forbes Publications
43 World Health Organization (1946) *Constitution of the WHO*, Reprinted in WHO Basic Documents, 37th edition, Geneva: WHO
44 Puska, P, Tuomilehto, J, Nissinen, A and Vartiainen, E (eds) (1995) *The North Karelia project: 20 year results and experiences*, Helsinki: National Public Health Institute of Finland
45 Commission on the Nutrition Challenges of the 21st Century (2001) 'Ending Malnutrition by 2020: An Agenda for Change in the Millennium', *Food and Nutrition Bulletin (United Nations)*, 21, 3, September pp14–16
46 Puska, P, Nissinen, A, Tuomilehto, J, Salonen, JT, Koskela, J et al (1996) 'The Community-based strategy to prevent coronary heart disease: conclusions from the 10 years of the North Karelia project' in Pan American Health Organization, *Health Promotion: an anthology*, Scientific Publication 557. Washington DC: World Health Organization, pp89–125
47 Rose, G (1992) *The Strategy of Preventive Medicine*, Oxford: Oxford University Press
48 Rose, G (1992) *The Strategy of Preventive Medicine*, Oxford: Oxford University Press
49 Wilkinson, RG (1996) *Unhealthy Societies: The Afflictions of Inequality*, London: Routledge
50 Marmot, M and Wilkinson, RG (eds) (1998) *The Social Determinants of Health*, Oxford: Oxford University Press
51 Rose, G (1985) 'Sick individuals and sick populations', *International Journal of Epidemiology*, 14, 1, pp32–38

52 Baum, R and Sanders, D (1995) 'Can health promotion and primary health care achieve health for All without a return to their more radical agenda?', *Health Promotion International*, 10, 2, pp149–160
53 Nestle, M (2002) *Food Politics: How the Food Industry Influences Nutrition and Health*, Berkeley CA: University of California Press
54 Cannon, G (1987) *The Politics of Food*, London: Century
55 WHO and FAO (2003) *Diet, Nutrition and the Prevention of Chronic Diseases. Technical Report Series 916*, Geneva: World Health Organization and Rome: Food and Agriculture Organization
56 WHO (2002) *World Health Report 2002*, Geneva: World Health Organization
57 Milio, N and Helsing E (eds) (1998) *European Food and Nutrition Policies in Action*, Copenhagen: World Health Organization WHO European Series 73
58 Truswell, AS (1987) 'Evolution of dietary recommendation, goals and guidelines', *American Journal of Clinical Nutrition*, 45, pp1060–1072
59 Egger, G and Swinburn, B (1997) An 'ecological' approach to the obesity pandemic. *British Medical Journal*, 315, pp477–480
60 Ministry of Health and Welfare (1999) *Dietary Guidelines for Adults in Greece*, Athens: Hellenic Ministry of Health, Supreme Scientific Health Council. http://www.nut.uoa.gr
61 Truswell, A (1987) 'Evolution of dietary recommendations, goals and guidelines', *American Journal of Clinical Nutrition*, 45, pp1060–1072
62 Truswell, A (1987) 'Evolution of dietary recommendations, goals and guidelines', *American Journal of Clinical Nutrition*, 45, pp1060–1072
63 Sims, L (1998) *The Politics of Fat: food and nutrition policy in America*, New York: M E Sharpe Inc.
64 Nestle, M (1999) 'Animal v. plant foods in human diets and health: is the historical record unequivocal?', *Proceedings of the Nutrition Society*, 58, pp211–218
65 Nestle, M (1993) 'Food lobbies, the food pyramid and US nutrition policy', *International Journal of Health Services*, 23, 3, pp483–496
66 Cannon, G (1987) *The Politics of Food*, London: Century
67 Walker, C and Cannon, G (1984) *The Food Scandal*, London: Ebury
68 Cannon, G (1987) *The Politics of Food*, London: Century
69 Norum, K (1997) 'Some aspects of Norwegian nutrition and food policy', in Shetty, P and McPherson, K (eds) *Diet, Nutrition and Chronic Disease: Lessons from contrasting worlds*, Chichester: J Wiley and Sons, p198
70 Scrimshaw, N (1990) 'Nutrition: prospects for the 1990s', *Annual Review of Public Health*, 11, pp53–68
71 Cannon, G (1993) 'The new public health', *British Food Journal*, 95, 5, pp4–11

72 Ludwig, DS (2002) 'The Glycemic Index: physiological mechanisms relating to obesity, diabetes, and cardiovascular disease', *Journal of the American Medical Association*, 287, 18, pp2414–2423
73 Nestle, M (2002) *The Politics of Food*, Berkley CA: University of California Press
74 Eurodiet, (2001) 'The Eurodiet Reports and Proceedings', *Public Health Nutrition*, Special Issue
75 WHO (2004)
76 Robertson, A, Tirado, C, Lobstein, T, Jermini, M, Khai, C, Jensen, JH, Ferro-Luzzi, A and James, WPT (2004) *Food and Health in Europe: A New Basis for Action*, WHO Regional Publications, European Series, No. 96, Copenhagen: WHO
77 Kickbusch, I (1989) 'Approaches to an ecological base for public health', *Health Promotion*, 4, 4, pp265–268
78 WHO (1990) *Diet, Nutrition and Chronic Disease, Technical Series 797*, Geneva: World Health Organization
79 WHO and FAO (2003) *Diet, Nutrition and Chronic Disease, Technical Series 916*, Geneva: World Health Organization and Food and Agriculture Organization
80 WHO (2003) Draft Global Strategy on Diet, Physical Activity and Health, Geneva, World Health Organization, 4 December, http://www.who.int/hpr/gs.strategy.document.shtml
81 WHO and FAO (2003) *Diet, Nutrition and Chronic Disease, Technical Series 916*, Geneva: World Health Organization and Food and Agriculture Organization
82 Jacobson, M and Nestle, M (2000) 'Halting the obesity epidemic: a public health approach', *Public Health Reports*, 115, Jan/Feb, pp12–24, http://www.cspinet.org/reports/obesity.pdf
83 Donaldson, L (2003) *Annual Report of the Chief Medical Officer for 2002*, London: Department of Health
84 Health Select Committee Inquiry into Obesity, Report published in May 2004; see Parliamentary website, http://www.parliament.uk/parliamentary_committees/health_committee.cfm
85 SEF (2003) 'Strategy March 2001', http://www.sef.no/assets/11000138/Strategy.pdf, accessed 25 February 2004
86 National Council on Nutrition and Physical Activity, Oslo: Norway, http://www.sef.no/
87 http://www.slv.se accessed 25 February 2004
88 Lang, T, Rayner, M, Rayner, G, Barling, D and Millstone, E (2004) *A New Policy Council on Food, Nutrition and Physical Activity for the UK?* London: City University Department of Health Management and Food Policy
89 US Surgeon General (2000) *Call to Action to Prevent and Decrease Overweight and Obesity*, Washington, DC: Office of the Surgeon General

90 Critzer, G (2003) *Fatland*, London: Penguin
91 Sims, L (1998) *The Politics of Fat: food and nutrition policy in America*, New York: M E Sharpe Inc
92 Daynard, R (2003) 'Lessons from tobacco control for the obesity control movement, *Journal of Public Health Policy*, 24, 3 & 4, pp291–295
93 Hulse, C (2004) 'Vote in House Offers a Shield for Restaurants in Obesity Suits', *New York Times*, 11 March 2004, http://www.nytimes.com/todaysheadlines, accessed 11 March 2004
94 Green, H (2003) 'Nestlé, nutrition and health', Presentation by Nestlé Research Centre to JP Morgan in London, 19 September 2003, Lausanne: Nestlé
95 International Sugar Organization (2004) Memo on 'Joint WHO/Technical Report 916', Memo (04) 05 17 February, London: ISO
96 Alden, E, Buckley, N and Mason, J (2004) 'Sweet deals: "big sugar" fights threats from free trade and a global drive to limit consumption', *Financial Times*, February

CHAPTER 4

1 McMichael, AJ (2001) *Human Frontiers, Environment and Disease*, Cambridge: Cambridge University Press
2 Kunast, R (2001) speech reported in *The Ecologist*, vol 31, no 3 pp48–49
3 Policy Commission on Farming and Food (2002) *Farming and Food – A Sustainable Future*, London: Cabinet Office
4 European Commission (2003) *Mid Term Review of the Common Agricultural Policy July 2002 Proposals: Impact Analyses*, Brussels: DG Agriculture, http:europa.eu.int/comm/agriculture/publi/reports/mtrimpact/rep_en.pdf
5 Heasman, M and Mellentin, J (2001) *The Functional Foods Revolution: Healthy People, Healthy Profits?* London: Earthscan
6 Jacobson, M and Silverglade, B (1999) 'Functional foods: health boon or quackery?', *British Medical Journal*, vol 319, pp205–206
7 Conway, G (1999) *The Doubly Green Revolution*, Harmondsworth: Penguin
8 Emerson, T (2001) 'Where's the beef?' *Newsweek*, Special Report, 26 February, pp16–21
9 Sancton, T (2001) 'Life without beef', *Time*, 26 February, pp22–27
10 Buzby, J (2001) 'Effects of food safety perceptions on food demand and global trade' in: Regmi, A (ed) *Changing structure of global food consumption and trade*, Washington DC: US Department of Agriculture. Agriculture and Trade Report WRS-01-1

11 Worldwatch Institute data: http://www.earth-policy.org
12 Worldwatch Institute data: http://www.earth-policy.org
13 Dyson, T (1996) *Population and Food: global trends and future prospects*, London: Routledge
14 Mannion, AM (1998) 'Future trends in agriculture: the role of biotechnology', *Outlook on Agriculture*, 27, 4, pp219–224
15 Mermelstein, NH (2002) 'A look into the future of food science and technology', *Food Technology*, 56, 1, pp46–55
16 Kurlansky, M (1997) *Cod: A biography of the fish that changed the world*, London: Jonathan Cape
17 Martin, MA (2001) 'The future of the world food system', *Outlook on Agriculture*, 30, 1, pp11–19
18 Delgado, C, Rosegrant, M, Steinfeld, H, Ehui, S and Courbois, C (1999) *Livestock to 2020: the next food revolution*, Food, Agriculture, and the Environment Discussion Paper 28. Washington DC: International Food Policy Research Institute.
19 Delgado, C, Rosegrant, M, Steinfeld, H, Ehui, S and Courbois, C (2001) 'Livestock 2020: the next food revolution', *Outlook on Agriculture*, 30, 1, pp27–29
20 Tansey, G and Worsley, T (1995) *The Food System*, London: Earthscan
21 Atkins, P and Bowler, I (2001) *Food in Society*, London: Hodder Headline
22 Goodman, D and Watts, MJ (1994) 'Reconfiguring the rural or fording the divide? Capitalist restructuring and the global agro-food system', *Journal of Peasant Studies*, 22, 1, pp1–49
23 Goodman, D, Sorj, B and Wilkinson, J (1987) *From Farming to Biotechnology*, Oxford: Blackwell
24 Jones, A (2001) *Eating Oil*, London: Sustain
25 Seth, A and Randall, G (1999) *The Grocers*, London: Kogan Page
26 Gabriel, Y (1988) *Working Lives in Catering*, London: Routledge and Kegan Paul
27 Thrupp, L-A (1995) *Bittersweet Harvests for Global Supermarkets*, Washington DC: World Resources Institute
28 Friedland, WH, Barton, AE and Thomas, RJ (1981) *Manufacturing Green Gold: Capital, Labour and Technology in the Lettuce Industry*, Cambridge: Cambridge University Press
29 Sims, L (1998) *The Politics of Fat: food and nutrition policy in America*, New York: M E Sharpe Inc.
30 Personal communication from Innova Market Insights, CNS Media, The Netherlands, http://www.win-food.com
31 Nestle, M (1999) 'Commentary', *Food Policy*, 24, p308
32 Goodman, D, Sorj, B and Wilkinson, J (1987) *From Farming to Biotechnology*, Oxford: Blackwell
33 Friedland, W, Busch, L, Buttel, F and Rudi, YA (eds) (1991) *Towards a new political economy of Agriculture*, Boulder Colorado: Westview

34 Pritchard, B and Burch, D (2003) *Agri-Food Globalization in Perspective: International Restructuring in the Processing Tomato Industry*, Aldershot: Ashgate
35 Buttel, F (2001) 'Some reflections on late twentieth century agrarian economy', *Sociologia Ruralis*, 41, 2, pp165–181
36 Fine, B, Heasman, M and Wright, J (1996) *Consumption in the Age of Affluence: world of food*, London: Routledge
37 Heffernan, W, Hendrickson, M and Gronski, R (1999) *Consolidation in the Food and Agriculture System: Report to the National Farmers Union*, University of Missouri: Department of Rural Sociology
38 Hendrickson, M, Heffernan, WD, Howard, PH and Heffernan, JB (2001) *Consolidation in Food Retailing and Dairy: Implications for Farmers and Consumers in a Global System: Report to National Farmers Union (USA)*, Columbia, Missouri: Dept. Rural Sociology University of Missouri
39 Heffernan, W, Hendrickson, M and Gronski, R (1999) *Consolidation in the Food and Agriculture System: Report to the National Farmers Union*, University of Missouri: Department of Rural Sociology
40 Hendrickson, M, Heffernan, WD, Howard, PH and Heffernan, JB (2001) *Consolidation in Food Retailing and Dairy: Implications for Farmers and Consumers in a Global System: Report to National Farmers Union (USA)*, Department of Rural Sociology University of Missouri, Columbia, Missouri
41 Heffernan, W, Hendrickson, M and Gronski, R (1999) *Consolidation in the Food and Agriculture System: Report to the National Farmers Union*, University of Missouri: Department of Rural Sociology p.13
42 McMichael, P (2000) 'The power of food', *Agriculture and Human Values*, 17, pp21–33
43 Lawrence, F (2004) *Not on the Label*, London: Penguin
44 Hendrickson, M, Heffernan, WD, Howard, PH and Heffernan, JB (2001) *Consolidation in Food Retailing and Dairy: Implications for Farmers and Consumers in a Global System: Report to National Farmers Union (USA)*, Columbia, Missouri: Dept Rural Sociology University of Missouri
45 Reuters (2003) 'Green Light for Global Study on Food Security', 4 August, http://www.planetark.org
46 Hendrickson, M, Heffernan, WD, Howard, PH and Heffernan, JB (2001) *Consolidation in Food Retailing and Dairy: Implications for Farmers and Consumers in a Global System: Report to National Farmers Union (USA)*, Columbia, Missouri: Dept Rural Sociology University of Missouri
47 Goodman, D and Redclift, M (1991) *Refashioning Nature: Food, Ecology and Culture*, London: Routledge
48 Magdoff, F, Buttel, FH and Foster, JB (eds) (1998) 'Hungry for Profit', *Monthly Review*, 50, 3, pp1–160

49 National Farmers' Union (2002) press release, 9 July 2002, London: National Farmers' Union
50 Lilliston, B and Ritchie, N (2000) 'Freedom to Fail', *Multinational Monitor*, July/August, pp9–12
51 Halweil, B (2000) 'Where have all the farmers gone?' *World-Watch* Sept/Oct vol 13, no 5, pp12–28
52 Watkins, K (2002) *Rigged Rules and Double Standards*, Oxford: Oxfam International
53 Friends of the Earth International (1999) *The world trade system: winners and losers*, London: FOE-I
54 *Agrow World Crop Protection News* (2002) 'Gap narrows between prospective agrochemical market leaders', 397, 1; and *Agrow*, passim
55 National Farmers' Union (2000) *The Farm Crisis, EU Subsidies, and Agribusiness Market Power*, Report presented to the Senate Standing Committee on Agriculture and Forestry, Ottawa, Ontario, 17 February
56 National Farmers' Union (2002) 'Free Trade: Is it working for farmers?', Saskatoon, Sask.: National Farmers' Union of Canada, http://www.nfu.ca
57 OECD (1999) *Agricultural Policies in OECD countries: Monitoring and Evaluation 1999*, Paris: OECD
58 Banana Link (2003) *The Banana 'Split'*. http://www.bananalink.org.uk/
59 Banana Link (2003) *Race to the Bottom*. http://www.bananalink.org.uk/
60 Wilson, N (1996) 'Supply Chain Management: A Case Study of a Dedicated Supply Chain for Bananas in the UK Grocery Market', *Supply Chain Management*, vol 1, no 1 pp39–46
61 Robbins, P (2003) *Stolen Fruit: the tropical commodities disaster*, London: Zed Press
62 Barrett, H, Ilbery, BW, Browne, AW and Binns, T (1999) Globalization and the changing networks of food supply: the importation of fresh horticultural produce from Kenya into the UK, *Trans Inst Br Geogr* NS 24 pp159–174
63 Data from Chen, Y (2001) *Analysis of the Competitiveness of China's Onion and Broccoli Exports to Japan*, Ehime University, Japan
64 Leatherhead Food RA (2001) *Food News*, vol 35, no 11, November, p2
65 Deboo, M (2002) 'Winners and Losers', *The Grocer*, 11 May, pp27–30
66 Data from CIAA, Brussels. http://www.ciaa.be
67 OECD (1981) *Food Policy*, Paris: Organisation for Economic Co-operation and Development
68 Longfield, J (1992) 'Information and advertising' in National Consumer Council, *Your Food, Whose Choice?* London: The Stationery Office
69 Source: http://www.productscan.com
70 From company reports

71 Cocaine, caffeine and alcohol, see http://www.lewrockwell.com/jarvis/jarvis17.html and http://www.foodreference.com/html/fcocacola.html
72 Teather, D (1999) 'Ice-cold Coke reacts calmly to crisis', *The Guardian*, 25 June
73 Martinson, J (2001) 'Why Coke's no longer the real thing', *The Guardian*, 14 August
74 Speeches and other Coca-Cola facts accessed from company website: http://www.cocoa-cola.com
75 Unilever (2000) 'Unilever plans faster growth', press release, 22 February, London: Unilever
76 *Fortune Magazine*, http://www.fortune.com, accessed 25 February 2003
77 Feeney, J (consumer products analyst at investment firm SunTrust Robinson Humphrey) quoted in Hopkins, J (2003) 'US economy follows the Wal-Mart way', *USA Today*, 3 February
78 IGD (2002) *Global Retailing*, Institute of Grocery Retailing, Letchmore Health, p113
79 IGD (2001) 'European Grocery Retailing', press release, 26 February, Institute of Grocery Distribution, Letchmore Heath
80 Vorley, W (2003) *Food, Inc: Corporate Concentration from Farm to Consumer*, London: UK Food Group, October, http://www.ukfg.org.uk
81 Grievink, J.-W. (2003) 'The Changing Face of the Global Food Supply Chain', paper presented at OECD Conference on Changing Dimensions of the Food Economy, The Hague, 6–7 February, http://webdomino1.oecd.org/comnet/agr/foodeco.nsf
82 Dobson, P, Clarke, R, Davies, S and Waterson, M (2001) 'Buyer Power and its impact on competition in the food retail distribution sector of the EU', *Journal of Industry, Competition and Trade*, 1, 3, pp247–281
83 Reardon, T, Timmer, P, Barrett, C and Berdegue, J (2003) 'The rise of supermarkets in Africa, Asia, and Latin America', *American Journal of Agricultural Economics*, 85, 5
84 Reardon, T and Berdegue, JA (2002) 'The rapid rise of supermarkets in Latin America: challenges and opportunities for development', *Development Policy Review*, 20, 4, 317–334
85 We are grateful to Professor Dinghua Hu of Kyoto University for this information
86 Reardon, T, Timmer, P, Barrett, C and Berdegue, J (2003) 'The rise of supermarkets in Africa, Asia, and Latin America', *American Journal of Agricultural Economics*, 85, 5
87 For further thoughts, particularly on the role of processing capital in this restructuring, see Wilkinson, J (2002) 'The Final Foods Industry and the Changing Face of the Global Agrofood System', *Sociologia Ruralis*, 42, pp329–347

88 Stanton, J (1999) 'Rethinking retailers' fees', *Food Processing*, 60, 8, pp32–34
89 Quoted in Hollinger, P (1999) 'Carrefour's revolutionary', *Financial Times*, 4 December
90 See Gabriel, Y and Lang, T (1999) *The Unmanageable Consumer*, London: Sage
91 Competition Commission (2000) *Report of the Competition Commission on the supply of Groceries from multiple stores in the United Kingdom*, London: Competition Commission (Cm 4842)
92 Competition Commission (2000) *Report of the Competition Commission on the supply of Groceries from multiple stores in the United Kingdom*, London: Competition Commission (Cm 4842)
93 IGD (2001) *Grocery Retailing 2001*, Institute of Grocery Distribution, Letchmore Heath, p43
94 Competition Commission (2003) *Safeway plc and Asda Group Limited (owned by Wal-Mart Stores Inc); Wm Morrison Supermarkets plc; J Sainsbury plc; and Tesco plc. A report on the mergers in contemplation*, London: Competition Commission, September
95 Euromonitor International (2001) *The world market for consumer foodservice 2001*, London: Euromonitor
96 Euromonitor International (2001) *The World Market for Consumer Foodservice*, London, Euromonitor, p52
97 ibid, p163
98 ibid, p52
99 Schlosser, E (2001) *Fast Food Nation*, Boston MA: Houghton Mifflin
100 Burger King, company data, supplied to authors, April 2002
101 Heasman, M (2003) 'Weighty matters loom large', *Innova*, 1, 1 CWS Media, The Netherlands
102 Biing-Hwan, L, Guthrie, J and Frazao, E (1998) 'Popularity of dining out present barrier to dietary improvements', *Food Review (USDA)*, May/August, pp2–10
103 Mason, J (2001) 'New farming techniques could "cut food crisis in south Asia"', *Financial Times*, 3 October, p12
104 Sloan, E (2002) The Natural & Organic Foods Marketplace, *Food Technology*, vol 56, no 1, pp27–37
105 NBJ (2001) *Organic foods report 2001*, San Diego: Nutrition Business Journal, http://www.nutritionbusiness.com
106 Data from www.organic-europe.net/europe_eu/statistics.asp
107 Yussefi, M and Willer, H (2003) *The World of Organic Agriculture*, Imsbach: International Federation of Organic Agriculture Movements
108 Compiled by ITC, January 2002, based on trade estimates, http://www.intracen.org/mds/sectors/organic/overview.pdf
109 Food Standards Agency (2001) *Statement on organics*, London: FSA, 16 October

110 Banks, J and Marsden, T (2001) 'The Nature of Rural Development: the organic potential', *Journal of Environmental Policy and Planning*, 3, 103–121
111 Data from http://www.isaaa.org
112 Persley, G (2003) *New Genetics, Food and Agriculture: Scientific Discoveries – Societal Dilemmas*, Paris: International Council for Science (ICSU). http://www.icsu.org
113 ETC Group (2002) *Ag Biotech Countdown, June 2002 Update*, http://www.etcgroup.org
114 Institute of Food Technologists (USA) (2000) IFT Expert Report on Biotechnology and Foods, Washington, DC, p54, http://www.ift.org/pdfs/biotech/report.pdf
115 Leaver, C (2001) *Food for thought*, 28th Bawden Memorial Lecture. Brighton: British Crop Protection Council Conference, http://www.bcpc.org
116 Persley, G (2003) *New Genetics, Food and Agriculture: Scientific Discoveries – Societal Dilemmas*, Paris: International Council for Science (ICSU), http://www.icsu.org
117 Royal Society (1998) *Genetically Modified Plants for Food Use*, London: Royal Society
118 Letter from the English Nature Chairman, Baroness Young of Old Scone, to the Prime Minister, the Rt Hon Tony Blair MP, 4 February 1999 (English Nature ref. PO/04.13 & 08.02/5361)
119 GM Science Review Panel (2003) *First Report*. London: Department of Environment, Food and Rural Affairs, http://www.gmsciencedebate.org.uk/report/
120 United Nations Development Programme (2001) *Human Development Report: Making new technologies work for human*, US: Oxford University Press (also available on http://www.undp.org/hdro)
121 ibid
122 Shiva, V (1993) *Monocultures of the Mind*, Penang: Third World Network
123 Shiva, V (1997), *Biopiracy: the plunder of nature and knowledge*, Boston MA: South End Press
124 Fowler, C and Mooney, C (1990) *Shattering: food, politics, and the loss of genetic diversity*, Tucson: University of Arizona Press
125 Altieri, M and Rosset, P (1999) 'Ten reasons why biotechnology will not ensure food security, protect the environment and reduce poverty in the developing world', *AgBio Forum*, 2, 3 & 4, http://www.agbioforum.org
126 de la Perriere, R and Seuret, F (2000) *Brave New Seeds: The threat of GM crop's to farmers*, London: Zed Books

Chapter 5

1 Goody, J (1982) *Cooking, cuisine and class*, Cambridge: Cambridge University Press
2 Murcott, A (1982) 'On the social significance of the "cooked dinner" in South Wales', *Social Science Information*, 21, pp677–695
3 Warde, A (1997) *Consumer, Food and Taste*, London: Sage
4 Mintz, S (1996) *Tasting Food, Tasting Freedom*, Boston MA: Beacon Press
5 Bourdieu, P (1984). *Distinction: A Social Critique of the Judgement of Taste*. London: Routledge
6 Latouche, S (1993) *In the wake of the affluent society*, London: Zed
7 Norberg Hodge, H (1991) *Ancient Futures*, San Francisco: Sierra Club Books
8 Levett, R (2003) *A Better Choice of Choice: Quality of Life, Consumption and Economic Growth*, London: Fabian Society
9 Jackson, T and Michaels, L (2003) Policies for Sustainable Consumptions, *A Report to the Sustainable Development Commission*, Guildford: University of Surrey/Oxford: Environment Change Institute
10 IEFS (1996) *A pan-EU survey of Consumer Attitudes to Food, Nutrition and Health*, Report number 1, Dublin: Institute of European Food Studies
11 IEFS (1996) *A pan-EU survey of Consumer Attitudes to Food, Nutrition and Health: influences on food choice and sources of information on healthy eating*, Report No. 2, Dublin: Institute of European Food Studies
12 IEFS (1996) *A pan-EU survey of Consumer Attitudes to Food, Nutrition and Health: definitions of healthy eating, barriers to healthy and benefits of healthy eating*, Report No. 3, Dublin: Institute of European Food Studies
13 IEFS (1996). *A pan-EU survey of Consumer Attitudes to Food, Nutrition and Health: dietary changes*, Report No. 4, Dublin: Institute of European Food Studies
14 Health Focus (2001) *Study of Public Attitudes and Actions Towards Shopping and Eating*, Atlanta: HealthFocus Inc., http://www.healthfocus.net
15 ibid
16 Food Science Australia (1999) *The future of food-related innovation. Report of Project Cassandra*, Sydney: Food Science Australia
17 Orbach, S (1986) *Hunger Strike: The Anorectic's Struggle as a Metaphor for our Age*, New York: W W Norton & Co
18 Lansley, S (1994) *After the Gold Rush: the trouble with affluence*, London: Century
19 Latouche, S (1993) *In the wake of Affluent Society*, London: Zed Press

20 Durning, AT (1992) *How much is enough?: the consumer society and the future of the earth*, London: Earthscan
21 Redclift, M (1996) *Wasted: Counting the Costs of Global Consumption*, London: Earthscan
22 See http://www.slowfood.com
23 Barling, D (2000) 'Regulating GM foods in the 1980s and 1990s' in Smith, DF and Phillips, J *Food, Science, Policy and Regulation in the 20th Century: international and comparative perspectives*, London: Routledge
24 Shiva, V (2000) *Stolen Harvest: the hijacking of the global food supply*, Cambridge MA: South End Press
25 Klein, N (2000) *No Logo*, London: Flamingo
26 Vidal, J (1997) *McLibel: burger culture on trial*, London: Pan
27 Salaman, R (1949) *The History and Social Influence of the Potato*, Cambridge: Cambridge University Press
28 Schlosser, E (2000) *Fast food nation*, London: Penguin
29 Hobhouse, H (1992) *Seeds of Change: five plants that transformed mankind*, London: Papermac, p214
30 Mintz, S (1985) *Sweetness and Power: the place of sugar in modern history*, Harmondsworth: Penguin
31 Hobhouse, H (1992) *Seeds of Change: five plants that transformed mankind*, London: Papermac, p43ff
32 Ritzer, G (1992) *McDonaldization of Society*, Thousand Oaks CA: Sage
33 Jakle, JA and Sculle, KA (1999) *Fast food: Roadside Restaurants in the Automobile Age*, Baltimore: Johns Hopkins University Press
34 Dickinson, R and Hollander, SC (1991) 'Consumer Votes', *Journal of Business Research*, 23, 1, pp9–20
35 Ewan, S (1992) 'From Citizen to Consumer?', *Intermedia*, 20, 3, May–June, p23
36 Baudrillard, J (1970/1998) *The Consumer Society: Myths and Structures*, London: Sage
37 Lang, T (1998) 'Towards a food democracy', in Griffiths, Sian (ed), *Consuming Passions*, Manchester: Manchester University Press, pp13–24
38 Adapted from Lang, T (1999) 'Diet, health and globalisation: five key questions', *Proceedings of the Nutrition Society*, vol 58, pp335–343
39 Kaeferstein, FK, Motarjemi, Y and Bettcher, DW (1997) 'Foodborne Disease Control: A Transnational Challenge', *Emerging Infectious Diseases*, 3: 503–510.
40 Galbraith, JK (1992) *Culture of Contentment*, Boston MA: Houghton Mifflin
41 Lang, T (1995) 'The contradictions of food labelling policy', *Information Design Journal*, 8, 1, pp3–16
42 Norberg-Hodge, H (1992) *Ancient Futures*, San Francisco: Sierra Club Books

43 Wackernagel, M and Rees, W (1996) *Our Ecological Footprint: reducing human impact on the earth*, Gabriola Island, BC/Philadelphia PA: New Society Publishers
44 Barlow, M (2000) *Blue Gold: the global water crisis and the commodification of the world's water supply*, San Francisco: International Forum on Globalisation
45 Barlow, M (2000) 'Commodification of water – wrong prescription', speech to 10th Stockholm Water Synmposium, 17 August
46 Ritzer, G (1999) *Enchanting a Disenchanted World*, Thousand Oaks: Pine Forge
47 Expanded from Durning, A T (1992) *How Much is Enough?*, Washington, DC, Worldwatch Institute; London, Earthscan
48 Appleton, J (1987) *Drought Relief in Ethiopia: planning and management of feeding programmes – a practical guide*, London: Save the Children
49 Walton, JK (1992) *Fish and Chips and the British Working Class, 1870–1940*, Leicester: Leicester University Press
50 Maslow, AH (1970) *Motivation and Personality*, New York: Harper and Row, 2nd edition
51 Maslow, AH (1973) *The Farther Reaches of Human Nature*, Harmondsworth: Penguin
52 Ingwerson, J (2000) 'US food industry targets biotech education', Reuters, 6 July
53 ibid
54 Heasman, M and Mellentin, J (2001) *Strategies in Functional Foods and Beverages 2002*, Brentford: New Nutrition Business, Dec/Jan
55 Dibb, S (1993) *Children: Advertisers' dream, Nutrition nightmare?* London: National Food Alliance
56 Young, B (1998) *Childrens' categorisation of food*, London: Food Advertising Unit
57 Barwise, P (1994) *Children, Advertising and Nutrition,* London: The Advertising Association
58 *Advertising Statistics Yearbook 2001*, Table 20.5, Unadjusted (at rate-card) ACNielson-MMS expenditure by product group, pp210–211
59 *Marketing*, http://www.marketing.haynet.com/feature00/big brands00/top50.htm (accessed 2001)
60 Adex International, Nielsen Media Research, 2002 (contact: adex.international@acnielsen.co.uk); http://www.mind-advertising.com/us/index.html
61 Perry, M (1994) *The Brand – Vehicle for Value in a Changing Marketplace*, Advertising Association, President's Lecture, London, 7 July 1994
62 Cecchini, P (1988) *The European Challenge*, London: Wildwood House
63 Consumers' Association (2001) 'The True Cost of Promotion in Schools', *Which?*, December, pp8–10
64 Rayner, G (2002) 'McDonald's and UNICEF: the odd couple?', Guardian Unlimited, http://www.guardianunlimited.co.uk, August

65 Wiseman, MJ (1990) 'Government: where does nutrition policy come from?', *Proceedings of the Nutrition Society*, 49, pp397–401
66 Lang, T (1995) 'The contradictions of food labelling policy', *Information Design Journal*, 8, 1, pp3–16
67 For example, in the press briefing to coincide with publication of four regulations on food safety by the European Commission ex parte Commissioner David Byrne, Health and Consumer Protection Commissioner, 17 July 2000
68 Department of Health (2003) *Health Check. Annual Report of the Chief Medical Officer 2002*, London: Department of Health
69 UBS Warburg (2002) *Absolute Risk of Obesity*, London: UBS Warburg, 27 November
70 Morgan, JP (2003) *Food Manufacture: Obesity Report*, London: JP Morgan, 16 April
71 ibid
72 Dalmeny, K (2003) 'Food marketing: the role of advertising in child health', *Consumer Policy Review*, vol 13, no 1, pp2–6
73 Dalmeny, K, Hanna, E and Lobstein, T (2003) *Broadcasting Bad Health*, London: International Association of Consumer Food Organizations, July
74 Centre for Social Marketing (2003) *Review of Research on the Effects of Food Promotion to Children*, Glasgow: the University of Strathclyde
75 Kraft press release, http://www.kraft.com accessed August 2003
76 Caraher, M, Lang, T and Dixon, P (2000) 'The influence of TV and celebrity chefs on public attitudes and behaviour among the English public', *Journal of the Association for the Study of Food and Society*, 4, 1, pp27–46
77 National Opinion Polls (1997) *Taste 2000*, Research carried out for Hammond Communications on Cooking (Geest Foods), London, UK: National Opinion Polls
78 National Food Alliance (1993) *Get Cooking!*, London: National Food Alliance, Department of Health & BBC Good Food
79 Fieldhouse, P (1995) *Food and Nutrition: Customs and culture*, London: Chapman & Hall
80 National Food Alliance (1993) *Get Cooking!*, London: National Food Alliance, Department of Health & BBC Good Food
81 Stitt, S, Jepson, M, Paulson-Box, E and Prisk, E (1997) 'Schooling for Capitalism: Cooking and the National Curriculum': in Köhler, BM, Feichtinger, E, Barlösius, E and Dowler, E (eds) *Poverty and Food In Welfare Societies*, Berlin: WZB
82 Lang, T (1998) 'Towards a food democracy', in Griffiths, S (ed) *Consuming Passions*, Manchester: Manchester University Press. pp13–24

83 Fieldhouse, P (1995) *Food and Nutrition: Customs and culture*, London: Chapman & Hall
84 Food and Drink Federation (2000) 'Out and about', *Feedback*, 36, summer, 8
85 Jakle, JA and Sculle, KA (1999) *Fast Food: Roadside Restaurants in the Automobile Age*, Baltimore: Johns Hopkins Press
86 Murcott, A (1995) 'It's such a pleasure to cook for him': food, mealtimes and gender in some South Wales households: in Jackson, S and Moores, S (eds) *The Politics of Domestic Consumption*, Hemel Hempstead, UK: Prentice Hall/Harvester Wheatsheaf
87 Gershuny, J and Fisher, K (2000) 'Leisure': in Halsey, AH with Webb, J (eds) *Twentieth-Century British Social Trends*, Basingstoke, Hampshire: Macmillan, pp620–649
88 Reisig, VMT and Hobbiss, A (2000) 'Food deserts and how to tackle them: a study from one city's approach', *Health Education Journal*, 59 (2), pp137–149
89 Robinson, N, Caraher, M and Lang, T (2000) 'Access to shops; the views of low income shoppers', *Health Education Journal*, 59 (2), pp121–136
90 Trager, J (1995) *The Food Chronology*, New York: Henry Holt & Co. p167
91 Raven, H and Lang, T (1995) *Off our Trolleys?: food retailing and the hypermarket revolution*, London: Institute of Public Policy Research
92 Department of Health (1996) *Low Income, Food, Nutrition and Health: Strategies for Improvement: Report by the Low Income Project Team for the Nutrition Taskforce*, London: Department of Health
93 Dowler, E, Blair, A, Donkin, A, Rex, D and Grundy, C (2001) *Measuring Access to Healthy Food in Sandwell*, Final Report, August, Sandwell: Sandwell Health Authority/Health Action Zone
94 Raghavan, C (1990) *Recolonisation: GATT, the Uruguay Round and the Third World*, Penang: Third World Network/London: Zed Books
95 Starr, A (2000) *Naming the enemy: anti-corporate movements confront globalization*, London: Zed and Pluto
96 Burbach, R (2001) *Globalization and postmodern politics*, London: Pluto
97 Lang, T (1996) 'Going public: food campaigns during the 1980s and 1990s', in Smith D (ed) *Nutrition Scientists and Nutrition Policy in the 20th Century*, London: Routledge
98 Millstone, E (1991) 'How to involve consumer organisations in the agricultural policy and international relations debate', Paper to conference: 'Comment Nourrir Le Monde? Les politiques alimentaires face à la libéralisation des économies et des échanges', Montpelier: SOLAGRAL, December
99 Smith, NC (1990) *Morality and the market: consumer pressure for corporate accountability*, London: Routledge

100 Witowski, TH (1989) 'Colonial consumers in revolt: buyer values and behavior during the nonimportation movement, 1764–76', *Journal of Consumer Research*, 16, 2, pp79–93

101 Allain, A (1991) 'Breastfeeding is politics: a personal view of the international baby milk campaign', *The Ecologist*, 21, 5, pp206–213

CHAPTER 6

1 Constance, DH and Heffernan, WH (1991) 'The global poultry agro-food complex', *International Journal of Sociology of Agriculture and Food*, 1, pp126–142

2 Bonnano, A and Constance, DH (2001) 'Corporate strategies in the Global Era: The Case of mega-Hog Farms in the Texas Panhandle region', *International Journal of Sociology of Agriculture and Food*, 9, pp5–28

3 Goodman, D and Watts, M (eds) (1997) *Globalising Food: Agrarian Questions and Global Restructuring*, London: Routledge

4 Burch, D, Lawrence, G and Goss, J (1999) *Restructuring Global and Regional Agricultures: Transformation in Australasian Economies and Spaces*, Aldershot, Hants: Avebury

5 Raynolds, LT (2000) 'Re-embedding Global Agriculture: the International Organic and Fair Trade Movements', *Journal of Agriculture and Human Values*, 17, 3, pp297–309

6 Draper, P (ed) (1991) *Health through Public Policy: the greening of public health*, London: Greenprint

7 McMichael, AJ (2001) *Human Frontiers, Environments and Disease*, Cambridge: Cambridge University Press

8 McMichael, AJ (1999) 'From Hazard to Habitat: Rethinking Environment and Health', *Epidemiology*, 10, 4, pp1–5

9 Lang, T, Barling, D and Caraher, M (2001) 'Food, Social Policy and the Environment: Towards a New Model', *Social Policy and Administration*, 35, 5, pp538–558

10 Meadows, DH, Meadows, DL, Randers, J and Behrens, W (1972) *The Limits to Growth*, London: Earth Island

11 Goldsmith, E, Allen, R, Allaby, M, Davoll, J and Lawrence, S (1972) 'Blueprint for Survival', *The Ecologist*, 2, 1, pp1–43

12 Carson, R (1962) *Silent Spring*, Boston: Houghton Mifflin Co

13 Lovelock, J (1979) *Gaia: a new look at life on earth*, Oxford: Oxford University Press

14 UNDP (1998) *World Development Report*, quoted in UNEP (2002) *Global Environment Outlook 3*. London: Earthscan/UNEP, p35

15 Durning, AT (1992) *How much is enough?* London: Earthscan

16 Brown, LR (1996) *Tough Choices: facing the challenge of food scarcity*, London: Earthscan
17 Dyson, T (1996) *Population and Food: global trends and future prospects*, London: Routledge
18 Ehrlich, PR, Ehrlich, AH and Daily, GC (1995) *The Stork and the Plough*, New York: Putnam Press
19 FAO *Agriculture: Towards 2015/2030, technical interim report April* 2000 Rome: Food and Agriculture Organization Economic & Social Department
20 Bowler, I (1992) 'The Industrialisation of Agriculture', in Bowler I, (ed), *The Geography of Agriculture in Developed Market Economies*, Harlow: Longman
21 Boyd Orr, J (1943) *Food and the People*, Targets for Tomorrow 3, London: Pilot Press
22 Mackintosh, J (1944) *The Nation's Health*, Targets for Tomorrow 4, London: Pilot Press
23 Goodman, D, Sorj, B and Wilkinson, J (1987) *From Farming to Biotechnology*, Oxford: Blackwell
24 Carson, R (1962) *Silent Spring*, Boston: Houghton Mifflin Co
25 Lang, T and Clutterbuck, C (1993) *P is for Pesticides*, London: Ebury
26 Colborn, T, Dumanoski, D and Peterson Myers, J (1996) *Our Stolen Future*, New York: Dutton
27 McGinn, AP (2000) 'Phasing out Persistent Organic Pollutants', in Brown, L R et al (eds) *State of the World 2000*, London: Earthscan
28 McMichael, AJ 'Dioxins in Belgian feed and food: chickens and eggs', *Journal of Epidemioliogy and Community Health* 1999; 53: pp742–743
29 McKee, M (1999) 'Trust me, I'm an expert', *European Journal of Public Health*, vol 9, pp161–162
30 Mackenzie, D (1999) 'Recipe for disaster', *New Scientist*, 12 June
31 McMichael, AJ (1999) 'Dioxins in Belgian feed and food: chickens and eggs', *Journal of Epidemiology & Community Health*, 53, pp742–743
32 Buzby, J (2001) 'Effects of food safety perceptions on and food demand and global trade'. In: Regmi, A (ed) *Changing structure of global food consumption and trade*, Washington DC: US Department of Agriculture, Agriculture and Trade Report WRS-01-1, p62
33 Lord Phillips of Worth Matravers, Mrs June Bridgeman and Professor Malcolm Ferguson-Smith (2000) *The BSE Inquiry: Report: evidence and supporting papers of the Inquiry into the emergence and identification of Bovine Spongiform Encephalopathy (BSE) and variant Creutzfeldt-Jakob Disease (vCJD) and the action taken in response to it up to 20 March 1996*, 16 volumes, London: The Stationery Office
34 Santer, J (1997) Speech by Jacques Santer, President of the European Commission at the Debate in the European Parliament on the report

into BSE by the Committee of Enquiry of the European Parliament, February 18 1997, Speech 97/39
35 Ghenremeskel, K and Crawford, MA (1994) 'Nutrition and health in relation to food production and processing', *Nutrition and Health*, 9, pp237–253
36 Ghenremeskel, K and Crawford, MA (1994) 'Nutrition and health in relation to food production and processing', *Nutrition and Health*, 9, pp237–253
37 http://www.euro.who.int/eprise/main/WHO/Progs/FOS/NewsEvents/20020520_1
38 Willett, W (1994) 'Diet and health: what should we eat', *Science*, 264, pp532–537
39 United Nations Environment Programme (2002) *Global Environment Outlook*, London: Earthscan/UNEP
40 McMichael, AJ, Bolin, B, Costanza, R, Daily, GC, Folke, C, Kindalh-Kiessling, K, Lindgren, E and Niklasson, B (1999) 'Globalization and the Sustainability of Human Health', *BioScience*, 49, 3, pp205–210
41 McMichael, AJ (1999) 'From Hazard to Habitat: Rethinking Environment and Health', *Epidemiology*, 10, 4, pp1–5
42 Wahlqvist, M and Specht, RL (1998) 'Food variety and biodiversity: Econutrition', *Asia Pacific Journal of Clinical Nutrition*, 7, 3–4, pp314–319
43 Gardner, G (1996) *Shrinking Fields: Cropland Loss in a World of Eight Billion*, Washington DC: Worldwatch Institute. Worldwatch Paper 131
44 Halweil, B (2000) 'Where have all the farmers gone?' *World Watch*, 13, 5, Sept–Oct, pp12–28
45 Ausubel, K (1994) *Seeds of Change*, San Francisco: HarperCollins
46 Henry, M (2001) 'Sow few, so trouble', *Green Futures*, 31, pp40–42
47 McMichael, P (2000) 'Power of Food', *Agriculture & Human Values*, 17, pp21–33
48 Henry, M (2001) 'Sow few, so trouble', *Green Futures*, 31, pp40–42
49 FAO (1996) *State of the World's Plant Genetic Resources*, Rome: Food and Agriculture Organization
50 Heffernan, W (1999) *Consolidation in the food and agriculture system*, Washington DC: National Farmers Union, p14
51 McMichael, AJ (2000) *Human frontiers, environments and disease*, Cambridge: Cambridge University Press, p310
52 United Nations Environment Programme (2002) *Global Environment Outlook*, London: Earthscan/UNEP
53 Pretty, J (2002) *Agri-Culture*, London: Earthscan
54 Funes, F, Garcia, L, Bourque, M, Perez, N and Rosset, P (eds) (2002) *Sustainable Agriculture and Resistance: transforming food production in Cuba*, Oakland CA: Food First Books
55 http://www.ukabc.org/iu2.htm

56 Curtis, V and Cairncross, S (2003) 'Water, sanitation, and hygiene at Kyoto', *British Medical Journal*, 327, pp3–4
57 Stockholm International Water Institute (2003) *General water statistics: World Water Week Symposium data sheets*, 10–16 August, Stockholm: Stockholm International Water Insitute, http://www.siwi.org/waterweek2003
58 ibid
59 United Nations Environment Programme (2002) *Global Environment Outlook*, London: Earthscan/UNEP
60 FAO (2003) *World Agriculture: towards 2015/2030: an FAO perspective*, Rome: Food and Agriculture Organisation/London: Earthscan p138
61 McMichael, AJ (1999) 'From Hazard to Habitat: Rethinking Environment and Health', *Epidemiology*, 10, 4, pp1–5
62 United Nations Environment Programme (2002) *Global Environment Outlook*, London: Earthscan/UNEP, p151
63 Gardner, G (1999) 'Irrigated Area Up', in Brown, LR, Renner, M and Halweil, B (eds) *Vital Signs 1999–2000*, London: Earthscan
64 de Moor, APG (1998) *Subsidizing Unsustainable Development*, The Hague: Institute for Research on Public Expenditure & Earth Council, S5.1–S5.5
65 Stockholm International Water Institute (2003) *General water statistics: World Water Week Symposium data sheets*, 10–16 August. Stockholm: Stockholm International Water Institute http://www.siwi.org/waterweek2003
66 Cosgrave, W, Vice-President of the World Water Council, quoted in Houlder, V (2003) 'World in drier straits', *Financial Times*, 11 August, p16
67 Postel, S (2000) 'Redesigning Irrigated Agriculture', in Brown, LR et al (eds) *State of the World 2000*, London: Earthscan. pp39–58
68 Shiva, V (2002) *Water Wars*, London: Pluto
69 Barlow, M (2000) *Blue Gold: the global water crisis and the commodification of the world's water supply*, San Francisco: International Forum on Globalization
70 Barlow, M (2000) 'Commodification of water – wrong prescription', speech to 10th Stockholm Water Synmposium, 17August
71 Carson, R (1962) *Silent Spring*, Boston: Houghton Mifflin Co
72 WHO 1997a, cited in UNEP (1999) *Global Environmental Outlook 2000*, Nairobi, UNEP, London, Earthscan
73 ISEC (2000) *Bringing the Food Economy Home: The Social, Ecological and Economic Benefits of Local Food*, Dartington: International Society for Ecology and Culture, p19
74 Lang, T and Clutterbuck, C (1994) *P is for Pesticide*, London: Ebury
75 Schafer, KS, Kegley, SE and Patton, S (2001) *Nowhere to Hide: Persistent Toxic Chemicals in the US Food Supply*, San Francisco: Pesticide Action Network/Bolinas: Commonweal

76 Orris, P, Chary, LK, Perry, K and Asbury, J (2000) *Persistent Organic Pollutants and Human Health*, Washington DC: World Federation of Public Health Associations
77 McGinn, AP (2000) 'Phasing out Persistent Organic Pollutants', in Brown, LR et al (eds) *State of the World 2000*, London: Earthscan
78 Baker, BP, Benbrook, CM, Groth, E and Benbrook, K (2002) 'Pesticide residues in conventional, integrated pest management (IPM)-grown and organic foods: insights from three US data sets', *Food Additives & Contaminants*, 2002; 19, pp427–446
79 World Cancer Research Fund/American Institute for Cancer Research (1997) *Food, nutrition and the prevention of cancer: a global perspective*, Washington DC: AICR
80 United Nations Environment Programme (2002) *Global Environment Outlook*, London: Earthscan/UNEP, p256
81 Commission of the European Communities (1994) *Directive on Packaging and Packaging Waste (94/62/EC)*, Brussels: European Commission
82 Murray, R (1999) *Creating wealth from waste*, London: Demos
83 INCPEN (2002) *Factsheet*, London: INCPEN
84 Miller, C (2001) 'Garbage by the numbers', *NSWMA Research Bulletin*, 01-02
85 Murray, R (1999) *Creating wealth from waste*, London: Demos
86 Smith, A (2002) 'Waste not', *Green Futures*, March/April, p12
87 INCPEN (2002) *Factsheet*, London: INCPEN
88 Roberts, M (2002) *Dual system: facts and figures*, http://www.gruener-punkt.de
89 United Nations Environment Programme (2002) *Global Environment Outlook*, London: Earthscan/UNEP
90 McMichael, AJ (2000) *Human frontiers, environments and disease*, Cambridge: Cambridge University Press
91 Oldeman, LR et al (1991) *World Map of the Status of Human-Induced Soil Degradation*, Wageningen, Netherlands and Nairobi: International Soil Reference and Information Centre, and UN Environment Programme
92 McMichael, AJ, Haines, A, Slooff, R and Kovats, RS (eds) *Climate Change and Human Health*, Geneva: World Health Organization, UN Environment Programme, World Meteorological Organization, 1996
93 United Nations Environment Programme (2002) *Global Environment Outlook*, London: Earthscan/UNEP
94 http://www.grida.no/climate/vital/36.htm
95 http://www.grida.no/eis-ssa/contry/kenya1.htm
96 UNEP (2001) 'Climate Change: Billions Across The Tropics Face Hunger And Starvation As Big Drop In Crop Yields Forecast. Soaring Temperatures Force Coffee and Tea Farmers to Abandon Traditional Plantations.' News Release 01/107, 8 November,

http://www.unep.org/documents/default.asp?DocumentID=225&ArticleID=2952

97 http://climatechange.unep.net

98 Stipp, D (2004) 'Climate collapse: The Pentagon's weather nightmare', *Fortune Magazine*, 26 January

99 Harrison, P (1992) *The Third Revolution: environment, population and a sustainable world*, New York: IB Tauris

100 See Millstone, E and Lang, T (2003) *Atlas of Food*, London: Earthscan/New York: Penguin pp50–51

101 Garnett, T (1999) *City Harvest: the feasibility of growing more food in London*, London: Sustain

102 FAO (1998) *The state of food and agriculture*, Rome: Food Agriculture Organization

103 ibid

104 Philippines National Statistics Office figures from NSO (2000) *2000 Census of Population and Housing*, information provided by Professor Maria Pedro and colleagues, Food and Nutrition Research Institute

105 Figures from Jimaima Tunidau Schultz, Secretariat of the Pacific Community

106 UN Population Division (2003). World Population Prospects 2002 revision, http://esa.un.org/unpp/p2k0data.asp accessed 16 October 2003

107 We are particularly grateful to Professor Prakash Shetty and colleagues from the FAO food and nutrition division's urbanization working party and to the participants at the FAO technical workshop on 'Globalization of food systems: impacts on food security and nutrition' (held in Rome, 8–10 October 2003) for thoughts on these processes. A report is due for publication in 2004

108 World Bank (1997) *The global burden of disease*, Washington DC: World Bank.

109 World Bank (1999) *World Development Report 1999–2000*, New York: World Bank

110 UNDP (1996) *Urban Agriculture: Food, Jobs and Sustainable Cities*, publication series for Habitat 2, New York: United Nations Development Programme, vol 1

111 WHO – Europe (1998) *Draft Urban Food and Nutrition Action Plan: elements for local action or local production for local consumption*, Copenhagen: World Health Organization Regional Office for Europe Programme for Nutrition Policy, Infant Feeding and Food Security together with the ETC Urban Agriculture Programme, Leusden, The Netherlands and the WHO Centre for Urban Health

112 Pretty, J (1998) *The Living Land*, London: Earthscan

113 Garnett, T (1999) *City Harvest: the feasibility of growing more food in London*, London: Sustain

114 Sandandreu, A, Gomez Perazzoli, A and Dubbeling, M (2002) 'Biodiversity, poverty and urban agriculture in Latin America', *Urban Agriculture*, 6, April, pp9–11
115 DETR (1998) *Sustainable business. Consultation paper on sustainable development and business in the UK*, London: Dept of the Environment, Transport and the Regions
116 Jones, A (2002) *Eating Oil*, London: Sustain
117 Heilig, GK (1993) 'Food, life-styles and energy', in van der Heij, DG, Loewik, MRH and Ockhuizen, TH (eds), *Food and Nutrition Policy in Europe*, Wageningen: Wageningen Pudoc Scientific Publishers/WHO Regional Office for Europe
118 Heilig, GK (1993) 'Food, life-styles and energy', in van der Heij, DG, Loewik, MRH and Ockhuizen, TH (eds), *Food and Nutrition Policy in Europe*, Wageningen: Wageningen Pudoc Scientific Publishers/WHO Regional Office for Europe
119 Koojiman, JM (1993) 'Environmental Assessment of Packaging', *Environmental Management*, vol 17, no 5, cited in Paxton, A (1994) *Food Miles*, London: SAFE Alliance, p17
120 ERR (1989) cited in Paxton, A (1994) *Food Miles*, London, SAFE Alliance, p18
121 Blaxter, K (1978) cited in Paxton, A (1994) *Food Miles*, London, SAFE Alliance, p21
122 Heilig, GK (1993) 'Food, life-styles and energy', in van der Heij, DG, Loewik, MRH and Ockhuizen, TH (eds), *Food and Nutrition Policy in Europe*, Wageningen, Wageningen Pudoc Scientific Publishers/WHO Regional Office for Europe
123 Jones, A (2001) *Eating Oil*, London, Sustain. Data for shipping and airfreight from *Guidelines for company reporting on greenhouse gas emissions*, Department of the Environment, Transport and the Regions: London, March 2001. Data for trucks is based on Whitelegg, J (1993) *Transport for a sustainable future: the case for Europe*, Belhaven Press, London and Gover, MP (1994). *UK petrol and diesel demand: energy and emission effects of a switch to diesel*, Report for the Department of Trade and Industry, The Stationery Office, London
124 Koojiman, JM (1993) Environmental Assessment of Packaging: Sense and Sensibility. *Environmental Management*, 17, 5
125 Boege, S (1993) *Road Transport of Goods and the Effects on the Spatial Environment*, Wupperthal: Wupperthal Institute
126 Pirog, R, Van Pelt, T, Enshayan, K and Cook, E (2001) *Food, fuel, and Freeways: an Iowa perspective on how far food travels, fuel usage, and greenhouse emissions*, Ames, Iowa: Leopold Center for Sustainable Agriculture, Iowa State University, http://www.leopold.iastate.edu/

127 Davidson, A (ed) (1999) *The Oxford Companion to Food*, Oxford: Oxford University Press; Masefield, GB, Wallis, B, Harrison, SG and Nicholson, BE (1969) *The Oxford Illustrated Book of Plant Foods*, Oxford: Oxford University Press; Bianchini, F, Corbetta, F and Pistoia, M (1975) *Fruits of the Earth*, London: Bloomsbury; Robinson, F, (ed) (1999) *The Oxford Companion to Wine*, Oxford: Oxford University Press
128 Paxton, A (1994) *The Food Miles Report*, London: Sustainable Agriculture, Food and Environment Alliance
129 Hoskins, R and Lobstein, T (1998) *The Pear Essentials*, London: Sustainable Agriculture, Food and Environment Alliance. Food Facts No. 3, Table 4, p11
130 DETR, 2000 *Focus on Ports*, Department of the Environment Transport and the Regions, The Stationery Office, London
131 DETR, 2000 *The Future of aviation: the Government's consultation document on air transport policy*, Department of Environment, Transport and the Regions, London
132 Raven, H, Lang, T with Dumonteil, C (1995) *Off our Trolleys?: food retailing and the hypermarket economy*, London: Institute for Public Policy Research
133 Whitelegg, J (1994) *Driven to shop*, London: Eco-logica/Sustainable Agriculture, Food and Environmental Alliance
134 Pretty, J, Brett, C, Gee, D, Hine, RE, Mason, CF et al (2000) 'An assessment of the total external costs of UK agriculture', *Agricultural Systems*, 65, pp113–136
135 Pretty, J, Brett, C, Gee, D, Hine, RE, Mason, CF et al (2001) 'Policy challenges and priorities for internalising the externalities of agriculture', *Journal of Environmental Planning and Management*, 44(2), pp263–283
136 Borgstrom, G (1973) *The Food and People Dilemma*, Pacific Grove, California: Duxbury Press
137 Wackernagel, M and Rees, W (1996) *Our Ecological Footprint*, Gabriola Island BC: New Society publishers
138 Borgstrom, G (1973) *The Food and People Dilemma*, California: Duxbury Press
139 Eurostat figures 1997–2000
140 MacLaren, D, Bullock, S and Yousuf, N (1998) *Tomorrow's World: Britain's share in a sustainable Future*, London: Earthscan
141 Durning, AT (1992) *How much is enough?*, London: Earthscan
142 Best Foot Forward (2002) *City Limits: a resource flow and ecological footprint analysis of Greater London*, London: Greater London Authority, http://www.citylimitslondon.com
143 OECD (1998) *Agricultural Policy Reform: Stocktaking of Achievements*, Discussion Paper prepared for ECD Agriculture Committee, 5–6

March 1998, Paris: Organisation for Economic Co-operation and Development

144 Hoskins, R and Lobstein, T (1998) *The pear essentials*, London: Sustainable Agriculture, Food and Environment Alliance, Food Facts No. 3

145 Hoskins, R and Lobstein, T (1998) *How green are our apples?* London: Sustainable Agriculture, Food and Environment Alliance, Food Facts No. 4

146 Jones, A (2001) *Eating Oil*, London: Sustain

147 Pretty, J, Brett, C, Gee, D, Hine, RE and Mason, CF et al (2001) 'Policy challenges and priorities for internalising the externalities of agriculture', *Journal of Environmental Planning and Management*, 44(2), pp263–283

148 McMichael, AJ, Bolin, B, Costanza, R, Daily, GC, Folke, C, Lindahl-Kiessling, K, Lindgren, E and Niklasson, B (1999) 'Globalization and the Sustainability of Human Health', *BioScience*, 49, 3, pp205–210

149 FAO (2001) *The State of World Fisheries and Acquaculture*, Rome: FAO

150 Pew Oceans Commission (2003) *America's Living Oceans: a report to the nation*, Arlington Virginia: Pew, http://www.pewoceans.org/

151 Delgado, C, Wada, N, Rosegrant, M, Meijer, S and Ahmed, M (2003) *The Future of Fish: Issues and Trends to 2020*, Washington DC: International Food Policy Research Institute

152 Kent, G (1997) 'Fisheries, Food Security, and the Poor', *Food Policy*, 22, 5, pp393–404

153 UNEP (2002) *Global Environment Outlook 3*, London: Earthscan/UNEP

154 FAO (2002) *The State of World Fisheries and Aquaculture 2002*, Rome: Food and Agriculture Organization

155 Environmental Justice Foundation and Action Aid (2003) *Smash and Grab: conflict, corruption and human rights abuses in the shrimp farming industry*, London: Environmental Justice Foundation and Action Aid

156 Lang, C (2001) *Vietnam: Shrimps, mangroves and the World Bank*, Bulletin 51, World Rainforest Movement, http://www.wrm.org.uy/bulletin/51/Vietnam.html

157 FAO (2002) *The State of World Fisheries and Aquaculture 2002*, Rome: Food and Agriculture Organization Figure 1, p5

158 Kurlansky, M (1998) *Cod: a biography of the fish that changed the world*, London: Jonathan Cape

159 WTO (1999) *Trade and Environment Bulletin*, PRESS/TE/029, 30 July, pp7–11

160 Schafer, KS, Kegley, SE and Patton, S (2001) *Nowhere to Hide: Persistent Toxic Chemicals in the US Food Supply*, San Francisco: Pesticide Action Network/Bolinas: Commonweal, p22

161 Food Standards Agency (2001) *Taskforce Report on the Burdens of Regulations on Small Food Businesses*, London: FSA, pp26–31, plus report on visits

162 McMichael, AJ (2001) *Human Frontiers, Environments and Disease*, Cambridge: Cambridge University Press

163 Davenport, J, Black, K, Burnell, G, Cross, T, Culloty, S et al (2003) *Acquaculture: the ecological issues*, Oxford: Blackwell/British Ecological Society

164 Davenport, J, Black, K, Burnell, G, Cross, T, Culloty, S et al (2003) *Acquaculture: the ecological issues*, Oxford: Blackwell/British Ecological Society

165 Delgado, C, Wada, N, Rosegrant, M, Meijer, S and Ahmed, M (2003) *Outlook for Fish to 2020: meeting global demand*, Washington DC: International Food Policy Research Institute

166 Rosegrant, MW, Leach, N and Gerpacio, RV (1999) 'Alternative futures for world cereal and meat consumption', *Proceedings of the Nutrition Society*, 58, pp219–234

167 Millward, DJ (1999) 'Meat or wheat for the next millennium?', *Proceedings of the Nutrition Society*, 58, pp209–210

168 Rosegrant, MW, Leach, N and Gerpacio, RV (1999) 'Alternative futures for world cereal and meat consumption', *Proceedings of the Nutrition Society*, 58, pp219–234

169 See the summary in Sanders, TAB (1999) 'The nutritional adequacy of plant-based diets', *Proceedings of the Nutrition Society*, 58, pp265–269

170 Cannon, G (1995) *Superbug: nature's revenge*, London: Virgin

171 Wallinga, D (2002) 'Antimicrobial Use in Animal Feed: an Ecological and Public Health Problem', *Minnesota Medicine*, 85, October, http://www.mmaonline.net/publications/MNMed2002/October/Wallinga.html

172 McMichael, AJ (1999) 'From Hazard to Habitat: Rethinking Environment and Health', *Epidemiology*, 10, 4, pp1–5

173 Young, R, Cowe, A, Nunan, C, Harvey, J and Mason, L (1999) *The Use and Misuse of Antibiotics in UK Agriculture: Part 2: Antibiotic Resistance and Human Health*, Bristol: Soil Association

174 GAO (1977) *Need to establish Safety and Effectiveness of Antibiotics Used in Animal Feeds*, Washington DC: General Accounting Office, GAO/HRD-77-81

175 GAO (1999) *The Agricultural Use of Antibiotics and Its Implications for Human Health*, Washington DC: General Accounting Office, GAO/RCED-99-74

176 GAO (1999) *The Agricultural Use of Antibiotics and Its Implications for Human Health*, Washington DC: General Accounting Office, GAO/RCED-99-74, pp18–19

177 Standing Medical Advisory Committee Sub-Group on Antimicrobial Resistance (1999) *Independent Review of the literature*, London: Department of Health

178 McMichael, AJ, Bolin, B, Costanza, R, Daily, GC, Folke, C et al (1999) 'Globalization and the Sustainability of Human Health', *BioScience*, 49, 3, pp205–210

179 House of Lords (1998) *Resistance to Antibiotics and Other Antimicrobial Agents. Seventh Report of the Select Committee on Science and Technology*, London: The Stationery Office

180 WHO (1997) *The Medical Impact of the Use of Antimicrobials in Food Animals: Report of a WHO meeting, Berlin, Germany, 13–17 October 1997*, Geneva: World Health Organization Division of Emerging and Other Communicable Diseases Surveillance and Control

181 World Health Organization (2002) *Impacts of antimicrobial growth promoter termination in Denmark*, Geneva: WHO

182 http://www.keepantibioticsworking.com (accessed 6 June 2003)

183 Lawrence, F (2003) 'So what if it's legal? It's disgusting', *The Guardian*, 23 May

184 Barza, M and Gorbach, SL (eds) (2002) 'The need to improve antimicrobial use in agriculture: ecological and human health consequences', *Clinical Infectious Diseases*, 34, supplement 3, S71–144

185 McDonald's (2003) *Global policy on antibiotics in food animals*, Policy statement, 19 June 2003, Chicago Ill: McDonald's Corporation, http://www.mcdonalds.com/corporate/social

186 Caraher, M and Anderson, A (2001) 'An apple a day', *Health Matters*, 46, pp12–14

187 World Health Organization (1990) *Diet, Nutrition and the Prevention of Chronic Diseases*, WHO Technical Report Series 1990, 797. Geneva: World Health Organization

188 Khaw, KT et al (2001) 'Relation Between Plasma Ascorbic Acid and Mortality in Men and Women in EPIC-Norfolk Prospective Study: A Prospective Population Study', *The Lancet*, 357, pp657–663

189 DEFRA (2001) *Agriculture in the United Kingdom 2000*, London: National Statistics/DEFRA, Chapter 5, Table 5.11

190 Key Note (2001) *Fruit and Vegetables*, London: Key Note Publications, April

191 Rayner, M/BHF Health Promotion Research Group data, quoted in Lang, T and Rayner, G (eds) (2002) *Why Health is the Key to Farming and Food*, London: UK Public Health Association

192 Ministers' Foreword to The National Schools Fruit Scheme (2001), http://www.doh.gov.uk/schoolfruitscheme/minister.htm

193 Cancer Research Campaign (2001) 'Poll finds children's diets seriously short on fruit and veg', 6 November, http://www.mori.com/polls/2001/crc-veg.shtml

194 Stoll, S (1998) *The Fruits of Natural Advantage*, Berkeley: University of California Press

195 Feder, E (1977) *Strawberry Imperialism*, The Hague: Institute of Social Studies

196 Tudge, C (1998) *Neanderthals, Bandits and Farmers: How Agriculture Really Began*, London: Weidenfeld and Nicolson

197 Tudge, C (1998) *Neanderthals, Bandits and Farmers: How Agriculture Really Began*, London: Weidenfeld and Nicolson

198 Crawford, M and Marsh, D (1989) *The Driving Force: Food, Evolution and the Future*, London: Heinemann

199 Crawford, M and Marsh, D (1989) *The Driving Force: Food, Evolution and the Future*, London: Heinemann, p190

200 Crawford, M and Marsh, D (1989) *The Driving Force: Food, Evolution and the Future*, London: Heinemann, p194

201 Goldsmith, E (1991) *The Way*, London: Weidenfeld & Nicolson

202 Lovelock, J (1979) *Gaia: A New Look at Life on Earth*, Oxford: Oxford University Press

203 Goldsmith, E, Khor, M, Norberg-Hodge, H, Shiva, V et al (1995) *The Future of Progress: Reflections on Environment and Development*, Dartington: Green Books

204 Hawken, P (1993) *The Ecology of Commerce: How Business can Save the Planet*, New York: HarperCollins

205 von Weizsaecker, E, Lovins, A and Lovins, L (1997) *Factor Four: Doubling Wealth, Halving Resource Use*, London: Earthscan

206 Wahlquvist, ML and Specht, RL (1998) 'Food variety and biodiversity: Econutrition', *Asia Pacific Journal of Clinical Nutrition*, 7, 3 & 4, 314–319

207 Lawrence, M (2002) *Folate fortification: a case study of public health policy-making*, Geelong (Aust.): Deakin University Phd thesis

208 Susser, M and Susser, E (1996) 'Choosing a future for epidemiology: from black box to Chinese boxes and eco-epidemiology', *American Journal of Public Health*, 86, 5, pp674–677

209 Leakey, R and Lewin, R (1992) *Origins Reconsidered: In Search of What Makes Us Human*, London: Little Brown & Co.

210 Nestle, M (2000) 'Paleolithic diets: a sceptical view', *Nutrition Bulletin*, 25, pp43–47

Chapter 7

1 Richards, D and Smith, M (2002) *Governance and Public Policy in the United Kingdom*, Oxford: Oxford University Press

2 Wallace, H and Wallace, W (2000) *Policy-making in the European Union*, 4th edition, Oxford: Oxford University Press

3 Deacon, B, Ollila, E, Koivusalo, M and Stubbs, P (2003) *Global Social Governance*, Globalism and Social Policy Programme, Helsinki: STAKES
4 World Resources Institute (2003) *World Resources 2002–2004*, Washington D.C.: World Resources Institute
5 Nestle, M (2002) *Food Politics*, Berkeley: University of California Press
6 Sims, L (1999) *The Politics of Fat: food and nutrition policy in America*, New York: M E Sharpe Inc
7 Cannon, G (1987) *The Politics of Food*, London: Century
8 For example, Alliance for People's Action on Nutrition (2003) *Founding Document*, Chennai, India, APAN March, http://www.apanutrition.org
9 Vidal, J (1997) *McLibel: burger culture on trial*, London: Pan
10 Lamont, J (2002) 'Big oil groups top league for 'greenwash', *Financial Times*, 25 August, p2
11 Cockett, R (1995) *Thinking the Unthinkable: think-tanks and the economic counter-revolution 1931–1983*, London: HarperCollins
12 UN Committee on Economic, Social and Cultural Rights (1999) *Substantive Issues arising in the Implementation of the International Covenant on Economic, Social and Cultural Rights: General Comment 12: The Right to Adequate Food (article 11)*, 12/05/99 E/C.12/1999/5, 20th Session, Geneva: United Nations Economic and Social Council
13 See, for example, Brown, L (2001) *Eco-economy*, London: Earthscan
14 Goldsmith, E and Mander, J (1996) *The Case Against the Global Economy*, San Francisco: Sierra Club Books
15 Hertz, N (2001) *The Silent Takeover*, London: Heinemann
16 Boyd Orr, J (1966) *As I recall*, London: MacGibbon and Kee, pp160–201
17 Watkins, K (2002) *Rigged Rules and Double Standards: Trade, globalisation and the fight against poverty*, Oxford: Oxfam
18 Watkins, K and von Braun, J (2003) *Time to stop dumping on the world's poor*, 2002–2003 IFPRI Annual Report Essay. Washington DC: International Food Policy Research Institute
19 Action Aid (2002) *The WTO Agreement on Agriculture*, London: Action Aid
20 Turner, M (2003) 'Rich world's subsidies hitting African growth', *Financial Times*, 31 August, p9
21 Watkins, K and von Braun, J (2003) *Time to stop dumping on the world's poor*, 2002–2003 IFPRI Annual Report Essay. Washington DC: International Food Policy Research Institute
22 WHO and WTO (2002) *WTO Agreements and Public Health*, Geneva: World Health Organization and World Trade Organization
23 Avery, N, Drake, M and Lang, T (1994) *Cracking the Codex: an analysis of who sets world food standards*, London: National Food Alliance

24 For example, McCrea, D (2001) *Report of the analysis of Codex documentation to aid the consumer decision-making process: Codex Committee on General Principles*, London: D McCrea Consulting
25 Cosbey, A (2001) *A forced evolution? The Codex Alimentarius Commission and scientific uncertainty*, Winnipeg: International Institute for Sustainable Development
26 CAC (2002) Conclusions and recommendations of the Joint FAO/WHO Evaluation of the Codex Alimentarius and other FAO and WHO work on Food Standards, *Alinorm 03/25/3*, Rome: Codex Alimentarius Commission, December
27 CAC (2003) Codex Alimentarius Commission, Twenty-Fifth (Extraordinary) Session, Geneva (Switzerland), 13–15 February 2003, *Alinorm 03/25/5*, Rome: Codex Alimentarius Commission, February
28 Codex Secretariat (2002) *Capacity Building for Food Standards and Regulations: A report provided by the Secretariat of the Codex Alimentarius Commission based on information provided by FAO and WHO*, Rome: Codex Alimentarius Commission, October, 5
29 Gupta, D (2002) Capacity Building and Technical Assistance – New Approaches and Building Alliances. FAO/WHO Global Forum of Food Safety Regulators, Marrakesh, Morocco, 28–30 January, GF 01/12
30 Stiglitz, J (2002) *Globalisation and its Discontents*, London: Allen Lane
31 Putnam, RD (2000) *Bowling Alone*, New York: Touchstone
32 Cooper, H (1999) 'Will Human Chains and Zapatistas Greet The WTO in Seattle?', *Wall Street Journal*, 16 July
33 Henderson, D (1999) *The MAI Affair: a Story and Its Lessons*, London: Royal Institute of International Affairs
34 Brundtland, GH (2001) 'FAO/WHO call for more international collaboration to solve food safety and quality problems' WHO Press Release WHO/30, coinciding with Codex Alimentarius Commission meeting, 2–7 July, Geneva: World Health Organization
35 Beaglehole, R and Bonita, R (1998) 'Public health at the crossroads: which way forward?', *The Lancet*, 351, 21 February, pp590–592
36 Yach, D and Bettcher, D (1998) 'The Globalization of Public Health, I: Threats and Opportunities', *American Journal of Public Health*, 88, 5, pp735–737
37 Yach, D and Bettcher, D (1998) 'The Globalization of Public Health, II: The Convergence of Self-Interest and Altrusim', *American Journal of Public Health*, 88, 5, pp738–741
38 Neville-Rolfe, N (1984) *The Politics of Agriculture in the European Community*, London: Policy Studies Institute
39 Whitehead, M and Nordgren, P (eds) (1996) *Health Impact Assessment of the EU Common Agricultural Policy*, Stockholm: National Institute for Public Health

40 Lang, T (1999) 'Food and nutrition', in Weil, O, McKee, M, Brodin, M and Oberlé, D (eds) *Priorities for Public Health Action in the European Union*, Vandoeuvre-les-Nancy: Société Française de Santé Publique, pp138–156
41 Elinder, L (2003) *Public Health Aspects of the EU Common Agricultural Policy*, Stockholm: National Institute of Public Health, http://www.fhi.se/shop/material_pdf/eu_inlaga.pdf
42 EU (2002) *Decision No. 2002/EC of the European Parliament and of the Council adopting a programme of Community Action in the field of public health (2003–2008)*, PE-CONS 3627/02 Brussels: European Union, 15 May, 2002
43 EC (2002) *The Common Agricultural Policy – an evolving policy*, Brussels: Commission of the European Communities, 12 July
44 Agra Europe (2002) 'Fischler's mid-term revolution', *Agra Europe*, 12 July, pA/1-A/2
45 Grant, W (2002) 'Prospects for CAP reform', *Political Quarterly*, 74, 1, pp19–26
46 Lastikka, L (2002) 'Finnish food industry-excellent products for consumers', *NORDICUM*, 2, p62
47 Barling, D, Lang, T and Caraher, M (2002) 'Joined-up Food Policy? The trials of Governance, Public Policy and the Food System, *Social Policy & Administration*, 36, 6, pp556–575
48 Barling, D and Lang, T (2003) 'A Reluctant Food Policy?: The First Five Years of Food Policy under Labour', *Political Quarterly*, 74, 1, pp8–18
49 Lang, T and Rayner, G (2003) 'Food and Health Strategy in the UK: A Policy Impact Analysis', *Political Quarterly*, 74, 1, pp66–75
50 Lang, T, Rayner, G, Barling, D and Millstone, E (2004) *A New Policy Council on Food, Nutrition & Physical Activity for the UK?* A Briefing, London: City University Dept Health Management and Food Policy, April
51 OECD (1999) *Agricultural Policies in OECD countries: monitoring and evaluation*, Paris: Organisation for Economic Co-operation and Development
52 Williams, F (2003) 'UN urges rich to restart trade talks', *Financial Times*, 26 June, http://www.un.org/publications
53 Krebs, A (2002) *Agribusiness Examiner*, 182, 15 August, http://www.eal.com/CARP/
54 OECD, PSE/CSE database 2002
55 US Department of Agriculture, cited in the in the *Financial Times*, 23 July, 2001
56 OECD (1999) *Agricultural Policies in OECD countries: monitoring and evaluation*, Paris: Organisation for Economic Co-operation and Development

57 Action Aid (2003) *Farmgate: The developmental impact of agricultural subsidies*, London: Action Aid, p1
58 Podbury, T (2000) *US and EU Agricultural Support: Who Does it Benefit?* ABARE research programme on agricultural trade liberalisation, Canberra: Australian Bureau of Agriculture Research (ABARE)
59 Nuffield College Oxford (2003) *Identifying the Flow of Domestic and European Expenditure into the English Regions*, Oxford: Nuffield College/London: Office of the Deputy Prime Minister, http://www.nuff.ox.ac.uk/projects/odpm/
60 de Moor, APG (1998) *Subsidizing Unsustainable Development*, The Hague: Institute for Research on Public Expenditure & Earth Council
61 Nestle, M (2002) *The Politics of Food*, San Francisco: University of California Press
62 Sims, L (1998) *The Politics of Fat*, New York: M E Sharpe Inc
63 James, WPT, Ralph, A and Berlizi, M (1997) 'Nutrition Policies in Western Europe: National Policies in Belgium, the Netherlands, France, Ireland and the United Kingdom', *Nutrition Reviews*, 55, 11, (ll)S4-S20
64 Norum, K (1997) 'Some aspects of Norwegian nutrition and food policy', in Shetty, P and McPherson, K (eds) *Diet, Nutrition and Chronic Disease: Lessons from contrasting worlds*, Chichester: J Wiley and Sons
65 National Nutrition Council (1994) *The Norwegian Diet and Nutrition and Food Policy*, Oslo: National Nutrition Council
66 Helsing, E (1993) 'Trends in fat consumption in Europe and their influence on the Mediterranean diet', *European Journal of Clinical Nutrition*, 47, Suppl. 1, S4–S12
67 Helsing, E (1987) *Norwegian Nutrition Policy in 1987: what works and why?* Report from a research seminar, Vettre, Norway 27–28 April. Copenhagen: WHO Regional Office for Europe
68 RNMA (1975) *On Norwegian nutrition and food policy*, Report No. 32 to the Storting, Oslo: Royal Norwegian Ministry of Agriculture
69 Milio, N (1990) *An analysis of the implementation of Norwegian Nutrition Policy 1981–1987*, first European Conference on Food and Nutrition Policy, Budapest, 1–5 October 1990, Copenhagen: WHO Regional Office for Europe. EUR/ICP/NUT 133/BD/1
70 Oshaug, A (1992) *Towards Nutrition Security*, country Paper for Norway, International Conference on Nutrition, Oslo: Nordic School of Nutrition, University of Oslo
71 Pietinen, P (1996) 'Trends in Nutrition and its consequences in Europe: The Finnish Experience', in Pietinenm, P, Nishida, C and Khaltaev, N (eds) *Nutrition and Quality of Life: Health Issues for the 21st century*, Geneva: World Health Organization

72 Puska, P, Tuomilehto, J, Nissinen, A and Vartiainen, E (eds) (1995) *The North Karelia Project: 20 years results and experiences*, Helsinki: National Public Health Institute & World Health Organization Regional Office for Europe
73 National Nutrition Council (1992) *Nutrition Policy in Finland*, country paper prepared for the FAO/WHO International Conference on Nutrition, Rome
74 Pietinen, P and Vartiainen, E (1995) 'Dietary changes', in Puska, P, Tuomilehto, J, Nissinen, A and Vartiainen, E (eds) (1995) *The North Karelia Project: 20 years results and experiences*, Helsinki: National Public Health Institute & World Health Organization Regional Office for Europe, pp107–117
75 Vail, D (1994) 'Sweden's 1990 Food Policy Reform', in McMichael, P (ed) *The Global Restructuring of Agro-Food Systems*, Ithaca: Cornell University Press
76 Vail, D (1994) 'Sweden's 1990 Food Policy Reform', in McMichael, P (ed) *The Global Restructuring of Agro-Food Systems*, Ithaca: Cornell University Press p66
77 Commission on Environmental Health (1996) *Environment for Sustainable Health Development – an Action Plan for Sweden*, Stockholm: Ministry of Health and Social Affairs. Swedish Official Reports Series 1996: 124
78 von Weizacher, E, Lovins, AB and Lovins, LH (1997) *Factor Four: doubling wealth, halving resource use*, London: Earthscan
79 Carlsson-Kanyama, A (1998) 'Climate Change and Dietary Choices: how can emissions of greenhouse gases from food consumption be reduced?', *Food Policy*, 23, 3 June
80 McMichael, AJ, Haines, A, Slooff, R and Kovats, RS (eds) (1996) *Climate Change and Human Health*, Geneva: World Health Organization, UN Environment Programme, World Meteorological Organization
81 EASA (1995) *Survey on self-regulation for advertising and children in Europe*, Brussels: European Advertising Standards Alliance, October
82 Walt, G and Gilson, L (1994) 'Reforming the health sector in developing countries', *Health Policy and Planning*, 9, 4, pp353–370
83 Millstone, E and van Zwanenberg, P (2001) 'The Politics of Scientific Advice', *Science & Public Policy*, 28, 2, pp99–112
84 Millstone, E (2000) 'Recent developments in EU food policy: institutional adjustments or fundamental reforms?', *Zeitschrift fur das gesamte Lebenmittelrecht*, 27, 6, pp1–15
85 Commission of the European Communities (2000) *Food Safety White Paper*, COM(1999)719 final, 12 January, Brussels: European Commission
86 Food Standards Australia New Zealand (2003), http://www.foodstandards.gov.au/

87 Lang, T, Millstone, E and Rayner, M (1997) *Food Standards and the State*, discussion paper 4, London: Centre for Food Policy, Thames Valley University
88 Lang, T, Millstone, E and Rayner, M (1997) *Food Standards and the State: a fresh start*, discussion paper 3, London: Centre for Food Policy, April
89 Raghavan, C (1990) *Recolonization: GATT, the Uruguay Round and the Third World*, London: Zed Press
90 Watkins, K (1991) 'Agriculture and food security in the GATT Uruguay Round', *Review of African Political Economy*, 50, March
91 Lawrence, F (2002) 'Fowl play', *The Guardian*, G2: 2–5 July 8
92 Lawrence, F and Evans, R (2003) 'Letters expose chicken scandal complacency; News of Panorama/Guardian investigation prompted Brussels to act', *The Guardian*, 24 May, http://www.guardian.co.uk/uk_news/story/0,3604,962556,00.html
93 Lawrence, F (2003) 'Sausage Factory', *The Guardian*, 10 May, http://www.guardian.co.uk/food/focus/story/0,13296,951917,00.html
94 Longfield, J (1992) 'Information and Advertising', in National Consumer Council (ed), *Your Food: Whose Choice?*, London: The Stationery Office
95 Body, R (1991) *Our Food, Our land: why contemporary farming practices must change*, London: Rider
96 Watkins, K (2002) *Rigged rules and double standards: trade, globalisation, and the fight against poverty*, Oxford: Oxfam International, http://www.maketradefair.com
97 Norberg-Hodge, H, Merrifield, T and Gorelick, S (eds) (2000) *Bringing the Food Economy Home*, Dartington: International Society for Ecology and Culture

Chapter 8

1 Brunner, E, Rayner, M, Thorogood, M, Margetts, B, Hooper, L et al (2001) 'Making public health nutrition relevant to evidence-based action', *Public Health Nutrition*, 4 (6), pp1297–1299
2 See, for example, Eric Hobsbawm's magisterial historical reviews in *The Age of Revolution*, *The Age of Capital*, *The Age of Empire* and *The Age of Extremes*
3 Commission on Macroeconomics and Health (2001) *Macroeconomics and Health: Investing in health for economic development*, Report to the World Health Organization from the Commission on Macroeconomics and Health, Geneva: World Health Organization, http://www.cid.harvard.edu/cidcmh/CMHReport.pdf
4 Millstone, E and Lang, T (eds) (2003) *Atlas of Food*, London: Earthscan/New York: Penguin

5 von Clausewitz, C (1976 [1832]) *On War*, Howard, M and Paret, P (ed and trans). Princeton: Princeton University Press, p87
6 Millstone, E and van Zwnaenberg, PF (2000) 'Food safety and consumer protection in a globalised economy', *Swiss Political Science Review*, 6, 3, pp109–118
7 Buck, T (2003) 'EU "obeys WTO ban on beef hormones"', *Financial Times*, 16 October
8 Esserman, S and Howse, R (2003) 'The WTO on trial', *Foreign Affairs*, 82, 1, pp130–140
9 Elinder, L (2003) *Public Health Aspects of the EU Common Agricultural Policy*, Stockholm: National Institute of Public Health, http://www.fhi.se/shop/material_pdf/eu_inlaga.pdf
10 Popkin, BM and Nielsen, J (2003) 'The Sweetening of the World's Diet', *Obesity Research*, 11, 11, pp1–8
11 For example, Hume Hall, R (1974) *Food for Nought: the decline in nutrition*, New York: Harper and Row
12 For example, Lerza, C and Jacobson, M (eds) (1975) *Food for People not for Profit: a sourcebook on the Food Crisis*, New York: Ballantine.
13 George, S (1976) *How the Other Half Dies*, Harmondsworth: Penguin
14 Robbins, P (2003) *Stolen Fruit: the tropical commodities disaster*, London: Zed Press
15 Lang, T (1996) 'Going public: food campaigns during the 1980s and 1990s', in Smith, D (ed) (1996) *Nutrition Scientists and Nutrition Policy in the 20th Century*, London: Routledge, pp238–260
16 Altieri, M (1996) *Agroecology: the science of sustainable agriculture*, Boulder: Westview Press
17 Pretty, J (2002) *AgriCulture*, London: Earthscan
18 Pretty, J and Hine, R (2001) *Reducing Food Poverty with Sustainable Agriculture: A Summary of New Evidence*, Final Report of the SAFE-World Research Project, February, Colchester: University of Essex
19 Funes, F, Garcia, L, Bourque, M, Perez, N and Rosset, P (2002) *Sustainable Agriculture and Resistance: Transforming Food Production in Cuba*, Oakland CA: Food First Books
20 http://ns.rds.org.hn/via/
21 http://www.sustainweb.org
22 http://www.ibase.org.br/
23 http://vshiva.net
24 Furusawa, K (1994) 'Co-operative alternatives in Japan', in Conford, P (ed) *A Future for the Land*, Bideford: Resurgence Books
25 Burlton, B (2000) *Presidential Address 2000 to Co-operative Congress*, Manchester: Co-operative Congress
26 http://www.slowfood.com
27 Jackle, JJ and Scully, KA (1999) *Fast Food: Roadside Restaurants in the Automobile Age*, Baltimore: the Johns Hopkins University Press

28 Elliot, L (2003) 'Cleaning Agent: Interview with Niall Fitzgerald', *The Guardian*, 5 July
29 Boyle, D (2003) *Authenticity: brands, fakes, spin and the lust for real life*, London: Flamingo
30 Norberg-Hodge, H, Goering, P and Page, J (2001) *From the Ground Up: Rethinking Industrial Agriculture*, London: Zed
31 McDonald's (2003) *Global Policy on Antibiotics in Food Animals*, Policy statement, 19 June 2003. Chicago Ill: McDonalds Corporation, http://www.mcdonalds.com/corporate/social
32 Dinham, B and Hines, C (1982) *Agribusiness in Africa*, London: Earth Resources Research
33 Barratt Brown, M (1993) *Fair Trade*, London: Pluto
34 Fairtrade Foundation, http://www.fairtrade.org.uk/
35 Coote, B (1992) *The Trade Trap: poverty and the global commodity markets*, Oxford: Oxfam publications
36 'Race to the top' project, coordinated by International Institute for Environment and Development. www.racetothetop.org
37 Raven, H and Lang, T (1995) *Off our Trollies?* London: Institute for Public Policy Research
38 'IFAP adopts NFU proposals attacking industrial concentration in Agriculture', *Agribusiness Examiner*, 16 May 2003, issue 248
39 Eurepgap, the global partnership for standards on fresh produce and flowers. See: http://www.eurep.org/sites/index_e.html
40 Nestle, M (2002) *Food Politics*, Berkeley: University of California Press
41 Maxwell, S and Slater, R (eds) (2003) 'Themed Issue: Food Policy Old and New', *Development Policy Review*, 21, Nos 5–6, September–November, pp531–710

INDEX